Advances in Biotechnology

Indu Ravi · Mamta Baunthiyal
Jyoti Saxena
Editors

Advances in Biotechnology

Springer

Editors
Indu Ravi
Indira Gandhi National Open University
Regional Centre Jaipur
Mansarovar, Jaipur, Rajasthan
India

Mamta Baunthiyal
Department of Biotechnology
Govind Ballabh Pant Engineering College
Ghurdauri, Pauri, Uttarakhand
India

Jyoti Saxena
Department of Biochemical Engineering
Bipin Tripathi Kumaon Institute of Technology
Dwarahat, Uttarakhand
India

ISBN 978-81-322-2891-2 ISBN 978-81-322-1554-7 (eBook)
DOI 10.1007/978-81-322-1554-7
Springer New Delhi Heidelberg New York Dordrecht London

Softcover reprint of the hardcover 1st edition 2014

Printed on acid-free paper

Springer is part of Springer Science+Business Media (www.springer.com)

Foreword

The twenty-first century has been nicknamed as the era of biotechnology. It has grown and evolved to such an extent over the past few years that increasing numbers of professionals work in areas directly impacted by it. It has been turned into a high science topic to our everyday vocabulary over a short period of time.

It is quite remarkable to note how different branches of biotechnology have emerged to have both substantial academic and industrial impact in the not so distant future. The opportunities become wider and the hopes brighter. Modern biotechnology has opened up many opportunities in various sectors such as agriculture, food, forestry, waste treatment, medicine, and pharmaceutical production. Covering even the most important aspects of biotechnology in a single book that reaches readers ranging from students to active researchers in academia and industry is an enormous challenge. To prepare such a wide-ranging book on biotechnology, editors have harnessed their own knowledge and experience, gained in several departments and universities, and has assembled experts to write chapters covering a wide array of biotechnology topics, including the latest advances. *Advances in Biotechnology* is an important book that provides the information and insight to enable readers to participate in the biotechnology debate. This book is intended to serve both as a textbook for university courses as well as a reference for researchers. It is increasingly important that scientists and engineers, whatever their specialty, have a solid grounding in the fundamentals and potential applications of biotechnology. The editors and their team are to be warmly congratulated for bringing together this exclusive, timely, and useful biotechnology book.

D. S. Chauhan Vice Chancellor
Uttarakhand Technical University, Dehradun

Preface

The twenty-first century looks to Biotechnology as the world's fastest growing and most rapidly changing technology that can improve the human conditions. Modern biotechnology enables an organism to produce a totally new product which the organism does not or cannot produce in its normal course of life. The book *Advances in Biotechnology* is a collection of topics on recent advances in certain ongoing biotechnological applications. Fourteen authoritative chapters on current developments and future trends in biotechnology are empathized. The book aims to cover a wide range of topics under all specialized domains of microbial, plant, animal, and industrial biotechnology.

Chapter 1 provides a detailed account of high capacity vectors used for various applications of genetic engineering. Chapter 2 is devoted to the modern era DNA sequencing dealing with next generation sequencing. Up-to-date methodological approaches such as use of molecular markers (Chap. 3), DNA microarray technology (Chap. 6) and proteomics (Chap. 8) have revolutionized biotechnology with a wide array of applications in studies related to cancer biology, microbiology, plant science, environmental science, etc. Proteomics has recently been of interest to scientists because it gives a better understanding of an organism than genomics.

Chapters 4, 5, 9, 11, and 12 are focused on the crucial role of biotechnology in health care through gene therapy, gene silencing, stem cell technology, monoclonal antibodies, and edible vaccines. Gene therapy is being used for correcting defective genes that are responsible for disease development; RNAi is a valuable research tool not only for functional genomics, but also for gene-specific therapeutic activities. Monoclonal antibodies are widely used for immunodiagnostic, immunotherapy, and in biological and biochemical research. Key aspects of edible vaccines like host plants, mechanism of action, advantages, limitations, and different regulatory issues are contemplated upon.

In today's world where products of microbial origin have proved their utility in almost every sphere of life, metagenomic studies (Chap. 7) have become highly important as they give a clue to the hidden wealth of microbial world. Chapter 10 describes the utilization of biosensors in various industries for monitoring food quality control, medical research, clinical diagnosis, environmental monitoring, agriculture, bioprocesses, and control. Genes from microbes, plants, and animals are being used successfully to enhance the ability of plants. Though improvement of plants by genetic engineering opens up new possibilities to tolerate, remove, and degrade pollutants, it is still in its research and development phase with many technical issues needing to be addressed as explained in Chap. 13. Finally, in Chap. 14, the great market potential involved for biotechnological companies has been highlighted with suggestions that can be set up for harnessing the vast potential involved in biotech products all over the world.

This book is clearly a team effort, and many thanks are due. The authors of the individual chapters have been chosen for their recognized expertise and their contributions to the various fields of biotechnology. Their willingness to impart their knowledge to their colleagues forms the basis of this book and is gratefully acknowledged. Authors relied on various sources, which are identified in the individual chapters. The authors would also like to thank their colleague Ms. Shweta Ranghar whose help during the preparation of this book was commendable. Thanks are also due to Mr. Vikas Kumar for working on the illustrations. The editors wish to thank their respective head of the institutions/center for their encouragement and providing required ambience.

Moreover, this work would not have been brought to realization without the prudence and the constant and conscientious support of the publisher. We are grateful to Springer for publishing this book with their customary excellence. Finally, special thanks go to our families, who put up with longer hours, helpful suggestions, indispensable help, and encouragement.

May 2013

Indu Ravi
Mamta Baunthiyal
Jyoti Saxena

Contents

Contributors

Pavan Kumar Agrawal Govind Ballabh Pant Engineering College, Ghurdauri, Pauri, Uttarakhand 246194, India, e-mail: p_k_agarwal@rediffmail.com

Rachana Arya Electronics and Communication Engineering Department, Bipin Tripathi Kumaon Institute of Technology, Dwarahat, Uttarakhand 263653, India, e-mail: rachna009@gmail.com

Bhakti Bajpai Biotechnology, ARIBAS, New Vallabh Vidya Nagar, Gujarat 388121, India, e-mail: bbajpai@yahoo.com

Mamta Baunthiyal Govind Ballabh Pant Engineering College, Ghurdauri, Pauri, Uttarakhand 246194, India , e-mail: mamtabaunthiyal@yahoo.co.in

J. P. Bhatt Department of Zoology and Biotechnology, HNB Garhwal University, Srinagar (Garhwal), Uttarakhand, India, e-mail: profjpbhatt@gmail.com

Pradeep Bhatnagar Life Sciences, The IIS University, Mansarovar, Jaipur, India, e-mail: Pradeepbhatnagar1947@yahoo.com

Satpal Singh Bisht Department of Biotechnology, School of Life Sciences, Mizoram University (A Central University), Tanhril, Aizawl, Mizoram 796004, India, e-mail: sps.bisht@gmail.com

Ashwini M. Charpe Plant Pathology Section, College of Agriculture, Dr. Panjabrao Deshmukh Krishi Vidyapeeth, Akola, Maharashtra 444104, India, e-mail: ashwinicharpe@yahoo.com

Sunita Chauhan Kumarappa National Handmade Paper Institute, Jaipur, Rajasthan, India, e-mail: itsneeru@yahoo.com

G. K. Joshi Department of Zoology and Biotechnology, HNB Garhwal University, Srinagar (Garhwal), Uttarakhand, India, e-mail: gkjoshi@rediffmail.com

J. Jugran Department of Zoology and Biotechnology, HNB Garhwal University, Srinagar (Garhwal), Uttarakhand, India, e-mail: jyotijugran28@gmail.com

Shweta Kulshreshtha Amity Institute of Biotechnology, Amity University of Rajasthan, Jaipur, India, e-mail: shweta_kul17@rediffmail.com

B. D. Lakhchaura Govind Ballabh Pant University of Agriculture and Technology, Pantnagar, India, e-mail: lakhchaurabd@rediffmail.com

Mayank Biochemical Engineering Department, Bipin Tripathi Kumaon Institute of Technology, Dwarahat, Uttarakhand 263653, India, e-mail: findmayank12@yahoo.co.in

Amrita Kumari Panda Department of Biotechnology, Roland Institute of Pharmaceutical Sciences, Berhampur, Orissa 760010, India, e-mail: itu.linu@gmail.com

Indu Ravi IGNOU Regional Centre, Jaipur, Rajasthan, India, e-mail: induravi11@yahoo.com

Shweta Rawat Biochemical Engineering Department, Bipin Tripathi Kumaon Institute of Technology, Dwarahat, Uttarakhand 263653, India, e-mail: shweta.biotech24@gmail.com

K. Rohini Unit of Biochemistry, Faculty of Medicine, AIMST University, 3½ Bukit Air Nasi, Jalan Semeling, 08100 Bedong, Kedah Darul Aman, Malaysia, e-mail: rohinik23@gmail.com

Jyoti Saxena Biochemical Engineering Department, Bipin Tripathi Kumaon Institute of Technology, Dwarahat, Uttarakhand 263653, India, e-mail: saxenajyoti30@gmail.com

Rahul Shrivastava Department of Biotechnology, Jaypee University of Information Technology, Waknaghat, DumeharBani, Solan, Himachal Pradesh, India, e-mail: rahulmicro@rediffmail.com

1 High Capacity Vectors

Bhakti Bajpai

Abstract

Since the construction of the first generation of general cloning vectors in the early 1970s, a large number of cloning vectors have been developed. Despite the bewildering choice of commercial and other available vectors, the selection of cloning vector to be used can be decided by applying a small number of criteria: insert size, copy number, incompatibility, selectable marker cloning sites, and specialized vector functions. Several of these criteria are dependent on each other. Most general cloning plasmids can carry a DNA insert up to around 15 kb in size. Several types of vectors are available for cloning large fragments of DNA too. This chapter presents a consolidated account of some new generation of high-capacity vectors such as cosmid, yeast artificial chromosome (YAC), bacterial artificial chromosome (BAC), P1 phage artificial chromosome (PAC), and human artificial chromosome (HAC).

1.1 Introduction

A prime requisite for a gene cloning experiment is the selection of a suitable cloning vector, i.e., a DNA molecule that acts as a vehicle for carrying a foreign DNA fragment when inserted into it and transports it into a host cell, which is usually a bacterium, though other types of living cells can also be used. A wide variety of natural replicons exhibit the properties that allow them to act as cloning vectors, however, vectors may also be designed to possess certain minimum qualification to function as an efficient agent for transfer, maintenance, and amplification of target DNA.

An ideal cloning vehicle would have the following four properties:

- Low-molecular weight
- Ability to confer readily selectable phenotypic traits on host cells
- Single sites for a large number of restriction endonucleases, preferably in genes with a scorable phenotype
- Ability to replicate within the host cell, so that numerous copies of the recombinant DNA molecule can be produced and passed to daughter cells.

B. Bajpai (✉)
Biotechnology, ARIBAS, New Vallabh Vidya Nagar, Gujarat, 388121, India
e-mail: bbajpai@yahoo.com

I. Ravi et al. (eds.), *Advances in Biotechnology*,
DOI: 10.1007/978-81-322-1554-7_1, © Springer India 2014

In 1970s, when recombinant DNA technology was being first developed, only a limited number of vectors were available based on either high-copy number plasmids or phage λ. Later phage M13 was developed as a specialist vector to facilitate DNA sequencing; over a time a series of specialist vectors were constructed for specific purpose. The examples of naturally occurring or artificially constructed vectors include vectors based on *Escherichia coli* plasmids, bacteriophages (e.g., λ, M13, P1), viruses (e.g., animal viruses—retrovirus, adenovirus, adeno-associated virus, Herpes Simplex virus, *Vaccinia* virus, etc.; insect viruses—baculo virus; plant viruses—cauliflower mosaic virus, potato virus X, Gemini virus, etc.), *Agrobacterium tumefaciens* based vectors, chimeric plasmids (e.g., cosmid, phagemid, phasmid, and fosmid), artificial chromosomes [e.g., YAC, BAC, PAC, MAC and HAC], and non-*E. coli* vectors (e.g., *Bacillus* and *Pseudomonas* vectors etc.). Table 1.1 gives an idea about the size of the insert possible with different types of vectors.

In order to determine the choice of vector for a particular cloning experiment, various factors need to be considered such as:

1. **Insert size**: The insert size may vary for different types of vectors ranging from 5 to 25 kb for plasmid vectors to >2,000 kb for HACs.
2. **Vector size**: The vector size range varies from 5 kb plasmid vectors to 6–10 megabases HAC high-capacity vectors.
3. **Restriction sites**: The number of restriction sites found in vectors is highly variable. There may be a few restriction sites in small plasmid vectors but they may be increased by the insertion of multiple cloning sites in vectors.
4. **Copy number**: Different cloning vectors are maintained at different copy numbers, dependent on the replicon of the plasmid. However, a high-copy number vector is desirable. The origin of replication determines the vector copy number, which could be in the range of 25–50 copies/cell if the expression vector is derived from the low-copy number plasmid pBR322, or between 150 and 200 copies/cell, if derived from the high-copy number plasmid pUC.
5. **Cloning efficiency**: The ability to clone a DNA fragment inserted into a vector is known as the cloning efficiency of the vector.
6. **Ability to screen for inserts**: For selection of recombinants, certain selectable markers should be present in vectors in order to distinguish them from non-recombinants.
7. Types of downstream experiments required.

Table 1.1 Maximum DNA insert possible with different cloning vectors

Vector	Host	Insert size
Plasmid	*E.coli*	5–25 kb
λ phage	*E.coli*	35–45 kb
P1 phage	*E.coli*	70–100 kb
PAC_s	*E.coli*	100–300kb
BAC_s	*E.coli*	<300 kb
YAC_s	*Saccharomyces cerevisae*	200–2000kb
Human Artificial Chromosomes (HACs)	*Cultured Human Cells*	>2000kb

1.2 Vectors for Cloning Large Fragments of DNA

1.2.1 Cosmid Vectors

Cosmids are hybrids between a phage DNA molecule and a bacterial plasmid or are basically a plasmid that carries a *cos* site, the substrate for enzymes that package λ DNA molecule into phage coat proteins. The *in vitro* packaging reaction works not only with one genome but also with any DNA molecule that carries *cos* site separated by 37–52 kb of DNA. It also needs a selectable marker, such as ampicillin resistance gene, and a plasmid origin of replication, as cosmids lack all the λ genes, therefore do not produce plaques. Instead colonies are formed on selective media, just as with a plasmid vector. The loading capacity of cosmids varies depending on the size of the vector itself but usually lies around 40–45 kb—much more than

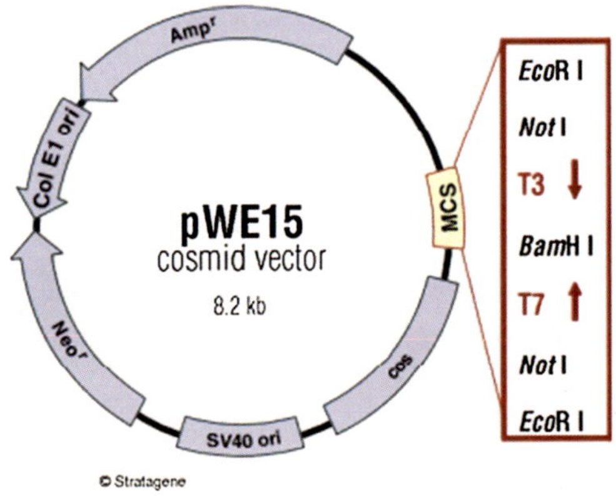

Fig. 1.1 Construct of a cosmid vector

a phage λ vector can accommodate. After packaging *in vitro*, the particle is used to infect suitable host. The recombinant cosmid DNA is injected into the cell where it circularizes like phage DNA but replicates as a normal plasmid without the expression of any phage functions. Transformed cells are selected on the basis of a vector drug resistance marker. The construct of a typical cosmid vector is shown in Fig. 1.1.

Cosmids provide an efficient means of cloning large pieces of DNA. Because of their capacity to carry large fragments of DNA, cosmids are particularly attractive for constructing libraries of eukaryotic genome fragments. Partial digestion with a restriction endonuclease provided suitably large fragments. However, there is potential problem associated with use of partial digests in this way. This is due to the possibility of two or more genome fragments joining together in the ligation reaction, hence creating a clone containing fragments that were not initially adjacent in the genome. The problem can be overcome by the size fractionation and dephosphorylation of the foreign DNA fragments so as to prevent their ligation together. But this method is very sensitive to the exact ratio of target-to-vector DNAs because vector-to-vector ligation can occur. Such difficulties have been overcome in a cosmid-cloning procedure devised by Ish-Horowicz and Burke (1981). By appropriate treatment of the cosmid vector pJB8, left-hand and right-hand vector ends are purified which are incapable of self-ligation but which accept dephosphorylated foreign DNA. Thus, the method eliminates the need to size the foreign DNA fragments and prevents formation of clones containing short foreign DNA or multiple vector sequences. Figure 1.2 describes the cosmid-cloning procedure devised by Ish-Horowicz and Burke (1981).

Problems associated with lambda and cosmid cloning:

1. Since repeats occur in eukaryotic DNA, rearrangements can occur *via* recombination of the repeats present on the DNA inserted into lambda or cosmid.
2. Cosmids are difficult to maintain in a bacterial cell because they are somewhat unstable.
3. Not easy to handle due to its very large size of approximately 50 kb.

1.2.2 Yeast Artificial Chromosomes

A YAC is a vector used to clone DNA fragments larger than 100 kb and up to 3,000 kb. YACs are useful for the physical mapping of complex genomes and for the cloning of large genes. First described in 1983 by Murray and Szostak, a YAC is an artificially constructed chromosome that contains a centromere (CEN), telomeres (TEL), and an autonomous replicating sequence (ARS) element which are required for replication and preservation of YAC in yeast cells. ARS elements are thought to act as replication origins. A YAC is built using an initial circular plasmid, which is typically broken into two linear molecules using restriction enzymes. DNA ligase is then used to ligate a sequence or gene of interest between the two linear molecules, forming a single large linear piece of DNA.

A plasmid-derived origin of replication (ori) and an antibiotic resistance gene allow the YAC vector to be amplified and selected for in *E. coli*. *TRP1* and *URA3* genes are included in the YAC vector to provide a selection system for identifying transformed yeast cells that include YAC by complementing recessive alleles *trp1* and *ura3* in yeast host cell. YAC vector cloning site for foreign DNA is located within

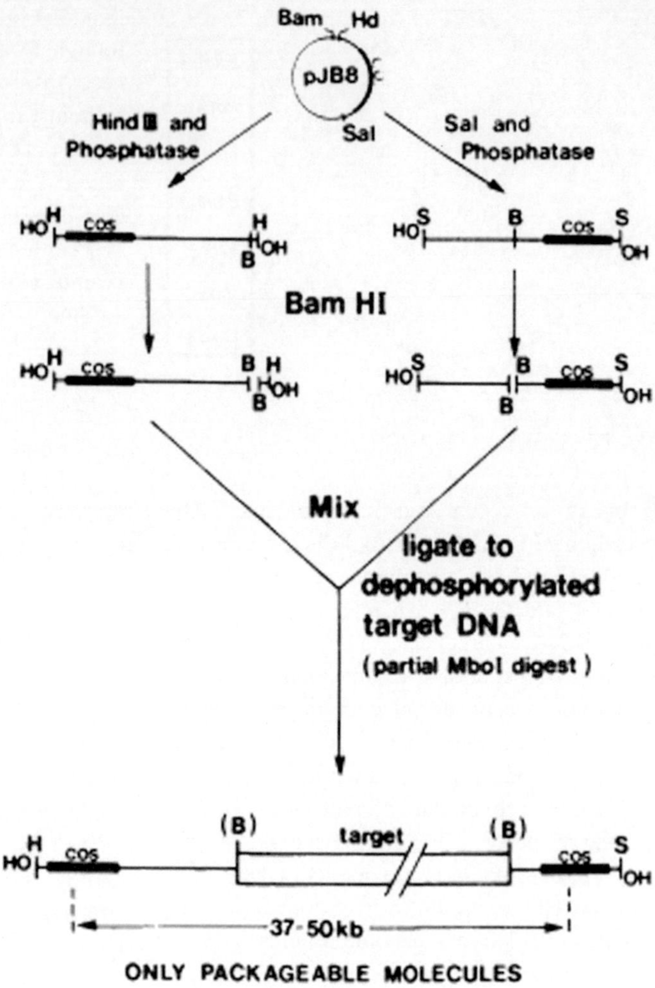

Fig. 1.2 Cosmid-cloning procedure (Ish-Horowicz and Burke 1981)

the *SUP4* gene. This gene compensates for a mutation in the yeast host cell that causes the accumulation of red pigment. The host cells are normally red, and those transformed with YAC only, will form colorless colonies. Cloning of a foreign DNA fragment into the YAC causes insertional inactivation, restoring the red color. Therefore, the colonies that contain the foreign DNA fragment are red.

1.2.2.1 Essential Components of YAC Vectors

1. Large DNA (>100 kb) is ligated between two arms. Each arm ends with a yeast telomere so that the product can be stabilized in the yeast cell. Interestingly, larger YACs are more stable than shorter ones, which favors cloning of large stretches of DNA (Fig. 1.3a, b).
2. One arm contains an autonomous replication sequence (ARS), a CEN, and TEL.
3. *amp*^{r} for selective amplification and markers such as TRP1 and URA3 for identifying cells containing the YAC vector.
4. Recognition sites of restriction enzymes (e.g., EcoRI and BamHI).
5. Insertion of DNA into the cloning site inactivates a mutant expressed in the vector DNA and red yeast colonies appear.
6. Transformants are identified as those red colonies which grow in a yeast cell that is mutant for *trp1* and *ura3*. This ensures that the cell has received an artificial chromosome

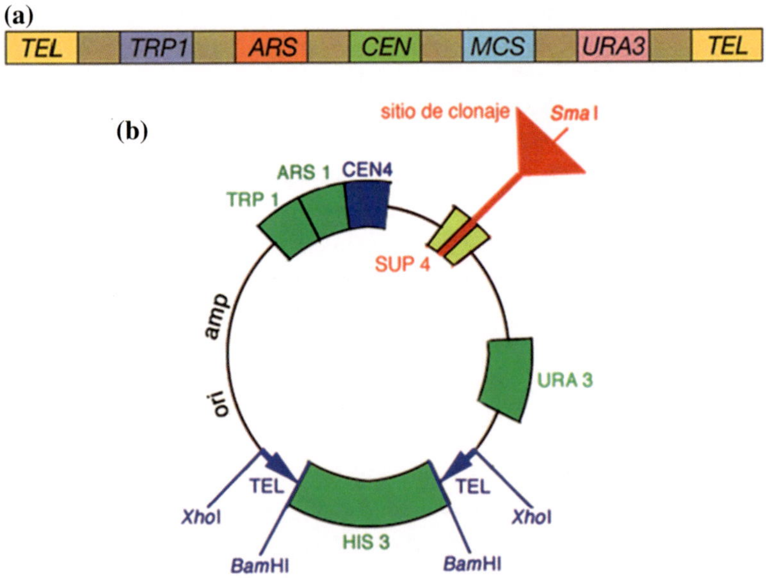

Fig. 1.3 **a** Linear form of yeast artificial chromosome. **b** Circular form of yeast artificial chromosome

with both TEL (because of complementation of the two mutants) and the artificial chromosome contains insert DNA (because the cell is red).

The procedure for cloning in YAC is as given below:

1. The target DNA is partially digested by EcoRI and the YAC vector is cleaved by EcoRI and BamHI.
2. The cleaved vector segment is ligated with a digested DNA fragment to form an artificial chromosome.
3. Yeast cells are transformed to make a large number of copies.

1.2.2.2 Advantages and Disadvantages

Yeast expression vectors, such as YACs, yeast integrating plasmids (YIps), and yeast episomal plasmids (YEps) have an advantage over bacterial vectors (BACs) in that they can be used to express eukaryotic proteins that require post-translational modification. However, YACs are significantly less stable than BACs, producing "chimeric effects": artifacts where the sequence of the cloned DNA actually corresponds not to a single genomic region but to multiple regions. Chimerism may be due to either coligation of multiple genomic segments into a single YAC, or recombination of two or more YACs transformed in the same host yeast cell. The incidence of chimerism may be as high as 50 %. Other artifacts are deletion of segments from a cloned region, and rearrangement of genomic segments (such as inversion). In all these cases, the sequence as determined from the YAC clone is different from the original, natural sequence, leading to inconsistent results, and errors in interpretation if the clone's information is relied upon. Due to these issues, the Human Genome Project ultimately abandoned the use of YACs and switched to BACs, where the incidence of these artifacts is very low.

1.2.3 Bacterial Artificial Chromosome

As the Human Genome Project was underway in the early 1990s, there was a need to create high-resolution physical map of each human chromosome, which would permit the isolation of short DNA fragments for direct sequencing and other manipulations. In response to this, the YAC system was developed. Although yeast can carry the DNA as large as one Mb, subsequent studies indicated that yeast system presented several difficulties in the creation of a human genome map. Additionally, yeast cells were not as familiar to molecular biologist as *E. coli*. To circumvent these difficulties, a bacterial cloning system based on the well-characterized *E. coli* F factor, a low-copy plasmid that exist in a supercoiled form was developed by Hiroaki Shizuya in 1992.

A BAC is a DNA construct, based on a functional fertility plasmid (or F-plasmid), used for transforming and cloning in bacteria, usually *E. coli.* F-plasmids play a crucial role because they contain partition genes that promote the even distribution of plasmids after bacterial cell division. The BAC's usual insert size is 150–350 kb. The replication of F factor is strictly controlled by the regulatory functions of *E. coli*; as a result F factor is maintained as a low-copy number (i.e., one or two copies per cell). This allows stable maintenance of large DNA inserts and reduces the potential for recombination between DNA fragments carried by the vector, which was a limitation observed with cosmid-cloning system. In addition to stable maintenance, the structural stability of F-factors allows complex genomic DNA inserts to be maintained with a great degree of structural stability in the *E. coli* host. The structure of a typical BAC is given in Fig. 1.4.

BACs have several advantages over YACs. It was observed that a large percentage of YACs carried chimeric inserts, making mapping efforts confusing and difficult. BACs in contrast are virtually free from chimerism. Another problem with YAC is that multiple YAC chromosomes may coexist in a single yeast cell, whereas in the BAC system the F factor encoded parA and parB gene are involved in exclusion of multiple F-factors, as a result multiple F-factors cannot coexist in a single cell.

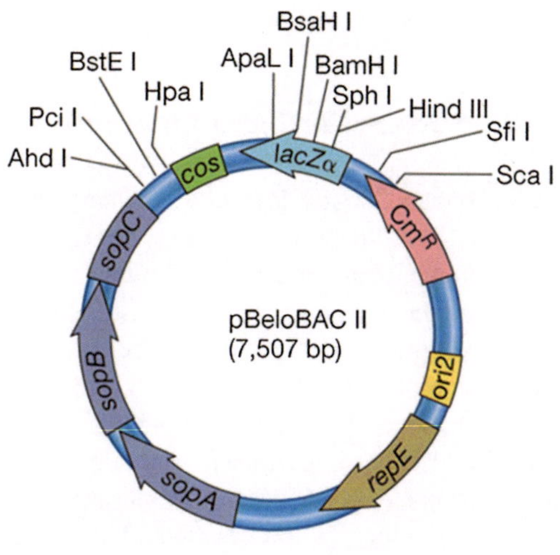

Fig. 1.4 Bacterial artificial chromosome

1.2.3.1 BAC Vector Cloning Site

The cloning segment of BAC vector includes (1) two bacteriophage markers lambda cosN and P1 loxP, (2) three restriction enzyme sites (*Eco*RI, *Hind*III, and *Bam*HI) for cloning, and (3) a GC- rich *Not*I restriction enzyme site for potential excision of inserts. The cosN site provides a fixed position for cleavage by bacteriophage lambda enzyme *terminase*, which allows the convenient generation of a linear form of the BAC DNA. The cosN site is also used to package approximately 50 kb DNA into bacteriophage lambda head as a particle. The method known as Fosmid for F-based cosmid system is extremely efficient, thus very useful when DNA is precious or available in small amounts. The P1 loxP site allows the retrofitting of additional components to BAC vector at a later stage. The loxP site is also utilized to linearize BACs through the P1 phage protein Cre, which catalyses strand exchange between two DNA strands at the loxP sites.

1.2.3.2 Uses

Inherited Disease

BACs are now being utilized to a greater extent in modeling genetic diseases, often alongside transgenic mice. BACs have been useful in this field as complex genes may have several regulatory sequences upstream of the encoding sequence, including various promoter sequences that will govern a gene's expression level. BACs have been used to some degree of success with mice while studying neurological diseases such as Alzheimer's disease or as in the case of aneuploidy associated with Down syndrome.

There have also been instances when they have been used to study specific oncogenes associated with cancers. They are transferred over to these genetic disease models by electroporation/transformation, transfection with a suitable virus or microinjection. BACs can also be utilized to detect genes or large sequences of interest and then used to map them onto the human chromosome using BAC arrays. BACs are preferred for these kinds of genetic studies because they accommodate much larger sequences without the risk of rearrangement, therefore more stable than other types of cloning vectors.

Infectious Diseases

The genomes of several large DNA and RNA viruses have been cloned as BACs. These constructs are referred to as "infectious clones," as transfection of the BAC construct into host cells is sufficient to initiate viral infection. The infectious property of these BACs has made the study of many viruses such as the herpes viruses, poxviruses, and coronaviruses more accessible. Molecular studies of these viruses can now be achieved using genetic approaches to mutate the BAC while it resides in bacteria. Such genetic approaches rely on either linear or circular targeting vectors to carry out homologous recombination.

Genome Sequencing

BACs are often used to sequence the genome of organisms in genome projects, for example the Human Genome Project. A short piece of the organism's DNA is amplified as an insert in BACs, and then sequenced. Finally, the sequenced parts are rearranged *in silico*, resulting in the genomic sequence of the organism.

1.2.4 P1 Phage Derived Artificial Chromosome

The P1-derived artificial chromosomes are DNA constructs derived from the DNA of P1 bacteriophage and BAC. They can carry large amounts (about 100–300 kb) of other sequences for a variety of bioengineering purposes. It is one type of vector used to clone DNA fragments (100- to 300-kb insert size; average, 150 kb) in *E. coli* cells. PACs have a low-copy number origin of replication based on P1 bacteriophage, which is used for propagation. Similar to BACs, PACs allow replication of the clones at one copy per cell and replicate clones across 60–100 generations. In contrast to BACs, PACs have a negative selection against non-recombinants. PACs also have an IPTG- inducible high-copy number origin of replication that can be utilized for DNA production. These can accommodate larger inserts of DNA than a plasmid or many other types of vectors. Sometimes, the number of inserts can be as high as 300 kb (Fig. 1.5).

1.2.4.1 Uniqueness of P1-bacteriophage

A P1 phage can exist in both lysogenic and lytic forms in the host cell, but its unique feature lies in its existence as an independent entity within the cell, rather than incorporating itself with the host chromosomes during the phase of 'lysogeny'. Thus, it acts like a plasmid during its existence and can replace the function of a plasmid during processes, which entails this feature. However, the scientists consider P1 derived chromosome to contain features of both plasmids and 'F' factor, which is a unique plasmid like DNA sequence used in creating BAC.

In comparison with YACs, PACs offer certain advantages: (1) these are bacterial systems that are easy to manipulate, (2) libraries are generated using bacterial hosts with well defined properties, (3) transformation efficiency is higher than that obtained by YACs, (4) PACs are nonchimeric, and (5) PACs have very stable inserts and do not delete sequences.

1.2.4.2 Construction of PACs Through Electroporation

During the construction of PACs, P1 phage containing cells will undergo a process known as 'electroporation', which will increase the permeability of the cell membrane and allow DNA material to enter the cell and couple with the

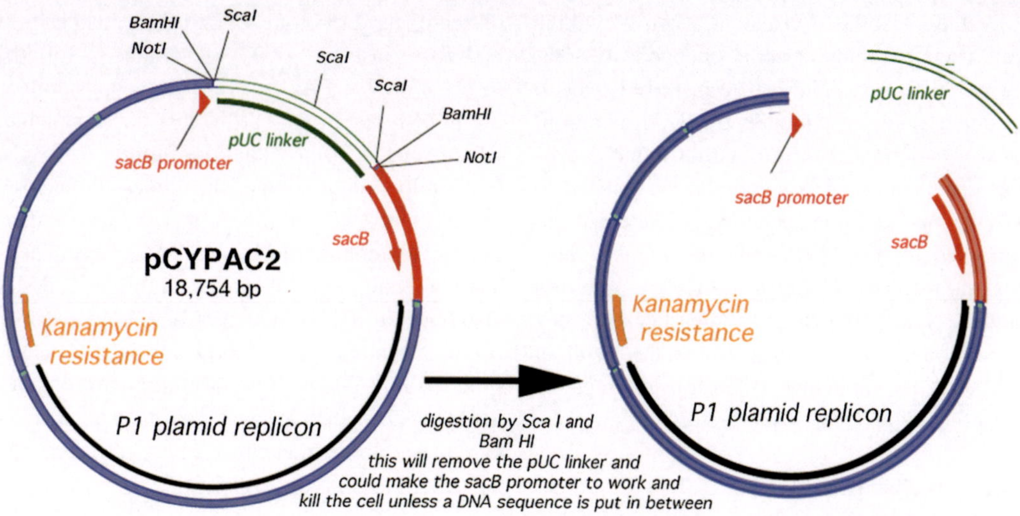

Fig. 1.5 Phage artificial chromosome

existing DNA. This process will give rise to PACs and from there onwards, the PACs can replicate within the cell through 'lysogeny', without destructing the cell or incorporating into rest of the chromosomes.

1.2.4.3 Uses of PACs

PACs are in high demand for cloning important biomedical sequences, which are essential for many scientific functions. One of its main uses is the genome analysis and map-based cloning of complex plants and animals, which requires isolation of large pieces of DNA rather than smaller segments. Furthermore, PAC-based cloning is useful in the study of 'phage therapy' and in scientific studies focusing on how antibiotics act on a particular bacteria.

Although there are other forms of artificial chromosomes which can accommodate more base pairs than PACs, relative user friendliness of these vectors makes them a popular choice among many biomedical researchers.

1.2.5 Human Artificial Chromosomes

The idea of using artificial chromosomes as potential vectors for gene therapy applications came as a consequence of the first studies involving chromosome manipulation, designed to understand human chromosome structure, and to identify elements necessary for their correct functioning. There are two approaches that can be used for the development of artificial chromosomes: *top-down* approach in which natural chromosomes are truncated by radiation or telomere-associated fragmentation; and the *bottom-up* approach in which a *de novo* chromosome is formed from the basic elements of CENs, TEL, and origins of replication.

The construction of YAC showed CENs, TEL, and origins of replication as the elements necessary for extrachromosomal retention and led to the development of mammalian chromosomes which are similar to yeast chromosomes. Many experiments designed to find mammalian origins of replication could not identify specific sequences responsible for mammalian genome replication. However, the structure of human TEL was soon described as an array. The confirmation came from the observation of newly formed TEL in the α globin gene, caused by insertion of $(TTAGGG)_n$ sequence. The discovery of the telomeric sequence as a tandemly repeated $(TTAGGG)_n$ sequence orientated $5'$-$3'$ toward the end of the chromosome and its role in telomere formation led to the development of telomere-mediated chromosome fragmentation (*top-down* approach), which

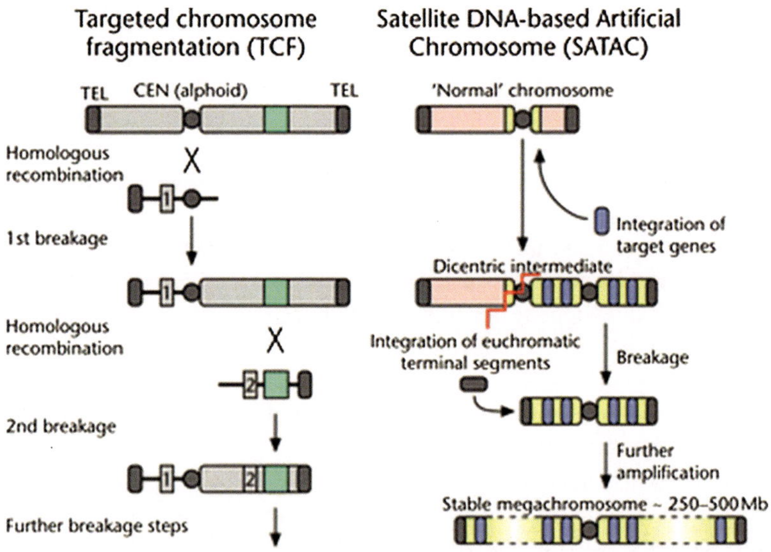

Fig. 1.6 Structural map of '*top-down*' engineered mammalian artificial chromosome systems

allowed the isolation of minichromosomes in somatic cell hybrid. The first "top down" approach or TACF (telomere-associated chromosome fragmentation) involved modifying natural chromosome into smaller defined minichromosomes in cultured cells. Following recombination and subsequent breakage between homologous sequences on the endogenous host chromosome and an incoming telomere containing the targeting vector, engineered minichromosomes as small as 450 kb in size in avian cells have been generated. The approach has been important for studying the structure, sequence organization, and size requirements of the human X and Y chromosomes.

Second approach, the "bottom up" or assembly approach involved generating HACs in human cells by introducing defined chromosomal sequences as naked DNA including human TEL, alpha satellite (alphoid) DNA and genomic fragmentation containing replication origins. The *de novo* HACs are generated following recombination and some amplification of the input DNA within the host cell. Together, the generation of minichromosomes and *de novo* HACs has identified alphoid DNA as the major sequence element of the CEN and determined the minimum size (~700–100 kb) required for CEN function and stability.

Second approach includes the generation of SATACs (satellite DNA-based artificial chromosomes) following integration of repetitive DNA into preexisting centromeric regions of host chromosomes and modifying small human marker chromosomes (minichromosomes derived from naturally occurring chromosomes). The two approaches are shown in Fig. 1.6.

The *de novo* HACs when introduced into the cell, undergo a process of recombination and amplification forming large (1–10 Mb) circular molecules (usually at one or two copies per cell) which are mitotically stable in the absence of any selection for 9 months in some cells. The efficiency of *de novo* HAC formation and stability depends on the presence of a CEN protein B-binding sequence (CENP-B box) and, to some extent, on the chromosomal origin of the alphoid template and the longer length of the alphoid array (>100 bp).

Established HACs can be either in a linear or a circular state. PAC-based constructs carrying ~70 kb of alphoid DNA array with or without telomeric sequences and in circular or linearized state were used to transfect by lipofection HT1080 cells. Circular alphoid DNA vectors established effectively as minichromosomes in any condition, demonstrating that TEL are not required for the circular conformation. However, capped TEL were essential for establishment of linear PAC vectors because these vectors showed poor chromosome formation in their absence.

1.2.5.1 Advantages and Uses of HAC

Human artificial chromosomes (HACs) represent another extrachromosomal gene delivery and gene expression vector system. Although this technology is less advanced than virus derived vectors, HACs have several potential advantages over currently used episomal viral vectors for gene therapy applications. The presence of a functional CEN provides a long-term stable maintenance of a HAC as a single copy episome without integration to the host chromosomes. There is no upper size limit to DNA that should be cloned in HAC that allows the use of complete genomic loci, including the upstream and downstream regulatory elements. Additionally, being solely human in origin, HAC vectors cannot evoke adverse host immunogenic responses or induce any risk of cellular transformation.

HAC-based vectors offer a promising system for delivery and expression of full-length human genes of any size.

1.3 Conclusion

A vector is a DNA molecule used as a vehicle to transfer foreign genetic material into another cell. All engineered vectors have an origin of replication, a multicloning site, and a selectable marker. Genome size varies among different organisms and the cloning vector must be selected accordingly. For a large genome, a vector with a large capacity is chosen so that a relatively small number of clones are sufficient for coverage of the entire genome. However, it is often more difficult to characterize an insert contained in a high capacity vector. The development of extrachromosomal large-capacity cloning vectors for mammalian cells represents a powerful tool for functional genomic studies. Further, the advances in genome library construction and DNA sequencing are mainly due to the development of high capacity vectors such as cosmids, BACs, PACs, YACs, and HACs.

References

Ish-Horowicz D, Burke JF (1981) Rapid and efficient cosmid cloning. Nucleic Acids Res 9(13):2989–2998

Murray AW, Szostak JW (1983) Construction of artificial chromosomes in yeast. Nature 305(5931):189–193

Shizuya H, Birren B, Kim UJ, Mancino V, Slepak T, Tachiiri Y, Simon M (1992) Cloning and stable maintenance of 300-kilobase-pair fragments of human DNA in *Escherichia coli* using an F-factor-based vector. Proc Natl Acad Sci 89:8794–8797

2 DNA Sequencing: Methods and Applications

Satpal Singh Bisht and Amrita Kumari Panda

Abstract

Determination of the precise order of nucleotides within a DNA molecule is popularly known as DNA sequencing. About three decades ago in the year 1977, Sanger and Maxam–Gilbert made a breakthrough that revolutionized the world of biological sciences by sequencing the 5,386-base bacteriophage φX174. From the year 1977 to till date DNA sequencing came across much advancement in terms of sequencing tools and techniques. The modern era DNA sequencing are dealing with Next generation sequencing and many other advancement are available to the researchers, practitioners, and academicians at a very reasonable cost with highest accuracy. The biological databases are being flooded with a huge flow of sequences coming out from various organisms across the world. Today the researchers and scientists across the various fields are utilizing these data for a variety of applications including food security by developing better crops and crop yields, livestock, improved diagnostics, prognostics, and therapies for many complex diseases.

2.1 Introduction

DNA is the blueprint of life consisting of chemical building blocks called nucleotides. These building blocks are made of three parts: phosphate, sugar group, and one of the four types of nitrogen bases viz Adenine (A), Thymine (T), Guanine (G), or Cytosine (C). To form a strand of DNA, nucleotides are linked into chains, with the phosphate and sugar groups alternately. The order or sequence of these bases determines what biological instructions are contained in a strand of DNA. For example, the sequence ATCGTT might instruct for blue eyes, while ATCGCT for brown. Each DNA sequence that contains instructions to make a protein is known as gene. The size of a gene may vary greatly, ranging from about 1,000 bases to 2300 kilo bases in humans. DNA has double helical structure in which two strands run in opposite directions. Each "rung" of the ladder is made up of two

S. S. Bisht (✉)
Biotechnology School of Life Sciences, Mizoram University, Tanhril, Mizoram, Aizawl, 796004, India
e-mail: sps.bisht@gmail.com

A. K. Panda
Biotechnology, Roland Institute of Pharmaceutical Sciences, Orissa, Berhampur, 760010, India
e-mail: itu.linu@gmail.com

I. Ravi et al. (eds.), *Advances in Biotechnology*,
DOI: 10.1007/978-81-322-1554-7_2, © Springer India 2014

nitrogen bases; paired together by hydrogen bonds, because of the highly specific nature of this type of chemical pairing, base A always pairs with base T, and likewise C with G. Therefore, if the sequence of the bases on one strand of a DNA double helix is known, it is simple to figure out the sequence of bases on the other strand.

The most significant advances in genetics during 1990s have come from complete sequencing of chromosomes. The first eukaryotic chromosome to be completely sequenced was chromosome III of *Saccharomyces cerevisiae*, published in 1992. This was followed by the first complete genome sequence for a free living organism, the bacterium *Haemophilus influenzae* in the year 1995 and the first complete sequence of an eukaryotic genome *S. cerevisiae* in 1996. Later the complete genomic sequences of important model organisms such as *Escherichia coli*, the nematode *Coenarhabditis elegans*, the fruit fly *Drosophila*, and the plant *Arabidopsis* became available. Genome projects for many organisms have either been completed or will be completed shortly, such as Palaeo-Eskimo, an ancient-human, Neanderthal *Homo neanderthalensis* (partial), Neanderthal genome project, Common Chimpanzee Pan troglodytes; Chimpanzee genome project, Domestic cow, Bovine genome, Honey-bee genome sequencing consortium, Human microbiome project, International grape genome program, International HapMap project including Human genome project which has now entered into functional genomics phase. The main objective of most genome projects is to determine the DNA sequence of the entire genome or of its large number of transcripts. This leads to the identification of all or most of the genes and to characterize various structural features of the genome. In many molecular biology laboratories DNA sequencing is chiefly used to characterize newly cloned cDNAs to confirm the identity of a clone or mutation, to check the fidelity of a newly created mutation, PCR products and screening tool to identify polymorphism. Now-a-days, by the advent of automated DNA sequencing and Next generation sequencing (NGS) complete genome sequencing data of many organisms are available for genetic studies.

2.2 Landmarks in DNA Sequencing

- 1953 Discovery of the structure of the DNA double helix.
- 1972 Development of recombinant DNA technology.
- 1977 The first complete genome of bacteriophage φX174 sequenced.
- 1977 Allan Maxam and Walter Gilbert publish "DNA sequencing by chemical degradation."
- 1984 Medical Research Council scientists decipher the complete DNA sequence of the Epstein-Barr virus, 170 kb.
- 1986 Leroy E. Hood's laboratory at the California Institute of Technology and Smith announced the first semi-automated DNA sequencing machine.
- 1987 Applied Biosystems marketed first automated sequencing machine, the model ABI 370.
- 1990 The U.S. National Institutes of Health (NIH) begins large-scale sequencing trials on *Mycoplasma capricolum*, *E. coli*, *C. elegans*, and *S. cerevisiae*.
- 1991 Sequencing of human expressed sequence tags begins in Craig Venter's lab.
- 1995 Craig Venter, Hamilton Smith, and colleagues at The Institute for Genomic Research (TIGR) published the first complete genome of a free-living organism, the bacterium *Haemophilus influenzae*.
- 1996 Pal Nyren and his student Mostafa Ronaghi at the Royal Institute of Technology in Stockholm published their method of pyrosequencing.
- 1998 Phil Green and Brent Ewing of the University of Washington publish "phred" for sequencer data analysis.
- 2000 Lynx Therapeutics publishes and markets "MPSS"—a parallelized, adapter/ligation-mediated, bead-based sequencing technology, launching "next-generation" sequencing.
- 2001 A draft sequence of the human genome published.
- 2004 454 Life Sciences markets a parallelized version of pyrosequencing. The first version of their machine reduced sequencing costs 6-fold compared to automated Sanger sequencing,

and was the second of a new generation of sequencing technologies, after MPSS.

- 2005 Solexa/ Illumina sequence analyzer which gave an output data of 10E+7 Kbp.
- 2010 Illumina Hi-seq 2000 was introduced which gave an output of 10E+8 Kbp.

2.3 Sequencing Methods

2.3.1 Maxam–Gilbert Method

Allan Maxam and Walter Gilbert developed a method for sequencing single-stranded DNA by a two-step catalytic process involving piperidine and two chemicals that selectively attack purines and pyrimidines (Maxam and Gilbert 1977). Purines react with dimethyl sulfate and pyrimidines react with hydrazine in such a way so as to break the glycoside bond between the ribose sugar and the base, displacing the base. Piperidine catalyzes cleavage of phosphodiester bonds where the base has been displaced. Moreover, dimethyl sulfate and piperidine alone selectively cleave guanine nucleotides but dimethyl sulfate and piperidine in formic acid cleave both guanine and adenine nucleotides. Similarly, hydrazine and piperidine cleave both thymine and cytosine nucleotides, whereas hydrazine and piperidine in 1.5 M NaCl only cleave cytosine nucleotides. The use of these selective reactions to DNA sequencing involves creating a single-stranded DNA substrate carrying a radioactive label on the 5′ end. This labeled substrate is subjected to four separate cleavage reactions, each of which creates a population of labeled cleavage products ending in known nucleotides. The reactions are loaded on high percentage polyacrylamide gels and the fragments are resolved by electrophoresis. The gel then is transferred to a light-proof X-ray film cassette, a piece of X-ray film placed over the gel, and the cassette placed in a freezer for several days. Wherever a labeled fragment stopped on the gel, the radioactive tag would expose the film due to particle decay (autoradiography).

The dark autoradiography bands on the film represent the 5′ to 3′ DNA sequence when read from bottom to top (Fig. 2.1). The process of base calling involves interpreting the banding pattern relative to the four chemical reactions. For example, a band in the lanes corresponding to the C only and the C + T reactions called a C. If the band present in the C + T reaction lane but not in the C reaction lane it is called as T. The same decision process can be obtained for the G only and the G + A reaction lanes. Sequences can be confirmed by running replicate reactions on the same gel and comparing the autoradiographic patterns between replicates.

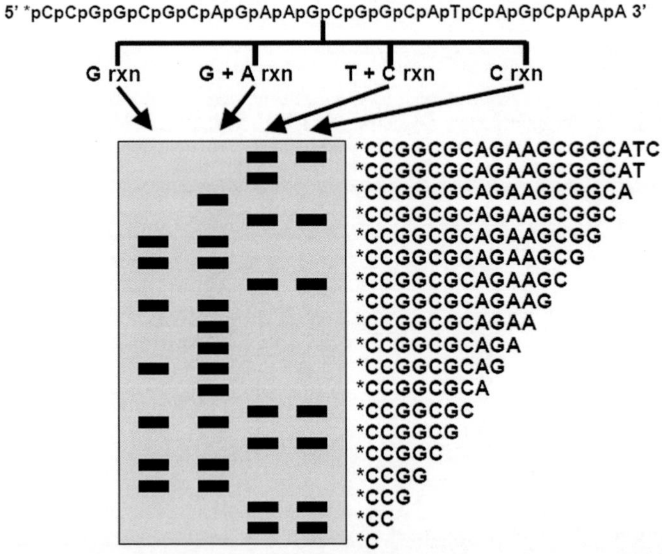

Fig. 2.1 The reading pattern of autoradiogram

2.3.2 Sanger Method

Frederick Sanger developed an alternative method, rather than using chemical cleavage reactions, Sanger opted for a method involving a third form of the ribose sugar (Sanger et al. 1977). Ribose has a hydroxyl group on both the 2′ and the 3′ carbons, whereas deoxyribose has only the one hydroxyl group on the 3′ carbon. There is a third form of ribose, dideoxyribose in which the hydroxyl group is missing from both the 2′ and the 3′ carbons (Fig. 2.2). Whenever a dideoxynucleotide incorporated into a polynucleotide, the chain irreversibly stops or terminates. The basic idea behind chain termination method developed in 1974 by Sanger was to generate all possible single-stranded DNA molecules complementary to a template that starts at a common 5′ base and extends up to 1 kilobase in the 3′ direction (Fig. 2.3). These single strands of DNA are labeled in such a way which allows the identity of the 3′-end base in each molecule. These molecules are separated according to size by electrophoresis and each band corresponding to a class of molecule differing in length by one nucleotide from the adjacent band (Fig. 2.4a, b).

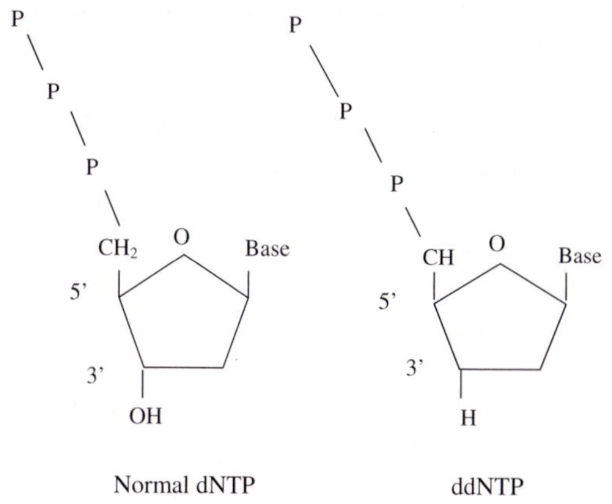

Fig. 2.2 Structural comparison of dNTP and ddNTP

2.3.3 Automated DNA Sequencing Methods

The principle of automated DNA sequencing is same as Sanger's method but the detection is different. In this automated method, the primer or the ddNTPs are labeled by incorporation of a fluorescent dye. Thus, rather than running the gel for a particular time and reading the results, the machine uses a laser to read the fluorescence of the dye as the bands pass a fixed point. Labeling of the ddNTPs is much more advantageous than the primer labeling because four ddNTPs each labeled with different dyes leads the sequencing reaction to run in a single tube and separated in a single lane, thus increasing the capacity of the machine. Automated DNA sequencers can sequence up to 384 DNA samples in a single batch and run up to 24 runs per day. DNA sequencers carry out capillary electrophoresis for size separation, detection and recording of dye fluorescence, and data output as fluorescent peak trace chromatograms. Since the capillary tubes have a high surface to volume ratio (25–100 mm diameter), it radiates heat readily, thus the samples do not over heat. Detection of the migrating molecules is

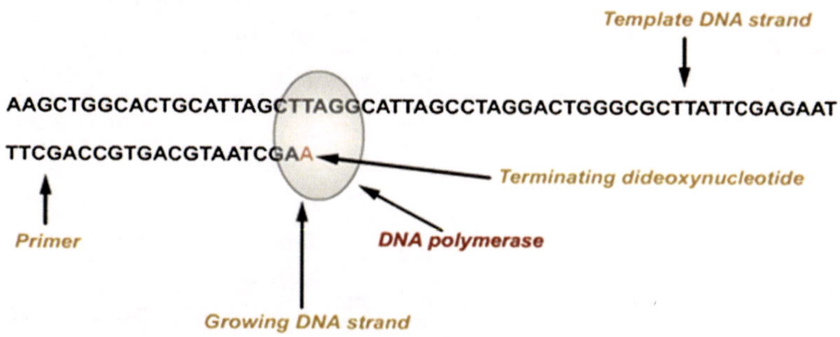

Fig. 2.3 Principle of Sanger sequencing

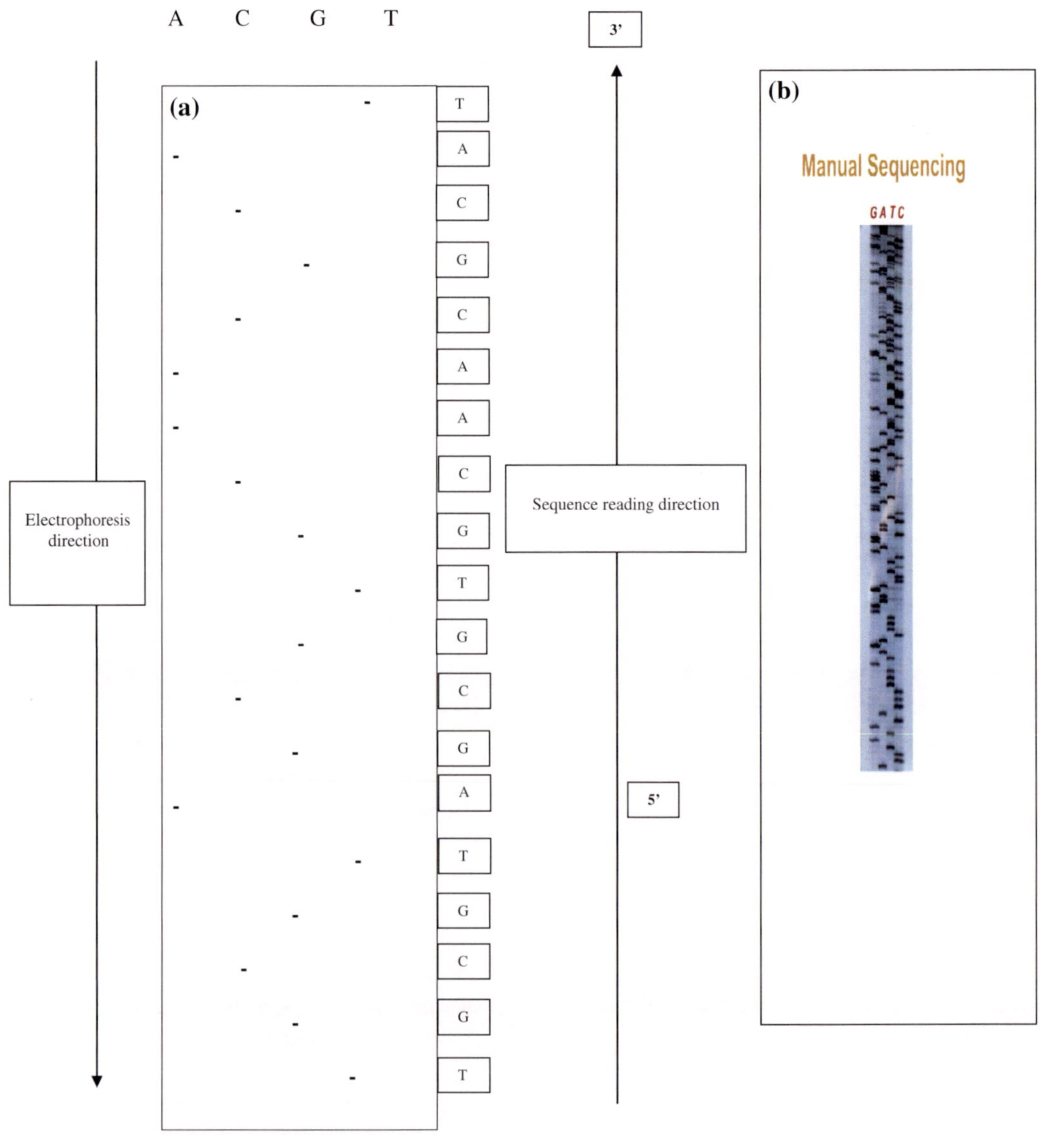

Fig. 2.4 Determination of DNA sequence by Sanger's dideoxy nucleotide method. **a** Depicted diagram. **b** Autoradiogram

accomplished by shining a light source through a portion of the tubing and detecting the light emitted from the other side (Fig. 2.5). In thermo cycling sequencing the reactions by thermo cycling, cleanup, and re-suspension in a buffer solution before loading onto the sequencer are performed separately. A number of commercial and non-commercial software packages can trim low-quality DNA traces automatically. These programs score the quality of each peak and remove low-quality base peaks (generally located at the ends of the sequence).

2.3.3.1 Base Calling

The raw sequence traces in automated sequencing can be read using automated softwares like *Phred programme* which convert traces into sequences that can be deposited in a database within seconds after sequencing run (Ewing et al. 1998).

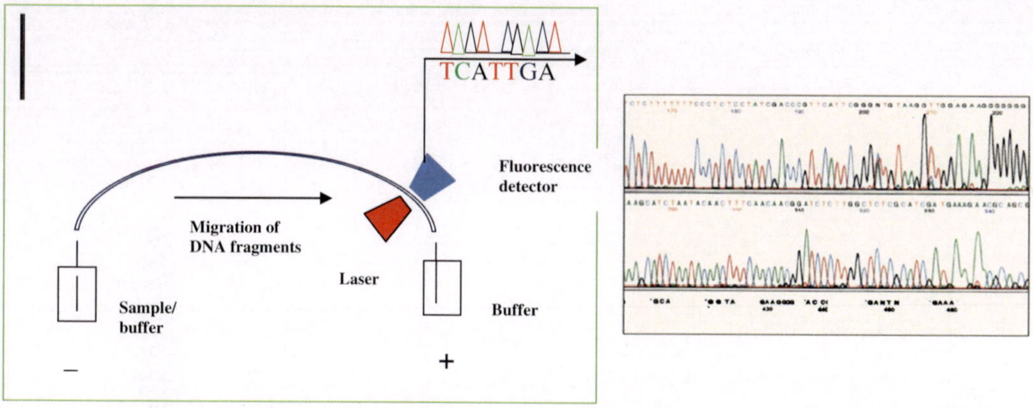

Fig. 2.5 Capillary electrophoresis and electropherogram with peaks representing the bands on the sequencing gel

The new techniques and equipment included in automated DNA sequencing are:

1. Four-color fluorescent dyes have replaced the radioactive label. Attachment of these dyes to the ddNTPs results in a fluorescent tag directly marking just the terminated DNA molecule, and consequently a single sequencing reaction spiked with all four ddNTPs is sufficient to sequence any template.
2. Rather than stopping the electrophoresis at a particular time the products are scanned for laser-induced fluorescence just before they run off the end of the electrophoresis medium. The sequence is collected as a set of four "trace files" that indicates the intensity of the four colors, a peak in the trace distribution implies that the particular base was the last one incorporated at the position.
3. Improvement in the chemistry of template purification and the sequencing reaction including use of bioengineered thermostable polymerases that can read through secondary structure with high fidelity extends the length of high quality sequence.
4. Slab gel electrophoresis gave way to capillary electrophoresis with the introduction in 1999 of Applied Biosystem's ABI Prism 3,700 automated sequencers. These sequencers give extremely high quality, long reads, save time and money by abolishing the laborious, and often frustrating step of gel pouring that add a new level of automation in which the capillaries are loaded by robot from 96-well plates rather than by hand. Each machine can handle six 96-well plates per day or approximately 0.5 Mb of sequence.
5. Matrix-assisted laser desorption/ionization, time-of-flight mass spectrometry (MALDI-TOF MS) was put forward as an alternative to the Sanger sequencing/capillary electrophoresis combination. It is the tool of choice in proteomics applications, while the full potential for DNA analysis was demonstrated in 1995 and for RNA in 1998. For MALDI-TOF MS analysis single-stranded nucleic acid molecules of 3–29 bp in length (1,000–8,600 Da range) need to be generated and deposited on a matrix (e.g., 3-hydroxy picolinic acid). The analyte/matrix molecules are then irradiated by a laser inducing their desorption and ionization, upon which the molecules pass through a flight tube connected to a detector on the other end. Separation occurs by the time of flight, which is proportional to the mass of the individual molecules. The main advantage of the method is that it directly measures an intrinsic physical property of the molecules i.e., mass and speed. Limitations lie in the size of the DNA molecules that can be detected intact to less than 100 bp (due to size-dependent fragmentation during the MALDI process); and that the analytes must be free from ion adducts which lead to mass distortion.

Compared to gel electrophoresis based sequencing systems, mass spectrometry produces very high resolution of sequencing fragments, rapid separation on microsecond time scales, and completely eliminates compressions associated with gel-based systems. While most of the research efforts have been focused on using mass spectrometers to analyze the DNA products from Sanger sequencing or enzymatic digestion reactions, the read lengths attainable are currently insufficient for large-scale *de novo* sequencing. The advantage of mass-spectrometry sequencing is that one can unambiguously identify frame shift mutations and heterozygous mutations making it an ideal choice for resequencing projects. In these applications, DNA sequencing fragments that are of the same length but with different base compositions are generated, which are challenging to consistently distinguish in gel-based sequencing systems. In contrast, MALDI-TOF MS produces mass spectra of these DNA sequencing fragments with nearly digital resolution, allowing accurate determination of the mixed bases. For these reasons, mass spectrometry based sequencing has mainly been focused on the detection of frameshift mutations and single nucleotide polymorphisms (SNPs). More recently, assays have been developed to indirectly sequence DNA by first converting it into RNA. These assays take advantage of the increased resolution and detection ability of MALDI-TOF MS for RNA.

6. For long oligonucleotides (>50 bases), e.g., microarray applications, Electrospray Ionization-Mass Spectrometry (ESI-MS) is used. The target molecules are ionized into multiple charge states producing a waveform that can be de-convoluted into parent peaks. As only the charge state will vary for the ions, oligonucleotides with high molecular weights can be analyzed using this method (Edwards et al. 2005). In addition, the inherently milder ionization conditions make this analytical technique a great tool for the analysis of labile compounds such as common quenchers, e.g., dabcyl, BHQ's, used in dual-labeled fluorogenic probes. The ESI-MS systems have mass resolution of approximately 0.03 %, i.e., resolution of ±3 Da on a 10 kDa oligonucleotide (Dale and Schantz 2007).

2.4 Genome Sequencing

The genome sequencing usually deals with large-scale sequencing, e.g., whole chromosomes, very long DNA pieces, etc. For longer targets, such as chromosomes, common approaches consisting of cutting (with restriction enzymes) and shearing (with mechanical forces) the large DNA fragments into shorter DNA fragments are used. The fragmented DNA is cloned into a DNA vector and amplified in *E. coli* or other suitable organisms. Short DNA fragments purified from individual clones and sequenced individually called shotgun sequencing, followed by electronic assembly into one long contiguous sequence. The overlapping fragments are joined together to form a contig; two or more contigs assembled to make draft sequence. This stage contains gaps in the assembled sequence which can be filled by primer walking and nested deletion strategies. The next stage is the finishing process which involves filling in the gaps and correcting the more obvious errors and uncertainties. The finished sequence does not contain gaps and is accurate to a defined level. The final stage is annotation which identifies the protein coding sequence. The Human genome project was completed by implementing two approaches: clone-by-clone sequencing and whole genome shotgun sequencing.

2.4.1 Clone-by-Clone Sequencing

In this approach the chromosomes were mapped and then split up into sections. A rough map was drawn for each section, and then the sections themselves were split into smaller bits, with plenty of overlap between each of the bits. Each of these smaller bits would be sequenced, and the overlapping bits would be used to put the genome back together again. First, by mapping the genome researchers produce at an early stage, a genetic resource that can be used to map genes. In addition, since every DNA sequence is derived from a known region, it was relatively easy to keep track of the project and to determine where gaps are in the sequence. Assembly of relatively short regions of DNA is an efficient step. However, mapping can be a time-consuming and costly process.

2.4.2 Whole Genome Shotgun Sequencing

The alternative to the clone-by-clone approach is the 'bottom-up' whole genome shotgun (WGS) sequencing. It was developed by Fred Sanger in 1982. First, DNA is broken into fragments followed by sequencing at random and assembling together the overlaps. Advantage of the whole genome shotgun is that it requires no prior mapping. Its disadvantage is that large genomes need computing power and sophisticated software to reassemble the genome from its fragments. Unlike the clone-by-clone approach, assemblies cannot be produced until the end of the project. Whole genome shotgun for large genomes is especially valuable if there is an existing 'scaffold' of organized sequences, localized to the genome, derived from other projects. When the whole genome shotgun data are laid on the 'scaffold' sequence, it is easier to resolve ambiguities. Today, whole genome shotgun is used for most bacterial genomes and as a 'top-up' of sequence data for many other genome projects.

2.5 Recent Advances in DNA Sequencing

During last 5 years many techniques, service agencies, and companies have come up scientifically with more advanced techniques and commercially viable services such as 454 life Sciences, Roche Applied Science, Microchip Biotechnologies, Agencourt Bioscience, and many more. The decade after the completion of the Human Genome Project, remarkable sequencing technology explosion has permitted a multitude of questions about the genome to be asked and answered at unprecedented speed and resolution. The advances in sequencing technology resulted to lower the DNA sequencing cost to become negligible for some applications and led far greater range of scientific experiments to be carried out, but also allow nonsensical uses of DNA sequencing and generate additional pressure on storage resources (Pettersson et al. 2009).

The prominent methods that are receiving sufficient attention from the researchers and scientists working on genome sequencing and related research areas of biomedical research are given below:

1. Illumina sequencing.
2. Roche 454 Genome Sequencing.
3. Pyro sequencing.
4. Solid sequencing.

Apart from these four sequencing methods few more are also in practice for various sequencing studies like Ion semiconductor sequencing, PacBio RS, DNA nanoball sequencing, Lynx Therapeutics' massively parallel signature sequencing (MPSS), Polony sequencing (Porreca 2010) etc.

2.5.1 Next-Generation DNA Sequencing

A new generation of non-Sanger sequencing technologies, i.e., next-generation DNA sequencing has the potential to dramatically accelerate biological and biomedical research by enabling the DNA sequencing at unprecedented speed. Comprehensive analysis of genomes, transcriptomes, and interactomes has come up with many advantages like being inexpensive, routine, and widespread rather than requiring significant production-scale efforts (Schuster 2008; Xiaoguang et al. 2010). First, DNA template library is constructed. DNA library fragments are prepared from either randomly sheared genomic DNA (10 s to 100 s bp in size) or alternatively pair-end fragments with controlled distance distribution. The double-stranded fragments are ligated with adaptor sequences at both ends and denatured. The resulting single-stranded template library is created and immobilized on a solid surface (either a planar surface or supporting beads) and clonally amplified by one of several means, e.g., bridge PCR, emulsion PCR, or *in situ* polonies. DNA clusters or amplified beads form an array of DNA clusters on a slide, which then undergo cyclic manipulation through enzyme such as polymerase or ligase. Optical events generated from the cyclic chain extension process are monitored by microscopic detection system, and images recorded through CCD camera. Sequential analysis of array image yields DNA fragment sequences, which are assembled into larger sequence contigs by computer algorithm. A comparison of the specifications of next generation sequencing technologies is given in Table 2.1.

2.5.1.1 Illumina Genome Analyzer

The amplification of single-stranded library fragments is carried out through a process called "bridge amplification." In Illumina/Solexa sequencing >50 million clusters/flow cells, each with 1,000 copies of the same template, 1 billion bases per run, and 1 % of the cost of capillary-based method makes it cost effective, highly accurate, straightforward sample preparation, well established open source software community and less time consuming method. Illumina Hi-seq produces approximately 1.6 billion short reads (18–150bp) per flow cell, whereas Illumina Mi-seq produces 7.5 Gbase/run producing 2 × 250 bp fragments.

The steps are as follows:

1 Prepare genomic DNA.
2 Attach DNA to surface.
3 Bridge amplification.
4 Fragment becomes double stranded.
5 Denature the double stranded molecules.
6 Complete amplification.
7 Determine first base.
8 Image first base.
9 Determine second base.
10 Image second base.
11 Sequence reads over multiple cycles.
12 Align data.

Table 2.1 Comparative specification of Next Generation Sequencing platforms (Quail et al. 2012; Liu et al. 2012; Loman et al. 2012)

Platform	454	Ion torrent	Illumina
Sequencing mechanism	Pyrosequencing	Sequencing by synthesis	Sequencing by synthesis
Sequence yield per run	0.7Gb	20-50 Mb on 314 chip100-200Mb on 316 chip1 Gb on 318 chip	600 Gb in Illumina HiSeq 20001.5-2 Gb in Illumina MiSeq
Sequence cost	$32/Mb	$1000 (318 chip)/Gb	$41/Gb in Illumina HiSeq 2000$502/Gb in Illumina MiSeq
Run time	8 hours	2 hours	11 days in Illumina HiSeq 2000 27 hours in Illumina MiSeq
Read length	700 bases	~200 bases	Up to 150 bases

2.5.1.2 Roche 454 Genome Sequencer

The 454 Sequencer utilizes emulsion PCR to yield amplicons used for the sequencing procedure. Tiny paramagnetic beads coated with DNA primers are mixed with single-stranded template DNA library together with components necessary for PCR reaction. Proportional amount of beads and library fragments are mixed to ensure most beads carry not more than one ssDNA molecule. The aqueous solution is mixed with oil to form emulsion where each water compartment forms an independent micro-reactor for subsequent PCR chemistry. After multiple rounds of thermo-cycling, each bead is coated with thousands of copies of DNA of the same sequence. Beads are further enriched, transferred, and deposited on a picotiter plate fabricated in organized array of tiny wells with each hole occupied by only one bead. The picotiter plate is engineered as part of flow cell for sequencing chemistry on one side and bounded with optic fibers as part of CCD-based optical detection system on the other.

2.5.1.3 Pyrosequencing

It is a method of DNA sequencing (determining the order of nucleotides in DNA) based on the "sequencing by synthesis" principle. It differs from Sanger sequencing, in that it relies on the detection of pyrophosphate release on nucleotide incorporation, rather than chain termination with dideoxynucleotides. The pyrosequencing usually relies on enzymes ATP sulfurylase and luciferase. Release of pyrophosphate, during nucleotide triphosphate incorporation into the DNA chain, triggers a cascade of biochemical reaction *via* ATP sulfurylase and luciferase, resulting in a burst of biochemiluminescent light being emitted. Apyrase, a nucleotide degrading enzyme, degrades unincorporated nucleotides, and ATP. After degradation another nucleotide is added. Sequencing is achieved by sequentially introducing each of the four dNTPs into the flow cell. Presence or absence of light burst of each picotiter well indicates the incorporation of corresponding nucleotide, therefore, reveals the identity of complementary base on the template DNA in that well (Fig. 2.6). Major advantages of pyrosequencing are its speed and read length-up to 500 bp. Unlike other next-generation technologies discussed, pyrosequencing does not need to carry out extra chemistries to the extending DNA chain beyond normal biochemical process by DNA polymerase, e.g., no need of removing label moiety or de-block terminators.

A limitation of the method is that the lengths of individual reads of DNA sequence are approximately 300–500 nucleotides, shorter than the 800–1,000 obtained with chain termination methods (e.g., Sanger sequencing method). This can make the process more difficult, particularly for sequences which contain large amount of repetitive DNA.

The reactions involved in pyrosequencing are:

$$(NA)_n + \text{Nucleotide} \xrightarrow{\text{Polymerase}} (NA)_{n+1} + \text{PPi}$$

$$\text{PPi} + \text{APS} \xrightarrow{\text{ATP sulfurylase}} \text{ATP} + SO_4^{2-}$$

$$\text{ATP} + \text{Luciferase} + O_2 \xrightarrow{\text{Luciferase}} \text{AMP} + \text{PPI} + \text{Oxyluciferase} + CO_2 + \text{Light}$$

2.5.1.4 Solid Sequencing Chemistry

Like 454, SOLiD system also employs emulsion PCR as a DNA template amplification scheme with paramagnetic beads. After breaking the emulsion, amplified beads are collected, enriched, and fixed on a flat glass substrate to create a disorder array.

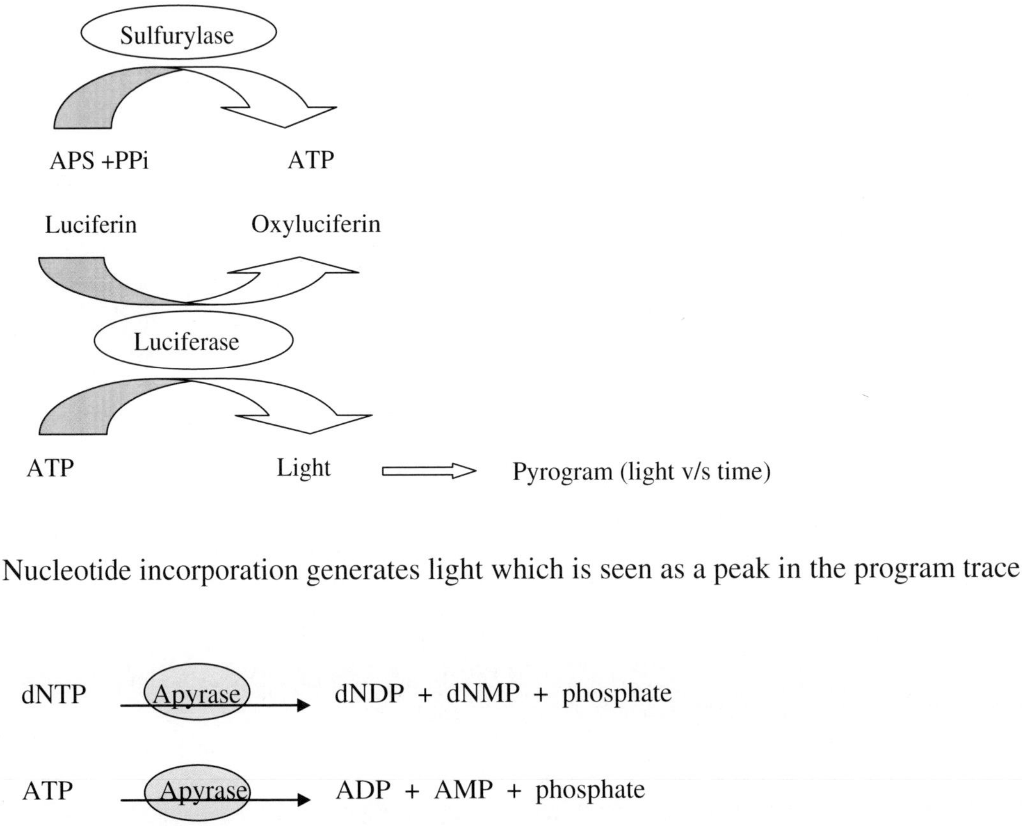

Fig. 2.6 Pyrosequencing

Its sequencing-by-synthesis is driven by ligation rather than polymerization. Primers hybridize to the P1 adapter sequence on the templated beads. A set of four fluorescently labeled di-base probes compete for ligation to the sequencing primer. Specificity of the di-base probe is achieved by interrogating every first and second base in each ligation reaction. Multiple cycles of ligation, detection, and cleavage are performed with the number of cycles determining the eventual read length. Following a series of ligation cycles, the extension product is removed and the template is reset with a primer complementary to the n-1 position for a second round of ligation cycles.

2.5.1.5 Error limitations

The 454 genome sequencer delivers the longest read length with lowest throughput (8 Mb/h during a 9-h run) and leads to errors in homopolymeric tracts, complex sample preparation and is relatively expensive). Ion torrent sequencer has 454-like chemistry without dye labeled nucleotides. It can read up to 400 bp reads (single end). In Ion torrent the output is dependent on chip-type. 318 chip produces about 1 G base in 3 hours. Ion Torrent run produces the shortest reads and performs poorly with homopolymers. However, it delivers the fastest throughput (80–100 Mb/h) with shortest run time ($\sim$3 h). Illumina genome analyzer cannot read missed base, has lowest rate of base substitution errors, difficult to resolve specific motifs (GGC motifs, inverted repeats) and uses longest-running time (Nakamura et al. 2011).

2.5.1.6 Quality and quantity requirements in sequencing

Illumine HiSeq lane typically produces about 150 million paired end reads under current run conditions (Haas et al. 2012). Thus, multiplexing 15–30 samples per lane will yield the 5–10 million reads per sample and will be sufficient for most applications of bacterial RNA-Seq but studies of differential gene expression need even significantly higher levels of multiplexing result. Approximately 12×10^4 transcripts will be covered by 45 million reads, for low abundance transcripts high coverage will be required that depends on whether it is a low or high abundance transcript.

As per ENCODE guidelines, with a reference genome 20 million reads per sample and more than 100 million reads per sample without a reference are required. Bacterial sample needs 2 to 10 million reads per sample.

2.6 Applications of DNA Sequencing

The Genome sequencing has revolutionized the understanding of treatment and prevention of human diseases at affordable price; genome sequencing has got many more significant applications (Drmanac et al. 2010).

1. DNA sequencing plays vital role in the field of agriculture. The mapping and sequencing of the whole genome of microorganisms has allowed the agriculturists to make them useful for the crops and food plants. For example, specific genes of bacteria have been used in some food plants to increase their resistance against insects and pests, as a result the productivity and nutritional value of the plants may increase.
2. In medical research, DNA sequencing can be used to detect the genes which are associated with some hereditary or acquired diseases.
3. In forensic science, DNA sequencing is used to identify the criminals by finding some proof from the criminal scene in the form of hair, nail, skin, or blood samples. DNA sequencing can also be used to determine the paternity of the child.
4. DNA sequencing information is important for planning the procedure and method of gene manipulation.
5. It is used for construction of restriction endonuclease maps.
6. It is used to find tandem repeats or inverted repeat for the possibility of hairpin formations.
7. The sequences can be used to find whether any open reading frame (ORF) coding for a polypeptide exists or not.
8. DNA sequences can be used to find a polypeptide sequence from the data bank or to compare with DNA sequences from other organisms for phylogenetic analysis.
9. DNA sequencing is used to construct the molecular evolution map.
10. Last but not least, it is useful in identifying exons and introns.
11. DNA sequencing discovers intra and inter species variations.
12. It can characterize the transcriptome of a cell.
13. It can identify DNA bindinig sites for proteins.
14. Metagenomic applications: It uses whole sample DNA/RNA with wide range of applications i.e phylogenetics/comparative/functional analysis in addition to environmental profiling.

2.7 Conclusion

DNA sequencing is a procedure for identifying the base sequence of a fragment of DNA. The most commonly used procedure is the Sanger method of DNA sequencing, which uses DNA polymerase I and a PCR reaction to repeatedly synthesize a new DNA strand using a segment of DNA as a template. Strand synthesis terminates when a dideoxynucleotide is incorporated instead of a normal nucleotide. Sanger sequencing can be done manually or by using an automated sequencer. In manual sequencing, four separate sequencing reactions are performed, one for each of the four bases. This creates a number of DNA fragments of different lengths which are electrophoresed on a sequencing gel, fragments differing by a single base in length will separate from each other and the base sequence of the DNA fragment can be determined. An automated sequencer uses a single reaction mixture in which each of the four terminating dideoxynucleoside triphosphates is labeled with a different fluorescent dye. The reaction is loaded in a single lane of a sequencing gel and a laser reads the dye color terminating each fragment band. Recently, a number of new DNA sequencing methods/sequencers have been developed which are fully automated requiring less time and money, e.g., Illumina Genome Analyzer, pyrosquencing, SOLiD Sequencing Chemistry, 454 Genome Sequencer, etc. DNA/genome sequencing has been applied to forensics, medicine, agriculture, and can be used to detect genes associated with heredity or acquired diseases, identification of microorganisms causing diseases, increase productivity, and resistance against pests in plants, etc.

References

Dale JW, Schantz MV (2007) DNA sequencing. In: from genes to genomes concepts and applications of DNA technology, 2nd edn. Wiley Ltd, West Sussex, England

Drmanac R et al (2010) Human genome sequencing using unchained base reads in self-assembling DNA nanoarrays. Science 327(5961):78–81. doi: 10.1126/science.1181498

Edwards JR, Ruparel H, Ju J (2005) Mass-spectrometry DNA sequencing. Mutat Res 573(1–2):3–12. doi: org/10.1016/j.mrfmmm.2004.07.021

Ewing B, Hillier L, Wendl MC, Green P (1998) Base-calling of automated sequencer races using Phred. I. Accuracy assessment. Genome Res 8:175–185. doi: 10.1101/gr.8.3.175

Haas BJ, Chin M, Nusbaum C, Birren BW, Livny J (2012) How deep is deep enough for RNA-Seq profiling of bacterial transcriptomes? BMC Genomics 13: 734

Liu L, Li Y, Li S, Hu N, He Y, Pong R, Lin D, Lu L, Law M (2012) Comparison of Next-Generation Sequencing Systems. J Biomed Biotechnol 2012:1–11. doi: 10.1155/2012/251364

Loman NJ, Misra RV, Dallman TJ, Constantinidou C, Gharbia SE, Wain J, Pallen MJ (2012) Performance comparison of benchtop high-throughput sequencing platforms. Nat Biotechnol 30: 434–439. doi:10.1038/nbt.2198

Maxam AM, Gilbert W (1977) A new method for sequencing DNA. Proc Nat Acad Sci 74(2):560–564. PMCID: PMC392330. doi:10.1073/pnas.74.2.560

Nakamura K, Oshima T, Morimoto T, Ikeda S, Yoshikawa H, Shiwa Y, Ishikawa S, Linak MC, Hirai A, Takahashi H, Altaf-Ul-Amin M, Ogasawara N, Kanaya S (2011) Sequence-specific error profile of Illumina sequencers. Nucleic Acids Res 39(13). doi: 10.1093/nar/gkr34410.1093/nar/gkr344

Pettersson E, Lundeberg J, Ahmadian A (2009) Generations of sequencing technologies. Genomics 93(2):105–11. doi.org/10.1016/j.ygeno.2008.10.003

Porreca JG (2010) Genome sequencing on nanoballs. Nat Biotechnol 28(1):43–44. doi:10.1038/nbt0110-43

Quail MA, Smith M, Coupland P, Otto TD, Harris SR, Connor TR, Bertoni A, Swerdlow HP, Gu Y (2012) A tale of three next generation sequencing platforms: comparison of Ion Torrent, Pacific Biosciences and Illumina MiSeq sequencers. BMC Genomics 13:341

Sanger F, Nicklen S, Coulson AR (1977) DNA sequencing with chain-terminating inhibitors. Proc Nat Acad Sci 74(12):54635467. doi:10.1073/pnas.74.12.5463

Schuster SC (2008) Next-generation sequencing transforms today's biology. Nat Methods 5(1):16–18. doi: 10.1038/nmeth1156

Xiaoguang Z, Lufeng R, Yuntao L, Meng Z, Yude Y, Jun Y (2010) The next-generation sequencing technology: a technology review and future perspective. Sci China Life Sci 53(1):44–57. doi:10.1007/s11427-010-0023-6

3 Molecular Markers

Pavan Kumar Agrawal and Rahul Shrivastava

Abstract

Detection and analysis of genetic variation help in understanding the molecular basis of various biological phenomena in eukaryotes. Since the entire eukaryotes cannot be covered under sequencing projects, molecular markers and their correlation to phenotypes provide with requisite landmarks for elucidation of genetic variation. There are different types of DNA-based molecular markers. These DNA-based markers are differentiated into two types; first nonPCR-based (RFLP) and second is PCR-based markers (RAPD, AFLP, SSR, SNP etc.). Amongst others, the microsatellite DNA marker has been the most widely used in ecological, evolutionary, taxonomical, phylogenetic, and genetic studies due to its easy use by simple PCR, followed by a denaturing gel electrophoresis for allele size determination and high degree of information provided by its large number of alleles per locus. Despite this, a new marker type, named SNP, for Single Nucleotide Polymorphism, is now on the scene and has gained high popularity, even though it is only a bi-allelic type of marker. Day by day development of such new and specific types of markers makes their importance in understanding the genomic variability and the diversity between the same as well as different species of the plants. In this chapter, we have discussed types of molecular markers, their advantages, disadvantages, and the applications.

3.1 Introduction

The concept of genetic markers is not a new one; Gregor Mendel used phenotype-based genetic markers in his experiment in the nineteenth century. Later, phenotype-based genetic markers for *Drosophila* led to the establishment of the theory of genetic linkage. The limitations of phenotype-based genetic markers directed the development of more general and useful direct

P. K. Agrawal
Govind Ballabh Pant Engineering College, Ghurdauri, Pauri, Uttrakhand 246194, India
e-mail: p_k_agarwal@rediffmail.com

R. Shrivastava (✉)
Department of Biotechnology and Bioinformatics, Jaypee University of Information Technology, Waknaghat, Solan, Himachal Pradesh, India
e-mail: rahulmicro@gmail.com

I. Ravi et al. (eds.), *Advances in Biotechnology*,
DOI: 10.1007/978-81-322-1554-7_3, © Springer India 2014

DNA-based markers that became known as molecular markers. A molecular marker is defined as a particular segment of DNA that is representative of the differences at the genome level. They may or may not correlate with phenotypic expression of a trait. Molecular markers offer numerous advantages over conventional phenotype-based alternatives as they are stable and detectable in all tissues regardless of growth, differentiation, development, or defense status of the cell. Further, these are not confounded by the environment, pleiotropic and epistatic effects.

Due to the rapid developments in the field of molecular genetics, a variety of techniques have emerged to analyze genetic variation during the last few decades. These genetic markers may differ with respect to important features, such as genomic abundance, level of polymorphism detected, locus specificity, reproducibility, technical requirements, and financial investment. No marker is superior to all others for a wide range of applications. The most appropriate genetic marker depends on the specific application, the presumed level of polymorphism, the presence of sufficient technical facilities and knowhow, time constraints, and financial limitations. The classifications of molecular marker technologies that have been widely applied during the last decades are summarized in Table 3.1.

Table 3.1 Some popular molecular markers, their characteristics and potential applications

Techniques	Marker type	Require prior molecular information	Mode of inheritance	Degree of polymorphism	Major applications	Discoverer
Non PCR-based techniques	Restriction fragment length polymorphism	Yes	Mendelian, codominant	Low	Linkage mapping, Physical mapping	Bostein et al. (1980)
PCR-based techniques	Random amplified polymorphic DNA	No	Mendelian, dominant	Intermediate	Fingerprinting for population study, Gene tagging, Hybrid identification	Williams et al. (1990)
	Amplified fragment length polymorphism	No	Mendelian, dominant	High	Linkage mapping, Gene tagging, population study	Vos et al. (1995)
	Microsatellite, SSR	Yes	Mendelian, codominant	High	Linkage mapping, population study, Genetic diversity, Paternity analysis	Litt and Lutty (1989)
	Minisatellite, VNTR	No	Mendelian, codominant	High	DNA fingerprinting for population study	Jeffrey (1985)
	Single nucleotide polymorphism	Yes	Mendelian, codominant	High	Linkage mapping, population study	Ching et al. (2002)

An ideal molecular marker technique should (1) be polymorphic and evenly distributed throughout the genome, (2) provide adequate resolution of genetic differences, (3) generate multiple, independent and reliable markers, (4) be simple, quick, and inexpensive, (5) need small amounts of tissue and DNA samples (6) have linkage to distinct phenotypes and (7) require no prior information about the genome of an organism.

3.2 Molecular Marker Techniques

A vast array of DNA-based genetic markers for the detection of DNA polymorphism have been discovered since 1980 and new marker types are developed every year. There are several DNA marker types, which may be classified broadly into two groups based on their detection method, (1) **Non PCR based** or those based on DNA–DNA hybridization and (2) **PCR based** or those based on amplification of DNA sequences using the polymerase chain reaction (PCR) (Table 3.1).

3.2.1 Non PCR Based or DNA–DNA Hybridization

3.2.1.1 Restriction Fragment Length Polymorphism

Restriction fragment length polymorphism markers were the first DNA-based genetic markers developed (Botstein et al. 1980). Eukaryotic genomes are very large and there was no simple way to observe genetic polymorphisms of individual genes or sequences. The property of complementary base pairing where small piece of DNA could be used as probe allowed for methods to be developed to reveal polymorphisms in sequences homologous to the probe. The genetic system derived using this approach is called restriction fragment length polymorphism. RFLP (commonly pronounced "rif-lip") refers to a difference between two or more samples of homologous DNA molecules arising from differing locations of restriction sites. If two organisms differ in the distance between sites of cleavage of particular *Restriction endonucleases*, the length of the fragments produced will differ when the DNA is digested with a restriction enzyme. The dissimilarity of the patterns generated can be used to differentiate species (and even strains) from one another. In RFLP analysis, the DNA sample broken into pieces (digested) by restriction enzymes that are separated according to their lengths on agarose gel electrophoresis. It is possible to visualize DNA within such a gel by staining it with ethidium bromide, however, due to typically so many restriction fragments of all possible sizes, discrete fragments cannot be seen. To overcome this problem, the fractionated DNA is transferred and chemically bound to a nylon membrane by a process called Southern blotting, named after its inventor E. M. Southern (1975). Specific DNA fragments are visualized by hybridizing the DNA fragments bound to the nylon membrane with a radioactively or fluorescently labeled DNA probe. Different sizes or lengths of restriction fragments are typically produced when different individuals are tested. Such a polymorphism can be used to distinguish plant species, genotypes and, in some cases, individual plants. Labeling of the probe may be performed with a radioactive isotope or with alternative non-radioactive stains, such as digoxigenin or fluorescein. These probes are mostly species-specific single locus probes of about 0.5–3.0 kb in size, obtained from a cDNA library or a genomic library. Though genomic library probes may exhibit greater variability than probes from cDNA libraries, a few studies reveal the converse (Miller and Tanksley 1990). In addition to genetic fingerprinting, RFLP is an important tool in genome mapping, localization of genes for genetic disorders, and determination of risk for disease and paternity testing.

Advantages: The RFLP markers are relatively highly polymorphic, codominantly inherited, and highly reproducible. Because of their presence throughout the eukaryotic genome, high heritability and locus specificity of the RFLP markers are considered superior. The method also provides opportunity to simultaneously screen numerous samples. DNA blots can be analyzed repeatedly by stripping and reprobing (usually 8 to 10 times) with different RFLP probes.

Disadvantages: The technique is not very widely used because it is time consuming, expensive, and only one out of several markers may be polymorphic, which is highly inconvenient especially for crosses between closely related species. The utility of RFLPs has been hampered due to the large quantities of purified, high molecular weight genomic DNA requirement for DNA digestion as well as Southern blotting. The use of radioactive isotope makes the analysis relatively expensive and hazardous. The requirement of prior sequence information for probe generation increases the complexity of the methodology. Their inability to detect single base changes restricts their use in detecting point mutations occurring within the regions at which they are detecting polymorphism.

Applications: Restriction Fragment length polymorphism is a powerful tool for the identification of organisms to the level of species, strains, varieties, or individuals. RFLPs have been widely used in population genetics to determine slight differences within populations, and in linkage studies to generate maps for quantitative trait locus (QTL) analysis. They also have been used to investigate relationships of closely related taxa or intraspecific level, as fingerprinting tools for diversity studies, and for studies of hybridization. They have been used in criminal and paternity tests, localization of genes for genetic disorders, and determination of risk for disease.

3.2.2 PCR-Based Marker

After the invention of polymerase chain reaction (PCR) technology (Mullis and Faloona 1987), a large number of approaches for generation of molecular markers based on PCR were detailed, primarily due to its apparent simplicity and high probability of success. Before PCR, the analysis of a specific DNA fragment generally required cloning of the fragment and amplification in a plasmid or compatible vector. PCR enables the production of a large amount of specific DNA sequence without cloning, starting with just a few molecules of the target sequence. Usage of random primers overcame the limitation of prior sequence knowledge for PCR analysis and facilitated the development of genetic markers for a variety of purposes.

3.2.2.1 Amplified Fragment Length Polymorphism

Amplified fragment length polymorphism PCR (or AFLP-PCR or just AFLP) is a PCR-based tool used in genetics research, DNA fingerprinting and in the practice of genetic engineering. Developed by Vos et al. (1995), AFLP is essentially an intermediate between RFLP and PCR. AFLP selectively amplifies a subset of restriction fragments from a complex mixture of DNA fragments obtained after digestion of genomic DNA with restriction endonucleases. Polymorphisms are detected from differences in the length of the amplified fragments by polyacrylamide gel electrophoresis (PAGE) or by capillary electrophoresis. AFLP uses restriction enzymes to cut genomic DNA, followed by ligation of adaptors to the sticky ends of the restriction fragments. A subset of the restriction fragments are then amplified using primers complementary to the adaptor and part of the restriction site fragments. The amplified fragments are visualized on denaturing polyacrylamide gels either through autoradiography or fluorescence methodologies (Vos et al. 1995). AFLPs are DNA fragments (80–500 bp) obtained from digestion with restriction enzymes, followed by ligation of oligonucleotide adapters to the digestion products and selective amplification by the PCR. AFLPs, therefore, involve both RFLP and PCR. The AFLP banding profiles are the result of variations in the restriction sites or in the intervening region. The AFLP technique simultaneously generates fragments from many genomic sites (usually 50–100 fragments per reaction) that are separated by polyacrylamide gel electrophoresis and that are generally scored as dominant markers.

Advantages: There are many advantages of AFLP, they are produced in greater amount and have higher reproducibility, resolution, and sensitivity at the whole genome level and also has the capability to amplify between 50 and 100 fragments at one time. In addition, no prior sequence information is needed for amplification. As a result, AFLP has become extremely beneficial in the study of taxa including bacteria, fungi, and plants, where much is still unknown about the genomic makeup of various organisms. AFLPs can be analyzed on automatic sequencers, but software problems concerning the scoring of AFLPs are encountered on some systems. The use of AFLP in genetic marker technologies has become the main tool due to its capability to disclose a high number of polymorphic markers by single reaction (Vos et al. 1995).

Disadvantages: Disadvantages of this technique are that alleles are not easily recognized, have medium reproducibility, labor intensive, and have high operational and developmental costs. AFLP needs purified, high molecular weight DNA, the dominance of alleles and the possible non-homology of co-migrating fragments belonging to different loci.

Applications: Most AFLP fragments correspond to unique positions on the genome and have the capability to detect various polymorphisms in different genomic regions simultaneously and hence can be exploited as landmarks in genetic and physical mapping. The technique can be used to distinguish closely related individuals at the subspecies level (Althoff et al. 2007) and also for gene mapping. Applications for AFLP in plant mapping include establishing linkage groups in crosses, saturating regions with markers for gene landing efforts (Yin et al. 1999), and assessing the degree of relatedness or variability among cultivars (Mian et al. 2002). Molecular markers are more reliable for genetic studies than morphological characteristics, because the environment does not affect them. AFLP is considered more applicable to intraspecific than to interspecific studies due to frequent null alleles. AFLP markers are useful in genetic studies, such as biodiversity evaluation, analysis of germplasm collections, genotyping of individuals, and genetic distance analysis. The availability of many different restriction enzymes and corresponding primer combinations provides a great deal of flexibility, enabling the direct manipulation of AFLP fragment generation for defined applications (e.g., polymorphism screening, QTL analysis, and genetic mapping).

3.2.2.2 Random Amplified Polymorphic DNA

The basis of randomly amplified polymorphic DNA technique is differential PCR amplification of genomic DNA. It deduces DNA polymorphisms produced by "rearrangements or deletions at or between oligonucleotide primer binding sites in the genome" using short random oligonucleotide sequences (mostly 10 bases long) (Williams et al. 1990). As the approach requires no prior knowledge of the genome that is being analyzed, it can be employed across species using universal primers. In the RAPD marker system, a PCR reaction is conducted using a very small amount of template DNA (even less than 10 ng is sufficient) and a single RAPD primer. Primers are usually just 10 base pairs long (10 mers) and are of random sequence. The analysis of RAPD is based on the PCR using short (about 10 bases) randomly chosen primers singly which anneal as reverted repeats to the complementary sites in the genome. The DNA between the two opposite sites with the primers as starting and end points is amplified by PCR. The amplification products are separated on agarose gels in the presence of ethidium bromide and viewed under ultraviolet light. The banding patterns distinguish organisms according to the presence or absence of bands (polymorphism). These polymorphisms are considered to be primarily due to variation in the primer annealing sites, but they can also be generated by length differences in the amplified sequence between primer annealing sites. Each product is derived from a region of the genome that contains two short segments in inverted orientation, on opposite strands that are complementary to the primer. It is a peculiarity of RAPD analysis that it discriminates at different taxonomical level, viz., isolates and species, depending on the organism investigated and the primer used

Advantages: The main advantage of RAPDs is that they are quick and easy to assay. Because PCR is involved, only low quantities of template DNA are required, usually 5–50 ng per reaction. Since random primers are commercially available, no sequence data for primer construction are needed. Moreover, RAPDs have a very high genomic abundance and are randomly distributed throughout the genome. They are dominant markers; hence have limitations in their use as markers for mapping, which can be overcome to some extent by selecting those markers that are linked in coupling (Williams et al. 1993). RAPD assay has been used by several groups as efficient tools for identification of markers linked to agronomically important traits, which are introgressed during the development of near isogenic lines. The RAPD analysis of NILs (nonisogenic lines) has been successful in identifying markers linked to disease resistance genes in tomato (*Lycopersicon* sp.), lettuce (*Lactuca* sp.), and common bean (*Phaseolus vulgaris*). Due to the speed and efficiency of RAPD analysis, high-density genetic mapping in many plant species was developed in a relatively short time.

Disadvantages: The major drawback of the method is that the profiling is dependent on the reaction conditions so it may vary within two different laboratories and as several discrete loci in the genome are amplified by each primer, profiles are not able to distinguish heterozygous from homozygous individuals (Bardakci 2001).

Applications: The application of RAPDs and their related modified markers in variability analysis and individual-specific genotyping has largely been carried out, but is less popular due to problems such as poor reproducibility, faint or fuzzy products, and difficulty in scoring bands, which lead to inappropriate inferences. RAPDs have been used for many purposes, ranging from studies at the individual level (e.g., genetic identity) to studies involving closely related species and determination of genetic diversity. RAPDs have also been applied in gene mapping studies to fill gaps not covered by other markers.

3.2.2.3 Sequence Characterized Amplified Region

Sequence characterized amplified regions (SCARs) markers were developed by Michelmore et al. (1991) and Martin et al. (1991). SCARs are DNA fragments amplified by the use of PCR using specific 20–30 bp primers. The primers are designed from terminal ends of a RAPD marker. RAPD marker fragments associated with a phenotypic condition of interest are cloned and sequenced. This nucleotide sequence is then used for designing of unique primers for specific amplification of a particular locus. Use of unique primers decreases site competition for specific regions among primers, making the results less sensitive to reaction conditions. Use of longer primers also increases the reproducibility of the results by increasing specificity of template binding by the primer. PCR products obtained after amplification are analyzed for the presence of length polymorphism by gel electrophoresis. Conversion of RAPDs into SCARs may have additional advantage of obtaining a codominant marker, although dominance may be exhibited by SCARs when one or both primers partially overlap the site of sequence variation.

Advantages: Due to the application of PCR, low amounts of genomic or template DNA is required (10–50 ng per reaction) for analysis of the marker. The major advantage of SCARs is the ease of their use, and requirement of relatively less time for the analysis. SCAR markers are locus specific and the results obtained from analysis of SCARs are highly reproducible.

Disadvantages: Major limitation lies in the fact that analysis of SCARs can be done only for organism, species, or DNA fragment with known sequence; as knowledge of sequence data is required for designing of the PCR primers.

Applications: Major application of SCAR makers has been found in gene mapping studies as they are locus specific. They are also used in marker assisted selection, where presence of a specific genotype and expression of its corresponding phenotype can be correlated to the presence or absence of a SCAR marker (Paran and Michelmore 1993).

3.2.2.4 Minisatellites, Variable Number of Tandem Repeats

The term 'minisatellite' was introduced by Jeffrey et al. (1985). These loci contain tandem repeats that vary in the number of repeat units between genotypes and are referred to as variable number of tandem repeats (VNTRs) (i.e., a single locus that contains variable number of tandem repeats between individuals) or hypervariable regions (HVRs) (i.e., numerous loci containing tandem repeats within a genome generating high levels of polymorphism between individuals). Minisatellites are a conceptually very different class of marker. They consist of chromosomal regions containing tandem repeat units of a 10–50 base motif, flanked by conserved DNA restriction sites. A minisatellite profile consisting of many bands, usually within a 4–20 kb size range, is generated by using common multi locus probes that are able to hybridize to minisatellite sequences in different species. Locus specific probes can be developed by molecular cloning of DNA restriction fragments, subsequent screening with a multi locus minisatellite probe and isolation of specific fragments. Variation in the number of repeat units, due to unequal crossing over or gene conversion, is considered to be the main cause of length polymorphisms. Due to the high mutation rate of minisatellites, the level of polymorphism is substantial, generally resulting in unique multilocus profiles for different individuals within a population.

Advantages: The main advantages of minisatellites are their high level of polymorphism and high reproducibility.

Disadvantages: Disadvantages of minisatellites are similar to RFLPs due to the high similarity in methodological procedures. If multilocus probes are used, highly informative profiles are generally observed due to the generation of many informative bands per reaction. In that case, band profiles cannot be interpreted in terms of loci and alleles and similar sized fragments may be nonhomologous. In addition, the random distribution of minisatellites across the genome has been questioned (Schlötterer 2004).

Applications: The term DNA fingerprinting was introduced for minisatellites, though DNA fingerprinting is now used in a more general way to refer to a DNA-based assay to uniquely identify individuals. Minisatellites are particularly useful in studies involving genetic identity, parentage, clonal growth and structure, and identification of varieties and cultivars and for population-level studies. Minisatellites are of reduced value for taxonomic studies because of hyper variability.

3.2.2.5 Simple Sequence Repeat or Microsatellites

The simple sequence repeats (SSR) are also referred to as microsatellites. 'Microsatellites', coined by Litt and Lutty (1989), consist of tandemly repeating units of DNA ubiquitous in prokaryotes and eukaryotes, scattered throughout most eukaryotic genomes (Powell et al. 1996). Microsatellites can comprise repetition of mono-, di-, tri-, tetra- or penta-nucleotide units. The variation in number of repeat units of a microsatellite resulting in length polymorphisms is mainly attributed to Polymerase slippage during DNA replication, or slipped strand mispairing. Function of such microsatellites is largely unknown, though they can occur in protein-coding as well as noncoding regions of the genome. Microsatellite sequences are especially exploited to differentiate closely related species or genotypes. Due to the high degree of variability present in microsatellites, they are preferred in population studies (Smith and Devey 1994) and to distinguish closely related plant cultivars.

Simple sequence repeat polymorphism is detected by the use of PCR, in conditions when nucleotide sequences of the flanking regions of the microsatellite are known, or by Southern hybridization using labeled probes. Unlike minisatellites, which also represent tandem repeats, microsatellites consist of short tandem repeat motifs of 1–6 base pair (bp) long units. Knowledge of nucleotide sequence is used to design specific primers (generally 20–25 bp) for amplification of the microsatellite region along with flanking sequences by PCR. Such amplified amplicon is then identified by construction of a small-insert genomic library, followed by screening the library with a synthetically labeled oligonucleotide probe. Positive clones thus obtained are sequenced for detection and confirmation of polymorphism.

Microsatellite polymorphism of organism whose genome is not sequenced may also be identified by screening sequence databases of closely related species for microsatellite sequence motifs. Primers designed from flanking regions after such comparative study are then used to amplify the SSR region followed by analysis.

Advantages: High number of alleles and their co-dominance, along with widespread distribution throughout the genome, are the most significant features of microsatellites which make them important molecular markers. The SSRs are present in high genomic abundance in eukaryotic organisms and are mostly localized in low-copy regions (Morgante et al. 2002). As the analysis of SSRs is PCR based, low quantities of template DNA (10–50 ng per reaction) are required. In comparison to other markers, such as RAPD, longer primers can be designed and used for microsatellites, thus, increasing reproducibility of the results obtained. The analysis of microsatellite can be semi-automated and does not necessitate the use of radioactive isotopes.

Microsatellite analysis allows identification of multiple alleles at a single locus. Although it is done mostly for single locus on the genome, in conditions where the size ranges of the alleles on different loci are markedly different, gel electrophoresis analysis or multiplex PCR can easily be used to simultaneously study multiple microsatellites (Ghislain et al. 2004). This would help in decreasing the analytical cost significantly.

Individual microsatellite loci can be converted into PCR-based markers for analysis as monolocus, codominant SSRs. In conditions where sequence of the microsatellite region is unknown, primers are designed from genomic regions adjacent to the SSR sequences. The amplified genomic fragment thus obtained is cloned and sequenced to obtain specific sequence of the microsatellite region. Primers are designed using this microsatellite sequence for specific amplification using PCR. This approach is termed as sequence-tagged microsatellite site (STMS).

Disadvantages: Principal drawback of microsatellites lies in their high development costs. Sequence information of the template or genomic DNA is required for designing and synthesis of primers. Thus, study or analysis of microsatellites is extremely difficult to apply for any unstudied group of species or organism of interest. Although SSRs are codominant in nature, mutations in the primer binding sites may result in false-negative result because of nonamplification of the intended PCR product (occurrence of null alleles), which may lead to errors in genotype analysis.

Such null alleles may lead to underestimation of heterozygosity and biased estimate of allelic and genotypic frequencies. Similarly, underestimation of genetic divergence may occur due to homoplasy at microsatellite loci, which is a result of different forward and backward mutations.

Interpretation of bands obtained on electrophoretic gel, after PCR amplification of microsatellite markers may be difficult sometimes. Appearances of artifacts in form of stutter bands have been seen to influence proper visualization and size determination of the fragments, thus interpretation of bands. Formation of such bands occurs due to DNA slippage during PCR amplification. Appropriate reference genotypes of known DNA fragments of specific band sizes can be run along with test samples to overcome the complications in interpretation of such results.

Applications: Microsatellites have high degree of mutability, hence are thought to play a significant role in genome evolution by creation and maintenance of quantitative genetic variation. Thus, high level of polymorphism shown by microsatellites makes them informative markers for population genetics studies, ranging from the individual level (e.g., strain identification) to that of closely related species level. Conversely, due to their high degree of mutability, microsatellites markers are not suitable for analysis and correlation studies of higher taxonomic groups.

SSR markers have been useful for genetic variation studies in germplasm collections and are considered to be ideal molecular markers for gene mapping studies (Hearne et al. 1992; Jarne and Lagoda 1996). SSRs are molecular markers used for recombination mapping studies, to verify parental relationships and population genetic studies. They are the only molecular marker to provide information about presence or absence of closely related alleles. Increase or decrease in the number of SSR repeats in genes of known function can be correlated with changes in biological functions or phenotypic variation in the organism (Ayers et al. 1997). SSRs have been shown to be useful for study of functional diversity among closely related species, in relation to adaptive changes (Eujay et al. 2001).

3.2.2.6 Inter-Simple Sequence Repeat

AFLP involves high cost of analysis, while RAPD is low in reproducibility; analysis of SSR markers also requires prior knowledge of DNA sequence of flanking regions for designing of species-specific primers. Thus, analysis of polymorphism by application of any of these techniques has its limitations, which pose major obstacle in regular use of these markers. These limitations can be overcome to some extent by the use of ISSR-PCR technique.

SSRs or microsatellites are short segments of DNA (1–4 base pairs) universally present in eukaryotic genomes. The number of repeats of SSRs varies in different organisms of same or different species amounting to polymorphism. ISSR is a PCR-based technique, in which SSRs are used as primers to primarily amplify the inter-SSR regions. The method involves PCR amplification of DNA fragments present at a suitable amplifiable distance in between two identical microsatellite regions. The ISSRs are regions between adjacent, oppositely oriented microsatellite sequences.

Primers are designed by using microsatellite repeat regions. As the repeat sequences flanking an inter-SSR region are same, a single primer of 17–25 bp length is generally used both as forward and reverse primers. The primers used for amplification of ISSRs may consist of di-, tri-, tetra-, or penta-nucleotide repeat units.

The designed primers are usually anchored at 3′ or 5′ end with 1–4 bases extended into the nonrepeat adjacent regions of SSRs (Zietkiewicz et al. 1994). Unanchored primers can also be used for amplification of ISSRs.

Using the primer, about 10–50 inter- SSR sequences of different sizes (about 200–3,000 bp) are amplified simultaneously from multiple genomic loci by a single PCR reaction. These amplicons are separated by gel electrophoresis and analyzed for the presence or absence of DNA fragments of particular length.

Some methods related to analysis of ISSRs are:

1. Single primer amplification reaction (SPAR)—that uses a single primer designed only from the core motif of microsatellite region (unanchored primer), and
2. Directed amplification of minisatellite region DNA (DAMD)—that employs a single primer designed only from the core motif of a minisatellite.

Advantages: Principle advantage of ISSRs is that no sequence data of the fragment to be amplified is required for designing of the primers, as primers are designed from the repeat units of flanking microsatellite region. The analysis requires very small quantities of template DNA (10–50 ng per reaction). ISSRs are mostly exhibited as dominant marker, though it might also be present in codominant state. In comparison to RAPD, which is also a PCR-based technique using a single primer, use of longer primers and corresponding higher annealing temperature (45–60 °C) leads to higher stringency of PCR amplification. Thus, better reproducibility in results is obtained from analysis of ISSRs as molecular markers.

Disadvantages: As ISSR is a multilocus technique, fragments with the same size and mobility originating from nonhomologous regions, may lead to misinterpretation of genetic similarity estimates. Further, under certain conditions ISSRs can also show low reproducibility of results similar to RAPDs.

Applications: ISSR analysis leads to multilocus fingerprinting profiles, thus it is useful in areas of study of genetic diversity, construction and study of phylogenetic tree, and evolutionary biology studies in a wide range of species. ISSR analysis can be applied in studies involving parentage establishment, clone and strain identification, and taxonomic studies of closely related species. ISSRs are also considered useful in genome mapping studies. Though ISSRs are present mostly as dominant markers, in some cases they may also segregate as codominant markers enabling distinction between homozygotes and heterozygotes.

3.2.2.7 Cleaved Amplified Polymorphic Sequence

Similar to RFLP, CAPS is based on the principle that any change or mutation in the genome of an individual could lead to creation or deletion of restriction sites. CAPS markers are generally developed by sequencing of RFLP probes used for hybridization. Moreover, 20–25 bp specific primers are designed for amplification of 800–2,000 bp DNA fragments. To increase the chance of finding a region containing polymorphism introns or 3′ untranslated regions are generally used for the amplification. Sequences of DNA from target genotypes are amplified using PCR and the resulting amplicons are digested individually with one or more restriction enzymes. The digested products after gel electrophoresis show readily distinguishable pattern. Thus, length polymorphisms resulting from variation in the occurrence of restriction sites are identified. CAPS have also been referred to as PCR-Restriction fragment length polymorphism (PCR–RFLP).

Advantages: Being a PCR-based technique, analysis of CAPS marker requires very low quantities of template DNA (50–100 ng per reaction). Compared to RFLPs, analysis of CAPS genotypes does not require the technically demanding and time taking steps of Southern hybridization and use of radioactive isotopes. CAPS markers are codominant in nature and locus specific. Most CAPS markers are easily scored and interpreted. Results obtained are highly reproducible.

Disadvantages: Sequence data of the genome to be analyzed are required for synthesis of primers. A limited size of DNA fragment can be amplified by PCR reaction, which would be used for restriction digestion. Hence, in comparison to RFLP analysis, specific regions with CAPS polymorphisms are more difficult to locate.

Applications: CAPS markers have been applied predominantly in gene mapping studies, also helpful in identification of mutation in the genome of an organism.

3.2.2.8 Single Nucleotide Polymorphism

A novel class of DNA markers namely single nucleotide polymorphism in genome (SNPs) has recently become highly applicable in genomic studies. SNP describes polymorphisms caused by point mutations that give rise to different alleles containing alternative bases at a given nucleotide position within a locus. Such sequence differences due to base substitutions have been well characterized since the beginning of DNA sequencing in 1977, but the ability to genotype SNPs rapidly in large numbers of samples was not possible until the application of gene chip technology in the late 1990s. SNPs are again becoming a focal point in molecular marker development since they are the most abundant polymorphism in any organism, adaptable to automation, and reveal hidden polymorphism not detected with other markers and methods. Theoretically, a SNP within a locus can produce as many as four alleles, each containing one of four bases at the SNP site: A, T, C, and G. Practically, however, most SNPs are usually restricted to one of two alleles (most often either the two pyrimidines C/T or the two purines A/G) and have been regarded as bi-allelic.

Analytical procedures require sequence information for the design of allele specific PCR primers or oligonucleotide probes. SNPs and flanking sequences can be found by library construction and sequencing or through the screening of readily available sequence databases. Once the location of SNPs is identified and appropriate primers designed, one of the advantages they offer is the possibility of high-throughput automation. To achieve high sample throughput, multiplex PCR and hybridization to oligonucleotide microarrays or analysis on automated sequencers are often used to interrogate the presence of SNPs. SNP analysis may be useful for cultivar discrimination in crops where it is difficult to find polymorphisms, such as in the cultivated tomato. SNPs may also be used to saturate linkage maps in order to locate relevant traits in the genome. For instance, in *Arabidopsis thaliana* a high density linkage map for easy to score DNA markers was lacking until SNPs became available (Cho et al. 2000; Ching et al. 2002). To date, SNP markers are not yet routinely applied in gene banks, in particular because of the high costs involved.

SNP markers are inherited as codominant markers. Several approaches have been used for SNP discovery including single-strand conformational polymorphism assays analysis, heteroduplex analysis, and direct DNA sequencing. DNA sequencing has been the most accurate and most used approach for SNP discovery. Random shotgun sequencing, amplicon sequencing using PCR, and comparative EST analysis are among the most popular sequencing methods for SNP discovery.

Advantages: Despite technological advances, SNP genotyping is still a challenging endeavor and requires specialized equipment. Traditional methods available for SNP genotyping include: direct sequencing, single base sequencing, allele specific oligonucleotide, denaturing gradient gel electrophoresis (DGGE), SSCP, and ligation chain reaction (LCR). Each approach has its advantages and limitations, but all are still useful for SNP genotyping, especially in small laboratories limited by budget and labor constraints. Large-scale analysis of SNP markers, however, depends on the availability of expensive, cutting-edge equipment. Several options are available for efficient genotyping using state-of-the-art equipment.

Particularly, popular are methods involving matrix-assisted laser desorption ionization time of flight (MALDI-TOF) mass spectrometry, pyrosequencing, Taqman allelic discrimination, real-time (quantitative) PCR, and the use of microarray or gene chips. Mass spectrometry and microarray technologies require a large investment in equipment.

Applications: In plants, SNPs are rapidly replacing simple sequence repeats as the DNA marker of choice for applications in plant breeding and genetics because they are more abundant, stable, amenable to automation, efficient, and increasingly cost-effective (Duran et al. 2009). Generally, SNPs are the most abundant form of genetic variation in eukaryotic genomes. Moreover, they occur in both coding and noncoding regions of nuclear and plastid DNA. As in the case of human genome, SNP-based resources are being developed and made publicly available for broad application in rice research. These resources include large SNP datasets, tools for identifying informative SNPs for targeted applications, and a suite of custom-designed SNP assays for use in marker assisted and genomic selection. SNPs are widely used in breeding programs for several applications such as, (1) marker assisted and genomic selection, (2) association and QTL mapping, positional cloning, (3) haplotype and pedigree analysis, (4) seed purity testing, (5) variety identification, and (6) monitoring the combinations of alleles that perform well in target environments (Kim et al. 2010).

The advantage and disadvantage of common molecular marker techniques are summarized in Table 3.2.

Table 3.2 Advantages and disadvantage of commonly used genetic markers

Marker	Advantages	Disadvantages
Amplified fragment length polymorphism	Sequence information not required	Very tricky due to changes in patterns with respect to materials used
	Can be used across species	
	Work with smaller RFLP fragments	Low reproducibility
	Useful in preparing contig maps	Need to have very good primers
	High genomic abundance	
	High polymorphism	
Restriction fragment length polymorphism	Sequence information not required	Large amount of good quality DNA required
	High genomic abundance	
	Codominant markers	
	High reproducibility	Labour intensive (in comparison to RAPD)
	Can use filters many times	
	Good genome coverage	Difficult to automate
	Can be used across species	Need radioactive labeling
	Needed for map-based cloning	Cloning and characterization of probe are required
Randomly amplified polymorphic DNA	Sequence information not required	No probe or primer information
		Dominant markers
		Very low reproducibility
	Good genome coverage	Cannot be used across species
	Ideal for automation	Not well established
	Less amount and poor quality DNA acceptable	
	No radioactive labeling	
	Relatively faster	
	High genomic abundance	

3.3 Array-Based Platforms

Several different types of molecular markers have been developed over the past three decades (Gupta and Rustgi 2004), motivated by requirements for increased throughput, decreased cost per data point, and greater map resolution. Recently, oligonucleotide-based gene expression microarrays have been used to identify DNA sequence polymorphisms using genomic DNA as the target.

3.3.1 Diversity Arrays Technology

Diversity arrays technology (DArT) is microarray hybridization-based technique that permits simultaneous screening of thousands of polymorphic loci without any prior sequence information. The DArT methodology offers a high multiplexing level, being able to simultaneously type several thousand loci per assay, while being independent of sequence information. DArT assays generate whole genome fingerprints by scoring the presence versus absence of DNA fragments in genomic representations generated from genomic DNA samples through the process of complexity reduction. DArT has been developed as a hybridization-based alternative to the majority of gel-based marker technologies currently in use. It can provide hundreds to tens of thousands of highly reliable markers for any species as it does not require any precise information about the genome sequence

(Jaccoud et al. 2001). Moreover, DArT was found to provide good genome coverage in wheat and barley (Akbari et al. 2006). An important step of this technology is a step called "genome complexity reduction" which increases genomic representation by reducing repetitive sequences that are abundant in eukaryotes. With DArT platform, comprehensive genome profiles are becoming affordable for virtually any crop, genome profiles which can be used in management of bio-diversity, for example in germplasm collections. DArT genome profiles enable breeders to map QTL in 1 week.

DArT profiles accelerate the introgression of a selected genomic region into an elite genetic background (for example, by marker assisted backcrossing). In addition, DArT profiles can be used to guide the assembly of many different regions into improved varieties (marker assisted breeding). The number of markers DArT detects is determined primarily by the level of DNA sequence variation in the material subjected to analysis and by the complexity reduction method deployed. Another advantage of DArT markers is that their sequence is easily accessible compared to amplified fragment length polymorphisms making DArT a method of choice for non-model species (James et al. 2008). DArT has also been applied to a number of animal species and microorganisms.

3.3.2 Restriction Site-Associated DNA

Another high-throughput method is restriction site-associated DNA (RAD) procedure which involves digestion of DNA with a particular restriction enzyme, ligating biotinylated adapters to the overhangs, randomly shearing the DNA into fragments much smaller than the average distance between restriction sites, and isolating the biotinylated fragments using streptavidin beads (Miller et al. 2007a). RAD specifically isolates DNA tags directly flanking the restriction sites of a particular restriction enzyme throughout the genome. More recently, the RAD tag isolation procedure has been modified for use with high-throughput sequencing on the Illumina platform. In addition, Miller et al. (2007b) demonstrated that RAD markers, using microarray platform, allowed high-throughput, high-resolution genotyping in both model and non-model systems.

3.3.3 Single Feature Polymorphism

A third high-throughput method is single feature polymorphism (SFP) which is done by labeling genomic DNA (target) and hybridizing to arrayed oligonucleotide probes that are complementary to insert-delete (INDEL) loci. The SFPs can be discovered through sequence alignments or by hybridization of genomic DNA with whole genome microarrays. Each SFP is scored by the presence or absence of a hybridization signal with its corresponding oligonucleotide probe on the array. Both spotted oligonucleotides and Affymetrix-type arrays have been used in SFP. Borevitz et al. (2003) coined the term "single feature polymorphism" and demonstrated that this approach can be applied to organisms with somewhat larger genomes, specifically *A. thaliana* with a genome size of 140 Mb. Similarly, whole genome DNA-based SFP detection has been accomplished in rice (Kumar et al. 2007), with a genome size of 440 Mb, barley, which has a 5,300 Mb genome composed of more than 90 % repetitive DNA (Cui et al. 2005). Thus, SFPs have become an attractive marker system for various applications including parental polymorphism discovery. The development of DNA-based technologies such as SFP, DArT, and RAD which are based on microarray have the merits of SNP without going through sequencing. These technologies have provided us platforms for medium- to ultra-high-throughput genotyping to discover regions of the genome at a low cost, and have been shown to be particularly useful for genomes, where the level of polymorphism is low (Gupta et al. 2008). These array-based technologies are expected to play an important role in crop improvement and will be used for a variety of studies including the development of high-density molecular maps, which may then be used for QTL interval mapping and for functional and evolutionary studies.

3.4 Conclusion

DNA-based molecular markers have acted as versatile tools and have found their own position in various fields like taxonomy, physiology, embryology, genetic engineering, etc. They are no longer looked upon as simple DNA fingerprinting markers in variability studies or as mere forensic tools. Ever since their development, they are constantly being modified to enhance their utility and to bring about automation in the process of genome analysis. The discovery of PCR was a landmark in this effort and proved to be a unique process that brought about a new class of DNA profiling markers. This facilitated the development of marker based gene tags, map-based cloning of agronomically important genes, variability studies, phylogenetic analysis, synteny mapping, marker assisted selection of desirable genotypes, etc. Thus, molecular marker provides new dimensions to concerted efforts of breeding and marker aided selection that can reduce the time span of developing new and better varieties and will make the dream of super varieties come true. These DNA markers offer several advantages over traditional phenotypic markers, as they provide data that can be analyzed objectively. With the advent of molecular markers, it is now possible to make direct inferences about genetic diversity and interrelationships among organisms at the DNA level without the confounding effects of the environment and/or faulty pedigree records. Genetic analyses of plant and animal populations and species for taxonomic, evolutionary, and ecological studies tremendously benefited from the development of various molecular marker techniques. Each molecular marker technique is based on different principles but their application is to bring out the genome-wide variability.

References

Akbari M, Wenzl P, Caig V, Carling J, Xia L, Yang S, Uszynski G, Mohler V, Lehmensiek A, Kuchel H, Hayden M, Howes N, Sharp P, Vaughan P, Rathmell B, Huttner E, Kilian A (2006) Diversity arrays technology (DArT) for high-throughput profiling of the hexaploid wheat genome. Theor Appl Genet 113:1409–1420

Althoff DM, Gitzendanner MA, Segraves KA (2007) The utility of amplified fragment length polymorphisms in phylogenetics: a comparison of homology within and between genomes. Syst Biol 56:477–484

Ayers NM, McClung AM, Larkin PD, Bligh HFJ, Jones CA, Park WD (1997) Microsatellites and a single-nucleotide polymorphism differentiate apparent amylose classes in an extended pedigree of US rice germ plasm. Theor Appl Genet 94:773–781

Bardakci F (2001) Random amplified polymorphic DNA (RAPD) markers. Turk J Biol 25:185–196

Borevitz JO, Liang D, Plouffe D, Chang HS, Zhu T, Weigel D, Berry CC, Winzeler E, Chory J (2003) Large-scale identification of single-feature polymorphisms in complex genomes. Genome Res 13:513–523

Botstein D, White RL, Skolnick M, Davis RW (1980) Construction of a genetic linkage map in man using restriction fragment length polymorphisms. Am J Hum Genet 32(314):331

Ching ADA, Caldwell KS, Jung M, Dolan M, Smith OS, Tingey S, Morgante M, Rafalski A (2002) SNP frequency, haplotype structure and linkage disequilibrium in elite maize inbred lines. BMC Genet 3:19

Cho YG, Shii T, Temnyk S, Chen X, Lipovich L, McCouch SR, Park WD, Ayres N, Cartinhour S (2000) Diversity of microsatellites derived from genomic libraries and GenBank sequences in rice (Oryza sativa L.). Theoretical and Applied Genetics 100:713–722

Cui XP, Xu J, Asghar R, Condamine P, Svensson JT, Wanamaker S, Stein N, Roose M, Close TJ (2005) Detecting single-feature polymorphisms using oligonucleotide arrays and robustified projection pursuit. Bioinformatics 21:3852–3858

Duran CN, Appleby T, Clark D, Wood M, Batley IJ, Edwards D (2009) Auto SNPdb: an annotated single nucleotide polymorphism database for crop plants. Nucleic Acids Res 37:D951–D953

Eujay I, Sorrells M, Baum M, Woltersand P, Powell W (2001) Assessment of genotypic variation among cultivated durum wheat based on EST-SSRs and genomic SSRs. Euphytica 119:39–43

Ghislain M, Spooner DM, Rodríguez F, Villamon F, Núñez C, Vásquez C, Bonierbale M (2004) Selection of highly informative and user-friendly microsatellites (SSRs) for genotyping of cultivated potato. Theor Appl Genet 108:881–890

Gupta PK, Rustgi S (2004) Molecular markers from the transcribed/expressed region of the genome in higher plants. Funct Integr Genomics 4:139–162

Gupta PK, Rustgi S, Mir RR (2008) Array-based high-throughput DNA markers for crop improvement. Heredity 101:5–18

Hearne CM, Ghosh S, Todd JA (1992) Microsatellites for linkage analysis of genetic traits. Trend Genet 8:288–294

Jaccoud D, Peng K, Feinstein D, Kilian A (2001) Diversity arrays: a solid state technology for sequence information independent genotyping. Nucleic Acids Res 29:e25

James KE, Schneider H, Ansell SW, Evers M, Robba L, Uszynski G, Pedersen N, Newton AE, Russell SJ, Vogel JC, Kilian A (2008) Diversity arrays technology (DArT) for pan-genomic evolutionary studies of non-model organisms. PLoS ONE 3:e1682

Jarne P, Lagoda PJL (1996) Microsatellites, from molecules to populations and back. Trends Ecol Evol 11:424–429

Jeffreys AJ, Wilson V, Thein SL (1985) Hypervariable 'minisatellite' regions in human DNA. Nature 314:67–73

Kim JS, Ahn SG, Kim CK, Shim CK (2010) Screening of rice blast resistance genes from aromatic rice germplasms with SNP markers. Plant Pathol J 26:70–79

Kumar R, Qiu J, Joshi T, Valliyodan B, Xu D, Nguyen HT (2007) Single feature polymorphism discovery in rice. PLoS ONE 2(3):e284

Litt M, Luty JA (1989) A hypervariable microsatellite revealed by *in vitro* amplification of a dinucleotide repeat within the cardiac muscle actin gene. Amr J Hum Genet 44:397–401

Martin GB, Williams JGK, Tanksley SD (1991) Rapid identification of markers linked to a Pseudomonas resistance gene in tomato by using random primers and near-isogenic lines. Proc Natl Acad Sci 88:2336–2340

Mian MAR, Hopkins AA, Zwonitzer JC (2002) Determination of genetic diversity in tall fescue with AFLP markers. Crop Sci 42:944–950

Michelmore RW, Paran I, Kesseli RV (1991) Identification of markers linked to disease resistance genes by bulked segregant analysis: a rapid method to detect markers in specific genomic regions using segregating populations. Proc Natl Acad Sci USA 88:9828–9832

Miller JC, Tanksley SD (1990) RFLP analysis of phylogenetic relationships and genetic variation in the genus *Lycopersicon*. Theor Appl Genet 80:437–448

Miller MR, Dunham JP, Amores A, Cresko WA, Johnson EA (2007a) Rapid and cost effective polymorphism identification and genotyping using restriction site associated DNA (RAD) markers. Genome Res 17(2):240–248

Miller MR, Atwood TS, Eames BF, Eberhart JK, Yan YL, Postlethwait JH, Johnson EA (2007b) RAD marker microarrays enable rapid mapping of zebrafish mutations. Genome Biol 8(6):R105

Morgante M, Hanafey M, Powell W (2002) Microsatellites are preferentially present with non-repetitive DNA in plant genomes. Nat Genet 30:194–200

Mullis KB, Faloona F (1987) Specific synthesis of DNA *in vitro* via polymerase chain reaction. Methods Enzymol 155:350–355

Paran I, Michelmore RW (1993) Development of reliable PCR-based markers linked to downy mildew resistance genes in lettuce. Theor Appl Genet 85:985–993

Powell W, Machray GC, Provan J (1996) Polymorphism revealed by simple sequence repeats. Tren Plant Sci 1:215–222

Schlötterer C (2004) The evolution of molecular markers—just a matter of fashion? Nat Rev Genet 5:63–69

Smith DN, Devey ME (1994) Occurrence and inheritance of microsatellites in *Pinus radia*. Genome 37:977–983

Southern EM (1975) Detection of specific sequences among DNA fragments separated by gel electrophoresis. J Mol Biol 98:503

Vos P, Hogers R, Bleeker M, Reijans M, van de Lee T, Hornes M, Frijters A, Pot J, Peleman J, Kuiper M, Zabeau M (1995) AFLP: a new technique for DNA fingerprinting. Nucleic Acids Res 23:4407–4414

Williams JGK, Kubelik AR, Livak KJ, Rafalski JA, Tingey SV (1990) DNA polymorphisms amplified by arbitrary primers are useful as genetic markers. Nucleic Acids Res 18:6531–6535

Williams JGK, Hanafey MK, Rafalski JA, Tingey SV (1993) Genetic analysis using random amplified polymorphic DNA markers. Meth Enzymol 218:705–740

Yin X, Stam P, Dourleijn CJ, Kropff MJ (1999) AFLP mapping of quantitative trait loci for yield-determining physiological characters in spring barley. Theor Appl Genet 99:244–253

Zietkiewicz E, Rafalski A, Labuda D (1994) Genome fingerprinting by simple sequence repeat (SSR) anchored polymerase chain reaction amplification. Genomics 20:176–183

4 Gene Therapy

K. Rohini

Abstract

Genes are the *blueprints for proteins* and serve as building blocks for tissues. Genes are regulators of chemical reactions inside the living cells. Mutation in a gene causes a change in the expressed structure of protein leading to protein dysfunction. Genetic errors disrupt gene expression and result in diseases. Advancement in biotechnology has helped in understanding the genetic basis of inherited diseases. Gene therapy is an experimental technique that uses genes as medicine for correcting defective genes that are responsible for disease development. Gene therapy has the potential to eliminate a wide range of inherited and acquired human diseases. A number of clinical trials in gene therapy are being carried out throughout the world. Gene therapy approach involves the treatment of disease by introducing new genetic instructions into the tissues of patients to correct the defective or abnormal gene. In gene therapy, mostly somatic (nonsex) cells are targeted for treatment. Due to controversies associated with the involvement of germ cells, this therapy is not attempted. As somatic gene therapy is directed at the individual, it has no impact on future generations. Changes directed at somatic cells are not inherited. The problem of "gene delivery" i.e., the need to get replacement genes into the desired tissues is the most difficult part and needs intensive research. The majority of current gene therapy clinical trials involve the mechanisms of viruses as "vectors". These viral vectors are capable of integrating at random sites in the host's cells. To date, gene therapy is being used to treat several genetic disorders. A large number of clinical trials are underway to test gene therapy as a treatment of choice for many life-threatening diseases.

K. Rohini (✉)
Unit of Biochemistry, Faculty of Medicine, AIMST University, 3½ Bukit Air Nasi, Jalan Bedong Semeling, 08100 Bedong, Kedah Darul Aman, Malaysia
e-mail: rohinik23@gmail.com

I. Ravi et al. (eds.), *Advances in Biotechnology*,
DOI: 10.1007/978-81-322-1554-7_4, © Springer India 2014

4.1 Introduction

Genes are hereditary units, carried on a chromosome. Mutation of the gene causes a protein dysfunction due to the change in the codon, which changes the amino acid. Researches in biotechnology are progressing by leaps and bounds and gene therapy is one of the most rapidly advancing areas of research.

Gene therapy is the use of genes as medicines which is an experimental technique for correcting defective genes, thus altering the course of a medical disease. It primarily involves genetic manipulations in animals or humans to correct a disease and is applied in several fields such as cancer therapy, monogenic diseases, infectious disease, vascular diseases, autoimmune diseases etc. Genetic diseases can be inherited, occur at birth or to be randomly acquired due to environmental factors such as exposure to toxins. Single gene disorders, such as cystic fibrosis (CF), hemophilia, muscular dystrophy, and sickle cell anemia, are caused by a single gene malfunctioning and are usually inherited.

In the early 1900s adenosine deaminase (ADA) deficiency, an immuno-deficiency state was identified and was recorded as the first condition to be treated with gene therapy. Therapeutic gene transfer into stem cells in the clinical arena was attempted for the first time for this condition. A mile stone in medical history was the successful treatment of another form of immune deficiency called severe combined immune deficiency (SCID).

Since involvement of gene therapy evokes greater risk, hence there are several regulatory agencies whose permission must be sought before any work is initiated. Recombinant DNA Advisory Committee (RAC) is the supervisory body of the National Institute of Health, involving gene therapy. Numerous researches in genetic disorders and other diseases are currently at various stages of clinical trials. Some significant examples are listed in Table 4.1.

4.2 Approaches to Gene Therapy

The two main approaches to gene therapy are:

4.2.1 Somatic Gene Therapy

Somatic cells are the nonreproductive cells of an organism (other than sex cells), e.g., blood cells, bone marrow cells, intestinal cells etc. Somatic cell gene therapy involves the insertion of fully functional and expressible gene into a target somatic cell to correct genetic disease. In somatic gene therapy, the fully functional therapeutic gene is inserted into the target somatic cell of a patient to correct the genetic disease permanently. Somatic gene therapy will not be inherited by the patient's offspring or later generations and any effects will be restricted to the individual patient only.

4.2.2 Germ Line Gene Therapy

The reproductive (sex) cells, i.e., sperm or eggs of an organism constitute germ cell line. Germ line gene therapy is the introduction of functional genes, which are integrated into genomes. Hence, the change due to germ line therapy is heritable and would be passed to later generations. This approach should be highly effective in treatment of genetic and hereditary disorders, however, for safety, ethical and technical reasons, this therapy is generally not attempted.

4.3 Principles of Gene Therapy

Gene therapy is introducing a functional copy of the gene for curing the defect. If a child or foetus is diagnosed to carry a defective gene leading to disability, one may like to correct it by one of the three methods: (1) by replacing the defective gene by correct gene, (2) by correcting the defective gene through gene targeting or (3) by gene augmentation either through increasing the number of copies of the gene or through a higher

Table 4.1 Selected list of some important human gene therapy trials

Disease	Gene therapy
Severe combined immunodeficiency (SCID)	Adenosine deaminase (ADA)
Cystic fibrosis	Cystic fibrosis transmembrane regulator (CFTR)
Sickle cell anemia	β-Globin
Hemophilia	Factor IX
Thalassemia	α or β–Globin
Peripheral artery disease	Vascular endothelial growth factor (VEGF)
Familial hyper cholesterolemia	Low-density lipoprotein (LDL) receptor
Lesch–Nyhan syndrome	Hypoxanthine-guanine phosphoribosyl transferase (HGPRT)

level of expression of introduced gene. Maximum progress in the area of gene therapy has been achieved through the gene augmentation method where normal foreign gene sequences for defective gene are introduced involving vector-mediated DNA delivery system.

The steps involved in the gene therapy are as follows:

1. Identification and characterization of gene.
2. Cloning of gene.
3. Choice of vector.
4. Method of delivery.
5. Expression of gene.

The primary step is to identify the genetic defect and to clone a functional copy of the gene. The functional gene is then transferred to the patient, which involves choosing a vector with a suitable method of delivery. The vector/gene must be designed to allow proper expression of the gene inside the patient. There are two types of gene therapies:

4.3.1 Ex Vivo Gene Therapy

Ex vivo gene therapy involves the transfer of genes in cultured cells which are then reintroduced into the patient (therapy takes place outside the patient). This technique involves the following steps:

1. Isolation of cells (selected tissues e.g., bone marrow) with genetic defect from a patient.
2. Growing the cells in culture.
3. Introduction of the therapeutic gene to correct the defective gene.
4. Selection of the genetically corrected cells and their growth.
5. Transplantation of the modified cells to the patient.

This therapy involves the use of patient's own cells for culture and genetic correction and then transferred back to the patient. Hence, this therapy does not develop adverse immunological reactions and is quite effective (Fig. 4.1).

4.3.2 *In Vivo* Gene Therapy

In vivo gene therapy is the direct delivery of the genes into the cells of the particular tissue.

4.4 Vehicles of Gene Transfer

A gene cannot be directly inserted into a person's cells. It must be delivered to the cell using a carrier, or "vector" (Table 4.2). Gene transferring systems can be divided into two groups, viral and nonviral vectors. The examples of viral vectors are: (a) Retrovirus, (b) Adenovirus, (c) Adeno-associated virus (AAV), and the examples of nonviral vectors are: (a) Liposomes, (b) Ribozymes, and (c) Oligonucleotides.

4.4.1 Viral Vectors in Gene Delivery

4.4.1.1 Recombinant Retrovirus Vectors

The most common class of gene transfer vector is the recombinant retroviruses. Retroviruses contain single-stranded RNA genome of 7–10 kb in size. These genomes are complexed with vector-encoded proteins and surrounded

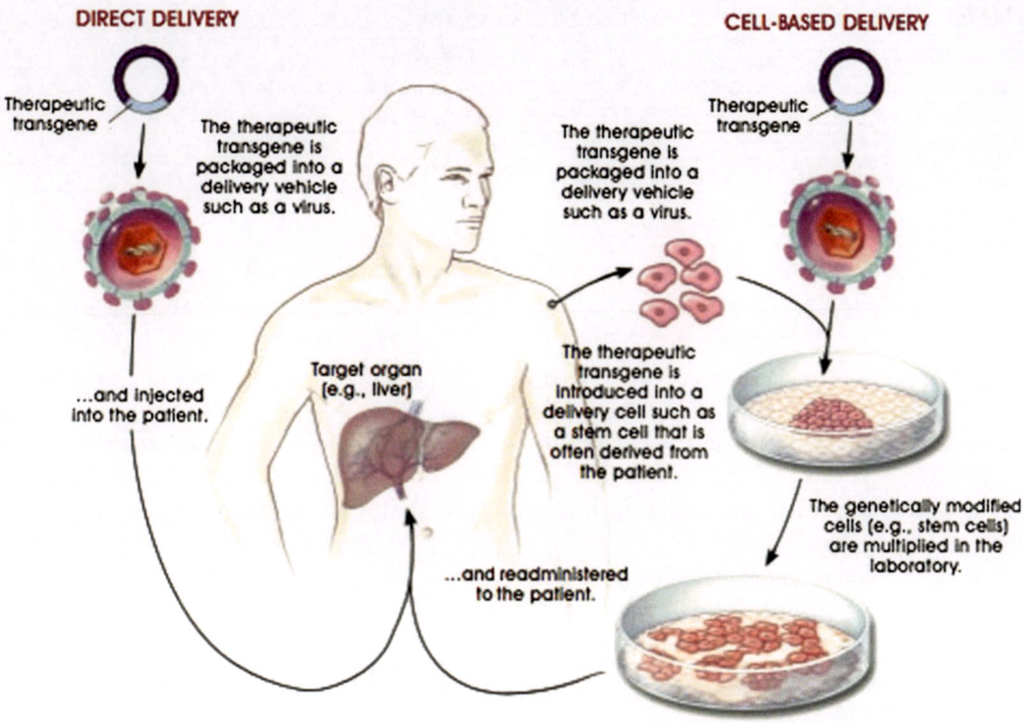

Fig. 4.1 Strategies for delivering therapeutic transgenes into patients; transfer of genes in cultured cells (*ex vivo*) and direct delivery of genes into the cells (*in vivo*) © 2001 Terese Winslow

Table 4.2 Gene transfer vehicles–vector delivery system include either recombinant viruses or various forms of plasmid DNA

Source	Vector	Genome	Advantages	Disadvantages
Viral	Retrovirus	ssRNA	Efficient transduction	Transduction requires cell division
			Stable gene transfer	Limits on transgene size/content
			Relatively safe	
	Adenovirus	dsDNA	Highly effective gene transfer	Transient transduction
			Higher titer	Immunological response
	Adeno-associated virus (AAV)	ssDNA	Potential stable transduction	Status of genome not fully elucidated
			Stable expression	
Nonviral	Plasmid/Liposomes	dsDNA	Absence of viral components	Low transfer and expression
			Potential for targeting	

with a lipid bilayer generated when the virus buds from the infected cell. Retrovirus genomes integrate into the target cell chromosomes as double-stranded DNA provirus generated by reverse transcription. This aids the virus genomes to be stably transferred to the targeted cells. Examples of retroviruses used as recombinant vectors include: oncoretroviruses, such as *Moloney Murine Leukemia virus* (MuLV) associated with oncogenesis in mice; lentiviruses, such as human immunodeficiency virus (HIV) associated with AIDS in humans; and spumaviruses, such as the human foamy virus, that has no known pathogenesis in humans (Buchschacher 2001; Bushman 2007).

The basic retrovirus genome consists of three genes (*gag, pol* and *env*) enclosed between two *long terminal repeats* (LTRs) (Fig. 4.2). The

Fig. 4.2 Retrovirus genome has packaging signal (Ψ) and gag, pol and env genes

LTR sequences are needed for integration of the DNA version of the virus genome into the host cell DNA. Between the upstream or 5′ LTR and the *gag* gene is the *packaging signal* (psi Ψ) which is essential for packaging the RNA into the budding virus particle, and the coding sequences for the reverse transcriptase and structural genes needed for the formation of mature virus particle.

The first step in generating replication-defective recombinant vectors is to introduce the coding sequences separately into cultured cells to generate packaging lines. The therapeutic gene is then combined with the remaining viral sequences, including the LTR and Ψ elements, in bacterial plasmids. This plasmid form of a vector is introduced into the packaging line, where the primary RNA transcripts are incorporated into the empty virus particles generated by the viral structural genes (Fig. 4.3). As a result, these virus particles contain the recombinant vector genomes, but not virus structural genes. After infection of the patient, the RNA inside the retroviral vector is reverse transcribed to give a DNA copy (although the retroviral vector does not carry a copy of the reverse transcriptase gene, a few molecules of reverse transcriptase enzyme are packaged in retrovirus particles). Ideally, the cloned gene, enclosed between the two LTR sequences, is then integrated into host cell DNA (Wolff and Budker 2005).

Recombinant retro viral vectors are advantageous because the integration of vectors into the target cell chromosomes is highly stable and do not cause an immune response. This class of vectors is also capable of highly efficient rates of gene transfer, depending on the choice of virus envelope. There are also several disadvantages for this class of recombinant virus vectors, which include a variable rate of

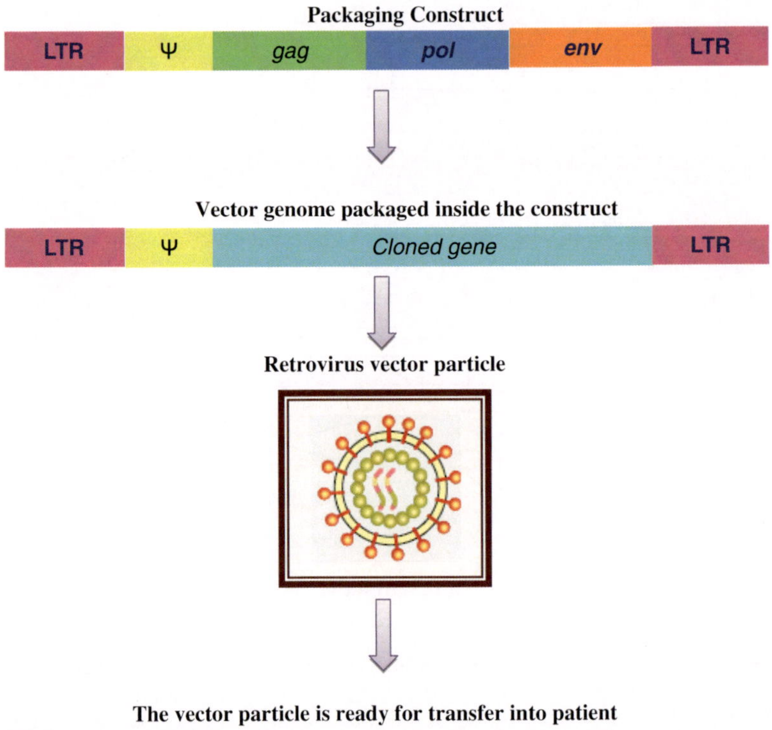

Fig. 4.3 Retrovirus vector system uses two virus constructs. The therapeutic vector carried the cloned gene and packaging signal. When both the constructs are present, the therapeutic vector and the cloned gene is packaged into the capsid (Mandel et al. 2006)

gene expression. They can carry only small amounts of DNA (about 8 kb) and cannot infect nondividing cells. These vectors are difficult to deliver in vivo because they are relatively fragile. Besides, there are growing safety concerns about potential problems of mutagenesis (Weidhaas et al. 2000).

4.4.1.2 Recombinant Adenovirus Vectors

Adenoviruses were among the first viruses chosen as vectors for use in human gene therapy and currently the second most common form of gene transfer vectors. Adenoviruses are relatively large 36 kb linear double-stranded DNA viruses

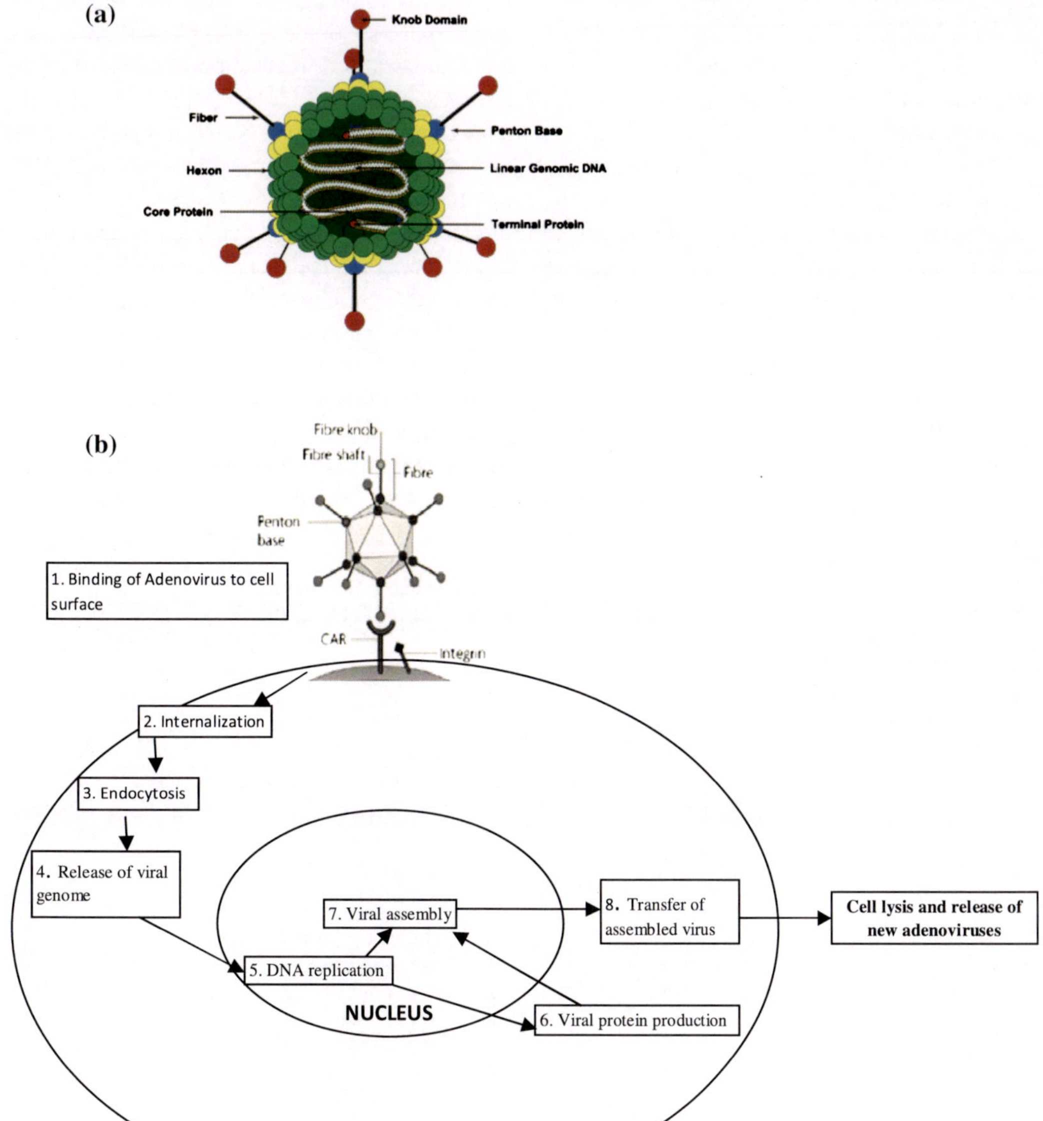

Fig. 4.4 **a** Adenovirus structure. **b** Adenovirus enters human cells by recognizing two receptors CAR and integrin. The virus is taken into the cell attached to the receptors and injects its DNA into the nucleus through a nuclear pore

that contain many regulatory and structural genes that are expressed at different times during the infection (Fig. 4.4a). These genomes are surrounded by glycoprotein capsids that are generated through lytic life cycle. The virus particle consists of a simple icosahedral shell, or capsid, containing a single linear dsDNA molecule of approximately 36,000 base pairs. A terminal protein protects each end of the DNA. The capsid is made of 240 hexons, each of which is a trimer of the hexon protein. Hexons are named for their 6-fold symmetry and are surrounded by six neighboring hexons. The hexon protein has loops that project outward from the virus surface. Adenoviruses are subdivided into about 50 serotypes by their response to antibody binding. This variation is largely due to variation in the loops of the hexon protein. The adenoviruses share the same receptor as B-group coxsackieviruses. This protein is therefore known as *coxsackievirus adenovirus receptor* (or CAR), but its normal physiological role is unknown (Wu et al. 2006).

Adenoviruses enter human cells by recognizing two receptors, CAR and integrin. The virus is taken into the cell attached to the receptors and surrounded by a membrane vesicle that dissociated in the cytoplasm. The adenovirus is injected into the nucleus through a nuclear pore. After the fiber tip binds to CAR, the penton base binds to integrins on the host cell surface (integrins are transmembrane proteins involved in adhesion) (Fig. 4.4b). The membrane puckers, move inward and form a vesicle that takes the adenovirus inside the cell. The virus is then released into the cytoplasm and travels toward the nucleus. The virion is disassembled outside the nucleus, and only the DNA (with its terminal proteins) enters the nucleus.

Adenoviruses have several advantages:

1. They are relatively harmless and cause mild infections of epithelial cells.
2. They are nononcogenic (i.e., they do not cause tumors).
3. They are relatively easy to culture and can be produced in large quantities.
4. The life cycle and biology of adenovirus are well understood.
5. The functions of most adenovirus genes are known. They include a highly efficient transfer associated with a large number of vector copies per cell.

However, adenoviruses do cause inflammation and can cause serious illness in patients with damaged immune systems. Therefore, when designing an adenovirus vector for gene therapy, the virus needs to be disarmed by crippling its replication system. This is done by deleting the gene for E1A protein, a virus protein made immediately on infection. E1A has two functions. First, it promotes transcription of other early virus genes. Second, it binds to host cell Rb protein, which normally prevents the cell from entering S-phase. This prompts the host cell to express genes for DNA synthesis, which the virus utilizes for its own replication. In the lab, crippled adenovirus is grown in genetically modified host cells that have the viral E1A gene integrated into host cell DNA. The virus particles generated by this approach cannot replicate in normal animal cells (St George 2003).

The main drawback of this class of vectors involves the transient nature of gene transfer and expression, which often lasts only a couple of months in nondividing cells and shorter time in dividing cells. Several safety concerns are also associated with these vectors. They can have lethal toxicity at high doses and are also highly infectious in humans.

4.4.1.3 Recombinant Adeno-Associated Virus (AAV) Vectors

AAVs are being used increasingly for clinical gene transfer and show considerable promise. AAVs are encapsidated parvoviruses with small (5 kb) linear single-stranded DNA genomes that encode only two gene products. AAV is a defective or "satellite" virus that depends on adenovirus (or some herpes viruses) to supply some necessary functions. AAV can replicate both as an episome and as an integrated provirus, depending on several factors including cell type.

AAVs are produced by first generating a plasmid form of the vector containing the therapeutic gene and critical cis-regulatory elements such as the inverted terminal repeats and packaging signal. This is combined with helper plasmids that express the AAV coding genes. These plasmids are co-transfected into a packaging cell line that contains the critical adenovirus E1 gene, so that the vector genomes can replicate. The transfected cells are eventually lysed and the virus particles are collected and run over a gradient or some other separation technique to purify the virus particles that contain only the vector genomes (Miller et al. 2002).

The advantages of using AAV are as follows:

1. It does not stimulate inflammation in the host.
2. It does not provoke antibody formation and can therefore be used for multiple treatments.
3. It infects a wide range of animals, as long as an appropriate helper virus is also present, hence can be cultured in many types of animal cells, including those from mice or monkeys.
4. It can enter nondividing cells of many different tissues, unlike adenovirus.
5. The unusually small size of the virus particle allows it to penetrate many tissues of the body effectively.
6. AAV integrates its DNA into a single site in the genome of animal cells (the AAV S1site on chromosome 19 in humans). This allows the therapeutic gene to be permanently integrated.

The main drawback of this class of vectors is that they can only accommodate small transgenes; transfer and expression is not always stable due to tissue-specific variations.

4.4.2 Nonviral Vectors in Gene Delivery

Although viral vectors are highly sophisticated and about 75 % of gene therapy trials have used these vectors only, nonviral vectors make attractive vehicle as they are inherently safer. A variety of alternative approaches investigated and found widely effective include:

1. Use of *naked nucleic acid* (DNA or less often RNA): Naked DNA is an attractive nonviral vector because of its inherent simplicity as it can easily be produced in bacteria and manipulated using standard recombinant DNA techniques. It shows very little dissemination and transfection at distant sites following delivery and can be re-administered multiple times into mammals (including primates) without inducing an antibody response against itself (i.e., no anti-DNA antibodies generated) (Li and Huang 2006). Direct *in vivo* gene transfer with naked DNA was first demonstrated when efficient transfection of myofibers was observed following injection of mRNA into skeletal muscle. Subsequent studies also found foreign gene expression after direct injection in other tissues such as heart, thyroid, skin, and liver. Since naked DNA is the simplest and safest gene delivery system, multiple approaches have been attempted for *in vivo* liver gene delivery, including intraportal injection, increased DNA retention time in the liver, electrogene transfer, hydrodynamic IV injection, and mechanical liver massage. Hydrodynamic transfection had been widely used because of its simplicity and high efficiency, and although nucleic acid delivery has been observed in all major organs, the best results are found in the liver (Niidome and Huang 2002).
2. *Particle bombardment*: In this technique, DNA is fired through the cell walls and membranes on metal particles. This method was originally developed to get DNA into plants. However, it has also been used to make transgenic animals and is occasionally used for humans.
3. *Receptor-mediated uptake*: DNA in this case is attached to a protein that is recognized by a cell surface receptor. When the protein enters the cell, the DNA is taken in with it.
4. *Polymer-complexed DNA*: Binding to a positively charged polymer, such as polyethyleneimine, protects the negatively charged DNA. Such complexes are often taken up by cells in culture and may in principle be used for ex vivo gene therapy.

5. *Encapsulated cells*: Whole cells engineered to express and secrete a needed protein may be encapsulated in a porous polymeric coat and injected locally. This approach may be of great value in treating conditions like Alzheimer's disease.

4.4.2.1 Liposomes in Gene Therapy

Nearly, a quarter of all gene therapy trials involve the use of plasmid DNA alone, which is typically very inefficient, or by complexing plasmid DNA with liposomes, which can improve the rate of gene transfer to varying degrees. Liposomes are hollow microscopic spheres of phospholipid, and can be filled with DNA or other molecules during assembly. The liposomes will merge with the membranes surrounding most animal cells and the contents of the liposome end up inside the cell, a process known as *Lipofection*. Liposomes, both cationic and anionic, are currently being considered as vectors for gene therapy and their accomplishment in transfection is being studied by research groups and pharmaceutical companies (Fig. 4.5a). Lipofection is a promising approach because "armed" liposomes can be injected directly into tumor tissue. In fact, liposomes are probably of more use in delivering proteins than DNA, something not feasible when using viruses as genetic engineering vectors. For example, toxic proteins such as tumor necrosis factor (TNF) can be packaged inside liposomes and injected into tumor tissue (Fig. 4.5b). The liposomes merge with the cancer cell membranes, and the lethal proteins are then released inside the cancer cells.

PEGylated liposome (PEGLip) technology is a new approach to improve the pharmacodynamic properties of therapeutic proteins. Instead of encapsulating the drug, PEGLips are used as carriers with the protein bound noncovalently but with high specificity to the outer surface. Unlike approaches such as mutagenesis, PEGylation, or fusion to carrier proteins, PEGLip technology does not involve changes to a protein's amino acid sequence and covalent attachment of stabilizing agents. PEGLip is a new platform technology that produces long-acting forms of coagulation factor VIII (FVIII) and activated FVII (FVIIa) and improves the efficacy of the hemophilia treatment (Suzuki et al. 2001).

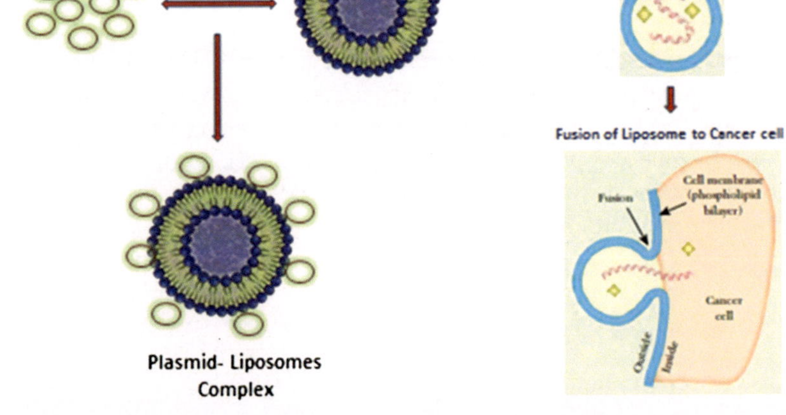

Fig. 4.5 **a** Cationic liposomes (positively charged) are complexed with DNA (plasmids); the liposome: DNA ratio is essential for optimal transfection. **b** Anticancer agent delivered by Lipofection. Liposomes can be filled with DNA/ proteins such as TNF; interacts and merges with cell membrane and delivers into the cell (http://www.azonano.com)

The development of liposomes as carriers for therapeutic molecules is an ever-growing research area. The possibility of modulating the technological characteristics of the vesicles makes them highly versatile both as carriers of several types of drugs (from conventional chemotherapeutics to proteins and peptides) and in therapeutic applications (from cancer therapy to vaccination).

4.4.2.2 Antisense Therapy

Antisense RNA binds to the corresponding RNA and prevents its translation by ribosome. Gene therapy for certain disorders may result in overproduction of some normal proteins. Hence, it is possible to treat these diseases by blocking transcription using single-stranded nucleotide sequence that hybridizes with the specific gene and this is called *antisense therapy*. Antisense therapy refers to the inhibition of translation by using a single-stranded nucleotide and also inhibits both transcription and translation by blocking the transcriptional factor responsible for gene expression (Aartsma-Rus and van Ommen 2007). Two main alternatives exist when using antisense RNA.

1. It is possible to use a full length anti-gene that is transcribed to give a full length antisense RNA. The anti-gene must be carried on a suitable vector and expressed in the target cells.
2. Shorter artificial RNA oligonucleotides may be used. An antisense RNA of 15–20 nucleotides is capable of binding specifically to part of the complementary mRNA and preventing translation.

Antisense oligonucleotides (oligos) can be designed to complement a region of a particular gene or messenger RNA (mRNA) and serve as potential blockers of transcription or translation through sequence-specific hybridization. Recent advances in the use of oligodeoxynucleotide and plasmid-derived RNA as antisense agents are of special relevance to cancer gene therapy which have lead to clinical trials and have proved promising results. Transformed cell lines bearing plasmids and viruses designed for the transcription of antisense RNA have the advantage that they can be characterized thoroughly and the effects of antisense RNA on target gene expression and phenotype can be studied easily *in vivo*. The importance of examination of antisense effects in syngeneic and immunocompetent hosts is illustrated by the studies of insulin-like growth factor and insulin-like growth factor receptor where tumor regression and protection against tumor formation have been observed in cell types.

Antisense oligonucleotides are being investigated as a potential therapeutic modality that takes direct advantage of molecular sequencing. A fundamental attraction of the antisense approach is that this method potentially may be applied to any gene product, in theory, for the treatment of malignant and nonmalignant diseases.

4.4.2.3 Ribozymes in Gene Therapy

Recent advances of biological drugs have broadened the scope of therapeutic targets and this holds true for dozens of RNA-based therapeutics currently under clinical investigation for diseases ranging from genetic disorders to HIV infection to various cancers.

RNA molecules that can serve as enzymes (catalytic ability) are referred to as *ribozyme*. Ribozymes have catalytic domains and recognize the substrate RNA by base paring and cleave RNA molecules. The naturally occurring ribozymes are to be modified, so that they can specifically hybridize with mRNA sequences and block protein synthesis. Thus, ribozymes might be engineered to recognize and destroy any target mRNA molecule (Pelletier et al. 2006).

A vector carrying a gene encoding of the ribozyme is used for delivery to the target cell. Transcription of this gene would result in production of the ribozyme RNA. The ribozyme would then bind and cleave the target mRNA. Genetically engineered ribozymes are a possibility for treating certain cancer and viral diseases.

4.4.2.4 RNA Interference

RNA interference (RNAi) is a mechanism to selectively degrade mRNA expression has emerged as a potential novel approach for drug target validation and the study of functional genomics. Like other biological drugs, RNAi-based therapeutics often requires a delivery vehicle to transport them to the targeted cells. Thus, the clinical advancement of numerous *small interfering RNAs* (siRNA) drugs has relied on the development of siRNA carriers, including biodegradable nanoparticles, lipids, bacteria, and attenuated viruses. Most therapies permit systemic delivery of the siRNA drug, while others use ex vivo delivery by autologous cell therapy. The most widely used siRNA delivery methodology consists of cationic lipids or cationic polymers that package siRNA into stable nanoparticles capable of translocation across the cellular membrane. Studies have shown that this targeted nanoparticle formulation with improved tissue specificity and delivery efficiency silenced the target gene in the metastasis effectively without any significant immunotoxicity and hence proved to be a potential tool for clinical cancer therapy.

4.5 Landmarks in Gene Therapy for Genetic Diseases

Clinical trials are becoming more increasingly prevalent. The most interesting is that gene therapy seems to have cured 8 of 10 children who had potentially fatal "bubble boy disease". Bubble boy disease is formally called severe combined immunodeficiency, or SCID, a genetic disorder diagnosed in about more than 40 babies each year in the United States. The name "bubble boy" is given after a Houston boy, David Vetter who was born with a genetic disorder leaving him no natural immunities against disease, became famous for living behind plastic barriers to protect him from germs. He died at the age of 12 in 1984.

Selected list of genetic diseases that are likely to be cured by gene therapy including SCID is Sickle cell anemia, Fanconi anemia, Thalassemia, Gausher's disease, Hunter disease, Hurler disease, Osteoporosis and Chronic granulomatous disease.

4.5.1 Therapy for Adenosine Deaminase Deficiency

The first human gene therapy was carried out to correct the deficiency of the enzyme ADA on September 14, 1990 at the National Institute of Health. Severe combined immunodeficiency (SCID) is a rare inherited immune disorder associated with T-lymphocytes and to a lesser extent B-lymphocytes dysfunction. According to Aiuti et al. (2009) in the deficiency of ADA, deoxyadenosine and its metabolites accumulated and destroyed T-lymphocytes. Both the lymphocytes participate in the defence mechanism of body, thus the patient of SCID (lacking ADA) suffers from infectious diseases. The technique for the treatment for ADA deficiency is as follows:

- A plasmid vector bearing proviral DNA is selected.
- A part of proviral DNA is replaced by the ADA gene and the gene (G 418) coding for antibiotic resistance and then cloned.
- Circulating lymphocytes are removed from the patient suffering from ADA deficiency.
- These cells are transfected with ADA gene by exposing to retrovirus carrying the said gene.
- The genetically modified lymphocytes are grown in cultures to confirm the expression of ADA gene and returned to the patient.

These lymphocytes persist in the circulation and synthesize ADA. Thus, the ability of the patient to produce antibodies is increased. The main drawback is that the lymphocytes have a short life span; hence the transfusions have to be carried out frequently.

4.5.2 Therapy for Lesch–Nyhan Syndrome

Lesch–Nyhan is an inborn error in purine metabolism due to the defect in a gene that encodes for the enzyme hypoxanthine-guanine phosphoribosyl transferase (HGPRT). In the absence of HGPRT, purine metabolism is disturbed and uric acid levels increases resulting in severe gout and kidney damage. Use of retroviral vectors, the HGPRT producing genes is successfully inserted into the bone marrow for treatment.

4.5.3 Therapy for Familial Hypercholesterolemia

The patients with familial hypercholesterolemia lack the low density lipoprotein (LDL) receptors on their liver cells. As a result, the LDL cholesterol is not metabolised in liver, leading to arterial blockage and heart diseases. The hepatocytes were transduced with retroviruses carrying genes for LDL receptors, which are infused into patient's liver. The hepatocytes establish and produce functional LDL receptors.

4.5.4 Therapy for Cystic Fibrosis

CF is a fatal genetic disease, characterized by the accumulation of sticky dehydrated mucus in the respiratory tract and lungs. Due to this, the patients of CF are highly susceptible to bacterial infections in their lungs. In normal persons, the chloride ions of the cells are pushed out through the participation of a protein called cystic fibrosis transmembrane regulator (CFTR). In patients of CF, the CFTR protein is not produced due to the gene defect. As a result, the respiratory tract and the lungs become dehydrated with sticky mucus, which serves as an ideal environment for bacterial infections. Adeno-associated virus vector system showed some encouraging results in the gene therapy for CF (Conese et al. 2011).

4.5.5 Therapy for Cancer

Cancer is a leading cause of death throughout the world and gene therapy is the latest and novel approach for cancer treatment.

4.5.5.1 Tumor Necrosis Factor

It is a protein produced by human macrophages which provide defense against cancer cells. A special type of immune cells, the *tumor infiltrating lymphocytes* (TILs), enhance the cancer fighting ability. These were transformed with the TNF gene (along with neomycin resistant gene) and used for the treatment for *malignant melanoma* (Woods and Bottero 2006).

4.5.5.2 Suicide Gene Therapy

Suicide gene therapy begins by delivering a therapeutic gene into the cancer cells. The gene encodes an enzyme that converts nontoxic prodrug into a toxic compound. The gene encoding the enzyme *thymidine kinase* (TK) is often referred to as suicide gene and is used for the treatment for certain cancers. TK phosphorylates nucleosides which are used for synthesis of DNA during cell division. The drug *ganciclovir* (GCV) bears a close structural resemblance to certain nucleosides (thymidine). The nontoxic prodrug, the nucleoside analog ganciclovir, is converted into its monophosphate by thymidine kinase, because only the cancer cells have thymidine kinase and all the noncancerous cells are unaffected. Normal cellular enzymes then convert the monophosphate to ganciclovir triphosphate (GCV-TP). This acts as a *DNA chain terminator.* DNA polymerase incorporates GCV-TP into the growing strands of DNA and kills the cancer cells (*bystander effect).* In suicide gene therapy, the vector used is the *herpes simplex virus* (HSV) with the gene for thymidine kinase inserted in its genome. Normal brain cells do not divide while the brain tumor cells undergo unchecked division. Studies by Deanna and Burmester (2006) have proved that by using GCV-HSVTK suicide gene therapy the proliferation of the cells can be reduced.

4.5.5.3 Gene Replacement Therapy

A gene named p53 codes for a protein with molecular weight of 53 kilo Daltons, hence the name p53. p53 is considered to be a *tumor suppressor gene,* since the protein it encodes binds with the DNA and inhibits replication. The tumor cells of several tissues were found to have altered genes of p53, and a main causative factor in tumor development. Researchers have found encouraging results by employing adenovirus vector system to replace the damaged p53 by a normal gene.

4.5.6 Therapy for AIDS

AIDS is a global disease and invariably a fatal one with no cure. Researches have made attempts to relieve its effects through gene therapy which are discussed below:

4.5.6.1 rev and env Genes

A mutant strain of HIV lacking *rev* and *env* genes has been developed. Due to the lack of these genes, the virus cannot replicate, hence used for therapeutic purposes. T-lymphocytes from HIV-infected patients are removed and mutant viruses are inserted into them. The modified T-lymphocytes are cultivated and injected into the patients. Due to the lack of the genes the viruses cannot multiply, but they can stimulate CD8 killer lymphocytes, which are proved to destroy HIV-infected cells (Cavazzana-Calvo et al. 2000).

4.5.6.2 Genes to Activate gp120

gp120 is a glycoprotein with molecular weight of 120 kilo Daltons, which is present in the envelope of HIV. It is essential for binding of virus to the host for replication. Researches have synthesized a gene *F105* to produce an antibody that can inactivate gp120. In the anti AIDS therapy, HIV cells were engineered to produce *anti-HIV antibodies* (Patterson et al. 2009).

4.6 Ethical Issues

One of the main concerns regarding gene therapy was that genetic material used to treat somatic cells would find its way into sperm and ova, thus affecting offspring of the patient. However, careful selection of vectors and target cells prevents the spread of genetic material to germ cells. Trials to define limits of safety and efficacy in somatic cell gene therapy have been underway for the better part of a decade. Beyond safety and efficacy issues, ethical concerns in somatic gene therapy research are familiar in many clinical settings, weighing potential harms and benefits, selection of patients for research, assurance that consents to experimental treatments is informed and voluntary, and protection of privacy and confidentiality of medical information. Ethical concerns about germ line therapy have been widely discussed, especially starting with the beginning of the Human Genome Project. Germ line gene therapy is far more technically difficult than somatic cell therapy. These concerns have raised queries about why research in germ line therapy should be pursued at all.

4.7 Conclusion

Gene therapy is moving on to higher gear for the treatment for a number of genetic diseases.

Gene therapy studies have shown promising results in the treatment of HIV and cancer. Ethical stem cells may be helpful to treat patients with osteogenesis; Red Orbit gene therapy may improve some form of blindness and successful clinical trials for Parkinson's disease are few examples of gene therapy success. Besides, researchers have engineered patients' own pathogen fighting T-cells to target a molecule on the surface of leukemia cells. The altered T-cells were grown outside the body and infused back to the patient suffering from late-stage chronic lymphocytic leukemia (CLL).

Theoretically, gene therapy is the permanent solution for genetic disease. It broadly involves isolation of a specific gene, making its copies, inserting them into target tissue cells to make the desired protein. The most significant factor is to ensure the gene is harmless to the patient and is expressed appropriately. Another important concern is the reaction of body's immune system to the foreign proteins.

It is true that gene therapy has several limitations, despite intensive research. But researchers are persistently working and waiting for a great breakthrough to revolutionize the world with the new medicine.

References

Aartsma-Rus A, van Ommen GJ (2007) Antisense-mediated exon skipping: a versatile tool with therapeutic and research applications. RNA 13:1609–1624

Aiuti A, Cattaneo F, Galimberti S, Benninghoff U, Cassani B, Callegaro L, Scaramuzza S, Andolfi G, Mirolo M, Brigida I (2009) Gene therapy for immunodeficiency due to adenosine deaminase deficiency. N Engl J Med 360(5):447–458

Buchschacher GL (2001) Introduction to retroviruses and retroviral vectors. Somat Cell Mol Genet 26:1–11

Bushman FD (2007) Retroviral integration and human gene therapy. J Clin Invest 117(8):2083–2086

Cavazzana-Calvo M, Hacein-Bey S, de Saint Basile G, Gross F, Yvon E, Nusbaum P, Selz F, Hue C, Certain S, Casanova JL (2000) Gene therapy of human severe combined immunodeficiency (SCID)-X1 disease. Science 288(5466):669–672

Conese M, Ascenzioni F, Boyd AC (2011) Gene and cell therapy for cystic fibrosis: from bench to bedside. J Cystic Fibr 10(2):S114–S128

Deanna C, Burmester JK (2006) Gene therapy for cancer treatment: past, present and future. Clin Med Res 4(3):218–227

Li SD, Huang L (2006) Gene therapy progress and prospects: non-viral gene therapy by systemic delivery. Gene Ther 13:1313–1319

Mandel RJ, Manfredsson FP, Foust KD, Rising A, Reimsnider S, Nash K, Burger C (2006) Recombinant adeno-associated viral vectors as therapeutic agents to treat neurological disorders. Mol Ther 13:463–483

Miller DG, Rutledge EA, Russell DW (2002) Chromosomal effects of adeno-associated virus vector integration. Nat Genet 30:147–148

Niidome T, Huang L (2002) Gene therapy progress and prospects: nonviral vectors. Gene Ther 9:1647–1652

Pelletier R, Caron SO, Puymirat J (2006) RNA based gene therapy for dominantly inherited diseases. Curr Gene Ther 6:131–146

Patterson S, Papagatsias T, Benlahrech A (2009) Use of adenovirus in vaccines for HIV. Handb Exp Pharmacol 275–293

St George JA (2003) Gene therapy progress and prospects: adenoviral vectors. Gene Ther 10:1135–1141

Suzuki S, Tadakuma T, Kunitomi M, Takayama E, Sato M, Asano T (2001) Liposome-mediated gene therapy using HSV-TK/Ganciclovir under the control of human PSA promoter in prostate cancer cells. Urol Int 67:216–223

Weidhaas JB, Angelichio EL, Fenner S, Coffin JM (2000) Relationship between retroviral DNA integration and gene expression. J Virol 74(18):8382–8389

Wolff JA, Budker V (2005) The mechanism of naked DNA uptake and expression. Adv Genet 54:3–20

Woods NB, Bottero V (2006) Gene therapy: therapeutic gene causing lymphoma. Nature 440(7088):1123

Wu Z, Asokan A, Samulski RJ (2006) Adeno-associated virus serotypes: Vector toolkit for human gene therapy. Mol Ther 14:316–327

5 RNA Interference and Its Applications

Jyoti Saxena

Abstract

RNA interference (RNAi) refers to a group of post-transcriptional or transcriptional gene silencing mechanisms conserved from fungi to mammals. It is a phenomenon primarily for the regulation of gene expression, self or nonself depending upon the surrounding factors or conditions. It is done in nature with the help of noncoding RNA molecules to control cellular metabolism and helps in maintaining genomic integrity by preventing the invasion of viruses and mobile genetic elements. It is a simple and rapid method of silencing gene expression in a range of organisms as a consequence of degradation of RNA into short RNAs that activate ribonucleases to target homologous mRNA. The process of RNAi can be mediated by either small interfering RNA (siRNA) or micro RNA (miRNA). The RNAi pathway is triggered by the presence of double-stranded RNA, which is cleaved by the ribonuclease-III domain-containing enzyme Dicer to generate 20–25 nucleotide long siRNA duplexes. siRNA is then loaded onto the RNA-induced silencing complex (RISC), in which an Argonaute (Ago)-family protein, guided by the siRNA, mediates the cleavage of homologous RNAs. Synthetic double-stranded RNA (dsRNA) introduced into cells can selectively and robustly induce suppression of specific genes of interest. Because of its exquisite specificity and efficiency, RNAi is being considered as valuable research tool, not only for functional genomics, but also for gene-specific therapeutic activities that target the mRNAs of disease-related genes.

5.1 Introduction

Gene silencing is a general term that describes epigenetic processes of gene regulation. It refers to 'switching off' of a gene by a mechanism other than genetic mutation. That is, a gene which would be expressed under normal circumstances

J. Saxena (✉)
Biochemical Engineering Department, BT Kumaon Institute of Technology, Dwarahat, Uttarakhand 263653, India
e-mail: saxenajyoti30@gmail.com

I. Ravi et al. (eds.), *Advances in Biotechnology*,
DOI: 10.1007/978-81-322-1554-7_5, © Springer India 2014

is switched off by machinery in the cell. Gene silencing may be a part of an ancient immune system protecting the genome of the organisms from infectious DNA elements such as transposons and viruses. RNA interference, commonly known as RNAi, regulates gene silencing mostly either at the transcriptional or post-transcriptional level. It is a sort of "Reverse Genetics."

RNAi is the mechanism in molecular biology where the presence of certain fragments of double-stranded RNA (dsRNA) interferes with the expression of a particular gene which shares a homologous sequence with the dsRNA. In other words, it is a process of using specific sequence of dsRNA to knock down the expression of complementary gene. This phenomenon is distinct from other gene silencing methods in that silencing can spread from cell to cell and generate heritable phenotypes in first generation. It only functions in eukaryotes as prokaryotic RNase III degrades shorter fragments of only 12 base pairs. RNAi is now known to operate in humans, mice and other mammals, as well as in frogs, fruit fly, hydra, trypanosomes, zebra fish, fungi, and plants.

In early 1990s horticultural researcher Richard Jorgensen and his team mates experimented on petunias by introducing numerous copies of a gene that encoded deep purple pigmentation in flowers. But to their surprise unexpected opposite result was achieved in the form of white or patchy flowers in place of purple colouration (Napoli et al. 1990). They concluded that somehow the introduced gene had silenced themselves as well as plants' endogenous purple pigment gene. They called this as co-suppression of gene expression but the molecular mechanism remained unknown.

A few years later, similar results of gene silencing were obtained in the experiments when genetically engineered RNA viruses containing plant gene were introduced into plants. Both the transgene and normal cellular copy of genes were turned off. This phenomenon was called VIGS, i.e., "virus induced gene silencing." At that time, these observations were puzzling until in 1998 Andrew Fire and Craig Mello at the Carnegie Institution of Washington and the University of Massachusetts Cancer Centre, respectively first described RNAi phenomenon in round worm, *Caenorhabditis elegans* by injecting dsRNA. They also coined the term RNAi (Fire et al. 1998). For their revolutionary work on RNAi they shared the Nobel Prize in medicine in 2006.

There are two classes of RNA molecules called small (short) interfering RNA (siRNA) and micro RNA (miRNA). siRNA can be defined as a class of dsRNAs consisting of approximately 18–23 nucleotides in length, generated from long dsRNAs by the action of a nuclease enzyme called Dicer. Its antisense strand functions in gene silencing by promoting the cleavage of mRNA. miRNA, on the other hand is a class of 19–25 nucleotides, single-stranded RNA encoded in genomes of most of the multicellular organisms. They vary from a few thousand to 40,000 molecules per cell. They are encoded by the genome but unlike mRNA, do not translate a protein, instead work for the regulation of the expression of mRNAs. Like siRNA, these are also produced by the cleavage of dsRNAs by Dicer.

RNAi has become a valuable research tool, both in *in vitro* studies and living organisms because synthetic dsRNA introduced into cells can selectively and robustly induce suppression of specific genes of interest. RNAi may be used for large-scale screens that systematically shut down each gene in the cell, which can help identify the components necessary for a particular cellular process or an event such as cell division. Besides, it has various applications in medicine, biotechnology, and functional genomics.

5.2 Small Interfering RNA

siRNAs are short double-stranded RNA molecules that induce sequence specific post-transcriptional gene silencing (PTGS). Their role was first discovered in 1999 by David Baulcombe's group in plants. siRNAs have a well-defined structure: two hybridized strands of 18–23 RNA bases (usually 21 bp), each with phosphorylated 5′ ends and hydroxylated 3′ ends with two overhanging DNA nucleotides at the 3′ terminus. The easiest way to achieve RNAi is the use of synthetic siRNA molecules. Once siRNA is transfected to a cell, it is incorporated into a nuclease complex called RNA-induced silencing complex (RISC). RISC targets and cleaves mRNA that is complementary to this siRNA, thus interrupting translation of targeted genes. The sequence specificity of siRNA molecules allows specific silencing of about 90 % of known genes.

siRNA plays many roles, but most notable is in the RNAi pathway, where it interferes with the expression of specific genes with complementary nucleotide sequence. It offers an efficient and easy method for post-transcriptional sequence specific silencing of genes and presents a powerful new tool to study gene function, target validation, and signal transduction. Besides, its applications are also in drug discovery and gene therapy.

There are many challenges encountered in siRNA research. Occasionally it has been found that nonspecific effects are triggered by the experimental introduction of a siRNA. When a mammalian cell comes across a double-stranded siRNA, it is mistaken for a viral by-product and mounts an immune response.

5.3 Micro RNA

miRNAs are noncoding genes present in cells of various eukaryotic organisms. These are a form of single-stranded RNA, typically 19–25 nucleotides long, and are thought to regulate the expression of other genes. These are RNA genes which are transcribed from DNA but do not form a protein. They were discovered in 1993 by Lee and Feinbaum during a study of the gene *lin-14* in *C. elegans* development. The DNA sequence that codes for miRNA is sometimes much longer than miRNA itself and includes not only the miRNA sequence but also an approximate reverse complement. Processing also involves components of the RNAi machinery. Once processed they seem to bind to their target mRNAs, however they are not 100 % complementary, a couple of base mismatches may be seen. The primary transcript known as a*pri-miRNA* is processed in the cell nucleus and forms a 70-nucleotide stem-loop structure called a *pre-miRNA*. This dsRNA hairpin loop or stem loop is formed due to pairing of single-stranded miRNA sequence and its reverse complement bases. In animals, RNAs III named as Drosha is used to cut the base of hairpin loop to release the pre-miRNA. This pre-miRNA is then transported into the cytoplasm by a carrier protein, Exportin 5. 20–25 nucleotides are removed by Dicer to release the mature miRNA. However, in plants (since Drosha has not been reported) the processing takes place through Dicer in nucleus.

The function of miRNAs appears to be in gene regulation. For that purpose, a miRNA is complementary to a part of one or more mRNAs. Gene silencing may occur either *via* mRNA degradation or by preventing mRNA from being translated. Since its discovery, miRNA research has revealed its multiple roles in negative regulation (transcript degradation and sequestering, translational suppression) and possible involvement in positive regulation (transcriptional and translational activation). By affecting gene regulation, miRNAs are likely to be involved in most biological processes. miRNAs occasionally also cause histone modification and DNA methylation of promoter sites, which affects the expression of target genes. The human genome may encode over 1,000 miRNAs, which may target about 60 % of mammalian genes.

The process of RNAi can be moderated by either siRNA or miRNA but there are subtle differences between the two. siRNA is considered exogenous double-stranded RNA that is taken up by cells or enters *via* vectors like viruses, while miRNA is single stranded and comes from endogenous noncoding RNA found within the introns of larger RNA molecules. Mature miRNAs are structurally similar to siRNAs, however they undergo extensive post transcriptional modifications before reaching maturity. Thus, miRNA and siRNA share the same cellular machinery downstream of their initial processing. The siRNAs derived from long dsRNA precursors also differ from miRNAs in that miRNAs, especially those in animals, typically have incomplete base pairing to a target and inhibit the translation of many different mRNAs with similar sequences. In plants, however, miRNA tends to have a more perfectly complimentary sequence which induces mRNA cleavage as opposed to just repression of translation. In contrast, siRNAs typically base-pair perfectly and induce mRNA cleavage only in a single, specific target. Both siRNA and miRNA can play a role in epigenetics through a process called RNA-induced transcriptional silencing (RITS). Likewise, both are important targets for therapeutic use because of the roles they play in controlling the gene expression.

5.4 Components of Gene Silencing

Many important enzymes and genes encoding them have been identified in various model organisms which have been identified by analysis of mutants. Some of the components identified serve as initiators, while others as effectors, amplifiers, and transmitters of the gene silencing process. These are described as follows.

5.4.1 RNase III

RNase III is divided into three classes on the basis of domain structure viz. bacterial RNase III, Drosha family nucleases, and Dicer enzyme. Dicer was identified in *Drosophila* extract which showed its involvement in the initiation of RNAi. It was named as Dicer (DCR) due to its ability to digest dsRNA into uniformly sized small RNAs (siRNA) by Hannon and a graduate student in his laboratory, Emily Bernstein in 2000 (Bernstein et al. 2001). In Drosophila, Dicer which is a large multidomain RNase III enzyme has been identified in existence into two forms: Dcr-1/Loquacious (Loqs)-PB (also known as R3D1-L[long]) that creates miRNA and Dcr-2/R2D2 that generates siRNA. Dcr-1 shares a structural homology with Dcr-2, despite that it displays different sets of properties such as ATP requirements and substrate specifications. Dcr-1 is an enzyme that shows ATP-independent functions and affinity toward stem-loop form of RNA and requires a dsRNA binding protein partner. Dcr-2 on the other hand shows ATP-dependent activity with substrate specificity to dsRNA. It also requires a dsRNA binding protein, namely R2D2, which function in association with specific RNase enzyme Dcr-2 forming a heterodimeric complex.

Dicer has four distinct domains such as an amino-terminal helicase domain, dual RNase III motifs, a dsRNA binding domain, and a PAZ domain. Cleavage by Dicer is thought to be catalyzed by its tandem RNase III domain. Complete digestion by RNase III enzyme results in dsRNA fragments of 12–15 bp, half the size of siRNAs. The RNase III enzyme acts as a dimmer, thus digests dsRNA with the help of two compound catalytic centers, whereas each monomer of the Dicer enzyme possesses two catalytic domains, with one of them deviating from the consensus catalytic sequences (Agarwal et al. 2003).

5.4.2 RNA-Induced Silencing Complex and Guide RNAs

The RISC is a multiprotein complex that incorporates one strand of a siRNA or miRNA. After recognizing the complementary mRNA, it activates RNase to cleave the RNA. The inability of cellular extracts treated with a Ca^{2+}-dependent nuclease (micrococcal nuclease, which can degrade both DNA and RNA) to degrade the cognate mRNAs and the absence of this effect with DNase I treatment showed that RNA was an essential component of the nuclease activity. Hammond et al. (2001) termed the sequence-specific nuclease activity observed in the cellular extracts responsible for ablating target mRNAs as RISC. The RNA endonuclease, Dicer plays a role in aiding RISC action by providing the initial RNA material to activate the complex as well as the first RNA substrate molecule. When the Dicer, which has endonuclease activity against dsRNAs and pre-miRNAs, cleaves a pre-miRNA stem-loop or a dsRNA, a 19–25-base pair dsRNA fragment is formed with a two-nucleotide 3′ overhang at each end.

Small RNAs are found to be associated with sequence-specific nuclease, which serves as guides to target-specific messages based upon sequence recognition. The multicomponent RNAi nuclease has been purified to homogeneity as a ribonucleoprotein complex of $\approx$500 kDa. One of the protein components of this complex belonged to Argonaute family of proteins and called as Argonaute2 (AGO2). AGO2 is homologous to RDE1, a protein required for dsRNA-mediated gene silencing in *C. elegans.* AGO2 is a $\approx$130 kDa protein containing polyglutamine residues, PAZ and PIWI domains characteristic of members of the Argonaute gene family. The Argonaute family members have been linked both to the gene-silencing phenomenon and to the control of development in diverse species.

5.4.3 RNA-Dependent RNA Polymerase

The effects of RNAi are potent and systemic in nature. This has led to a proposed mechanism in which RNA-dependent RNA polymerases (RdRPs) play a role in both triggering and amplifying the silencing effect. An excess amount of dsRNA is considered better for the formation of more 21–23 nt dsRNAs to survive dilution by cell division. This dsRNA signal is amplified in nature by an RdRP activity. RdRP probably converts an aberrant ssRNA population into dsRNA or repeatedly copy the dsRNAs so as to produce a population of ssRNAs that could then interact with target RNA.

In plants, the DNA-dependent RNA polymerase IV (Pol IV) is important for siRNA production and RNAi-directed transcriptional silencing. However, Pol IV homologues are not found in fungal or animal genomes. In yeast, mutants of the RNA polymerase II (Pol II) subunits show loss of centromeric siRNA and RNAi-dependent heterochromatin formation at the centromeric repeat regions, probably by coupling transcription with transcriptional silencing machinery. It is not known whether Pol II plays a similar role in post-transcriptional silencing and in the silencing of other chromosomal regions.

5.4.4 RNA and DNA Helicases

A diverse array of proteins specific for most of the eukaryotic organisms seems to carry out elimination of aberrant RNA. Broadly, they fall in the biochemically similar group of RNA-DNA helicases. A strain of *C. reinhardtii* carrying *mut-6* mutation was found to relieve silencing by a transgene, and by also activating the transposons. The deduced MUT6 protein contains 1,431 amino acids and is a member of the DEAH box RNA helicase family. In *Neurospora crassa*, three classes of quelling-defective mutants (*qde1*, *qde2*, and *qde3*) have been isolated. The *qde3* gene has been cloned, and the sequence encodes a 1,955 amino-acid protein. The protein shows homology with several polypeptides belonging to the family of RecQ DNA helicases, which includes the human proteins for Bloom's syndrome and Werner's syndrome. In addition, QDE3 protein is believed to be involved in the activation step of gene silencing.

Seven *smg* genes involved in nonsense-mediated decay have been identified from the mutants of *C. elegans*. The SMG proteins could unwind dsRNA to provide a template for amplification activity. In this way, the three SMG proteins might facilitate amplification of the silencing signal and cause persistence of the silenced state. Alternatively, SMG proteins could increase the number of dsRNA molecules by promoting endonucleolytic cleavage of existing dsRNA molecules, which has been observed in *Drosophila* flies. No SMG2 homologues have been identified in plants or fungi. However, a search of the *Arabidopsis thaliana* genome sequence database revealed a number of candidates with either helicase and/or RNase domains.

5.4.5 Trans-Membrane Protein (Channel or Receptor)

It is well established now that in a number of organisms the spread of gene silencing from one tissue to another is systemic in nature. A green fluorescent protein (GFP) was identified from a special transgenic strain of *C. elegans* HC57. Out of 106 *sid* mutants belonging to three complementation groups (*sid1*, *sid2*, and *sid3*), *sid1* mutant was characterized which failed to show systemic RNAi. The SID1 polypeptide is predicted to be a 776 amino-acid membrane protein consisting of a signal peptide and 11 putative trans-membrane domains. Based on the structure of SID1, it was suggested that it might act as a channel for the import or export of a systemic RNAi signal or might be necessary for endocytosis of the systemic RNAi signal, perhaps functioning as a receptor.

5.5 Mechanism

Our knowledge on RNAi has expanded dramatically in short time since its discovery. In the last few years, important insights have been gained in elucidating the mechanism of RNAi, although it seems to be very complicated. RNAi appears to be a highly potent and specific process, which starts when a dsRNA is introduced either naturally or artificially in a cell. An endoribonuclease enzyme cleaves the long dsRNA into small pieces of miRNA or siRNA depending upon the origin of long dsRNA, i.e., endogenous or exogenous, respectively. A dsRNA may be generated by either RNA-dependent RNA polymerase or bidirectional transcription of transposable elements.

Based on the *in vitro* and *in vivo* experimental results, a two-step mechanistic model for RNAi has been established (Fig. 5.1). The first step, referred to as the RNAi initiating step, involves the binding of RNA nucleases to a large dsRNA and its cleavage into discrete $\approx$18- to $\approx$23-nucleotide RNA fragments (siRNA). In the second step, these siRNAs join a multinuclease complex, RISC, which degrades the homologous single-stranded mRNAs.

5.5.1 Initiation

This stage is characterized by generation of siRNA mediated by type III endonuclease Dicer. It appears that the RNAi machinery, once it finds RNA duplexes, cuts it up into small molecules with a length of about 21 nt. These also have 2-nucleotide, 3′ overhangs, and 5′ phosphates. Of course, different organisms have different numbers of Dicer genes that process different sorts of dsRNAs. In Drosophila, Dicer-1 is seen to interact with the dsRBD protein Loquacious (Loqs). Immunoaffinity purification experiments have revealed that Loqs resides in a functional pre-miRNA processing complex, and stimulates and directs specific pre-miRNA processing activity. Dicer-2 requires R2D2, a dsRNA binding protein, which unlike Loq is supposed to be composed of two dsRNA binding domains. Both dsRNA binding domains of Dcr-2 and R2D2 are critical for Dcr-2/R2D2 complex to bind and load siRNA into siRISC complex.

In general, Dicer works by recognizing the ends of dsRNA with its PAZ domain (Dicer has PAZ domain at one end and dual RNase III domains on the opposing end) and then cutting the dsRNA with its RNase III domain. The distance between the PAZ and RNase III domain determines the length of siRNA produced by Dicer which varies by species. Long dsRNA and miRNA precursors are processed to siRNA/miRNA duplexes by the RNase-III-like enzyme Dicer. These short dsRNAs are subsequently unwound and assembled into effector complexes, RISCs, which can direct RNA cleavage, mediate translational repression or induce chromatin modification.

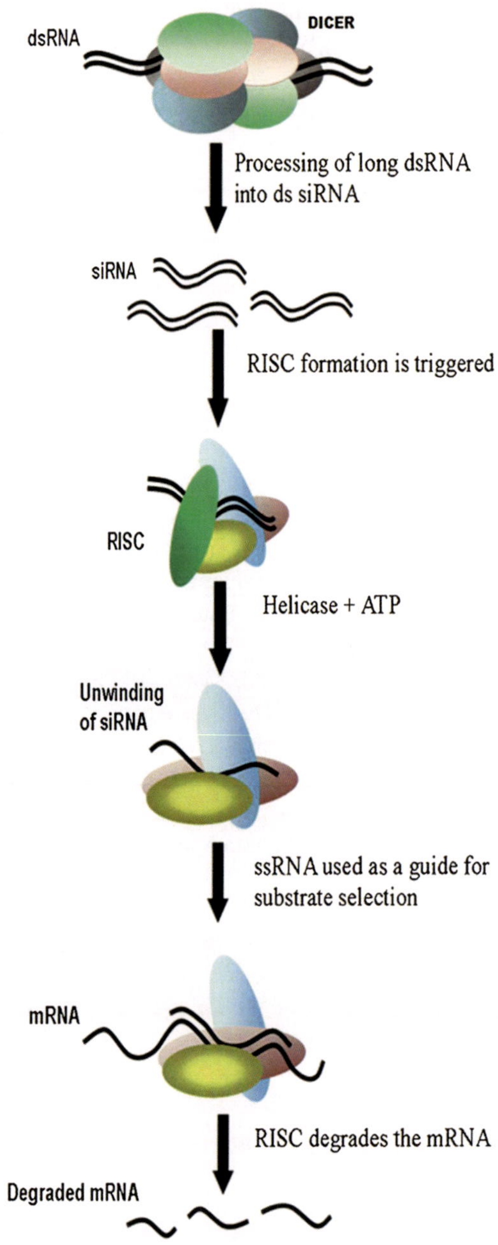

Fig. 5.1 Mechanism of RNA interference

5.5.2 Effector

The second step of RNAi mechanism is referred as effector step; the incorporation of guide strand. This is characterized by the formation of silencing complex. The siRNA and miRNA duplexes that contain ribonucleoprotein particles (RNPs) are now made into RISC. Generally, effector complexes containing siRNAs are known as a RISC, while those containing miRNAs are known as miRNPs.

It is thought that RISC undergoes an ATP-dependent step that activates the unwinding of the double-stranded siRNAs. RISC is composed of PPD proteins (PAZ PIWI Domain proteins), which are highly conserved super-family. Members of PPD proteins contain PAZ (100 amino acids) and C-Terminal located PIWI (300 amino acids). The Argonaute protein and one strand of the siRNA form the RISC. Once the RISC complex has been formed from 5′-phosphorylated siRNAs and endogenous Argonaute protein, the siRNAs in the RISC complex guide the degradation that is sequence-specific, of the complementary or near complementary mRNAs. RISC binds to the mRNA which is targeted by the single RNA strand within the complex and cleaves the mRNA or represses their translation by homology-dependent m-RNA degradation. mRNA is cleaved in the middle of its complementary region. This cleaved mRNA cannot be translated. These newly synthesized siRNAs destroy the target with the help of RNA-dependent RNA polymerase (RdRP). Afterwards RISC dissociates and is ready to cleave other mRNAs. By that even a few numbers of the RISC can lead to high-level gene silencing. The siRNA derived in RNAi pathway may have two different fates, either it may be converted into holo-RISC for the destruction of the target m-RNA, or it may contribute to the amplification step by the generation of secondary siRNA.

RISC and miRNP complexes work by catalyzing hydrolysis of the phosphodiester linkage of the target RNA. The mechanism by which repression of translation guided by miRNA works, is not as well understood. It has been suggested that miRNAs affect translation termination or elongation rather than actual initiation of the process. Besides, miRNAs can act as siRNAs and vice versa. Perhaps mRNA degradation and translational regulation guided by miRNAs are used as simultaneous mechanisms for natural regulation.

5.6 Delivery Methods

RNAi has rapidly advanced from a laboratory observation into a major area of research within biology and medicine. The RNAi treatments are entering various stages of clinical trials. However, development of RNAi-based agents has been hindered because siRNAs are unstable in serum and delivery across the cell membrane is highly inefficient. Numerous methods have been developed to facilitate delivery of RNAi in animals and patients, each with their own set of advantages and disadvantages. To improve the effect of RNAi-based therapy, the enzymatic stability and cellular uptake of siRNA should be significantly enhanced, while their immunoactivation should be decreased. The two most common approaches for RNAi delivery are lipid mediated transfection and viral mediated transduction. Typically, researchers strive to achieve the highest levels of transfection efficiency possible. Transfection efficiency describes the percentage of cells that have received the RNAi duplex or expression plasmid. For many disease models, the most desirable cell types to use are primary cultures. However, these cannot be transfected adequately with commercially available transfection reagents. A powerful alternative to cationic lipid-mediated transfection is viral delivery of vectors expressing RNAi sequences. This option is the best for delivery to primary, hard-to-transfect, and non-dividing cells. Viral delivery can also be used to create stable cell lines with inducible RNAi expression or to express RNAi sequences with tissue-specific promoters.

5.6.1 Viral Methods

Since viruses work by delivering genetic material into a cell, scientists have been viewing them as a potential method of delivering DNA or RNA into a cell since long. Lentiviruses, a family of retroviruses that includes the human immunodeficiency virus (HIV), use RNA as their genetic material and have been used to deliver RNA to the cells of live mammals (including humans) for RNAi. Lentiviruses are valued among retroviruses for this purpose because they are the only ones that can work in non-dividing cells. Other viruses may also be used for RNAi delivery. Adenoviruses and herpes simplex-1 virus, both of which use double-stranded DNA instead of RNA, have been studied as RNAi vectors. Jonge et al. (2006) utilized virosomes derived from influenza as a means of delivering siRNA *in vivo*. Virosomes, which are empty influenza virus envelopes were complexed with cationic lipids for protection from nuclease activity and, following intraperitoneal injection, were demonstrated to successfully encapsulate siRNA duplexes and deliver them to cells in peritoneal cavity of mice. Virosomes may also serve as an effective means to target cells of the respiratory epithelium and cells that possess Fc receptors, such as dendritic-cells, macrophages, and natural-killer cells. Viral vectors have also been used to suppress the expression of HIV human co-receptor, chemokine receptor 5 to prevent HIV infection to penetrate lymphocytes. Such technologies are undergoing tests in a clinical trial (Nguyen et al. 2008).

Despite initial promise as a vehicle for the delivery of RNAi into live animals, viral vectors no longer appear to be the tool of choice due to adverse effects observed in viral vector gene therapy trials, and RNAi-induced hepatotoxicity.

5.6.2 Liposomal Delivery

For many immortalized cell lines, transfection with a lipid- or amine-based reagents is the preferred option. A major focus in the development of RNAi therapeutics is liposomal formulations for siRNA delivery to improve efficacy and safety. Many commercial institutions as reported by Nguyen et al. (2008) such as Silence Therapeutics plc are in the process of improving liposomal delivery. The work is in progress on a novel class of lipid delivery vehicles using 'AtuPLEX,' a mixture of cationic and fusogenic lipids complexed with negatively charged siRNAs, to promote uptake of siRNAs into endothelial cells of blood vessels in the liver and tumors.

Protiva Biotherapeutics Inc and Alnylam have utilized stable nucleic acid lipid particle (SNALP)-formulated siRNAs. SNALPs are specialized lipid nanoparticles that encapsulate siRNAs and are coated with a diffusible PEG-lipid conjugate which stabilizes the particle during formulation by providing a neutral hydrophilic exterior and prevents rapid systemic clearance *in vivo* by shielding the cationic bilayer; therefore, SNALPs facilitate the cellular uptake and endosomal release of the particle's siRNA. In addition, Alnylam is exploring a library of 'lipidoids' or lipid-like molecules developed by Robert Langer and Daniel Anderson, to develop siRNA therapeutics for two different disease targets in the liver: hypercholesterolemia and liver cancer. Chemical modification and bioconjugation with lipids are known to improve the stability and cellular uptake of siRNA.

5.6.3 Chemical Modification

Chemical modifications to sugars, backbones, or bases of siRNAs have been shown to enhance their stability, prevent them from triggering an immune response, control their pharmacokinetic profiles and reduce nonspecific effects without affecting their biological activity. Most commonly used chemical modifications of siRNA include phosphorothiolation (P = S) of the non-bridging oxygen at the 3′-end, 2′-sugar modification (such as 2′-OMe or 2′-F) and locked nucleic acid (LNA). Chemical modifications of siRNA confer enhanced gene silencing at lower doses and reduced dosing frequency. In an experiment it was found that the placement of one, two or three ribonucleotides at the 5′ end of the modified siRNA of HBV263 and HBV1583 improved the median inhibitory concentration of the stabilized siRNAs by approximately fivefold.

5.6.4 Bioconjugation

A siRNA can be linked to a carrier molecule or covalently conjugated to a targeting ligand *via* a cleavable spacer to increase its cellular uptake or confer cell-specific targeting. To date, siRNA has been conjugated to lipids, polymers, peptides, and aptamers. Conjugation of siRNA to apatamer which can bind to prostate-specific membrane antigen (PSMA) has shown some promise for targeted delivery of siRNA to prostate cancer. siRNAs have also been conjugated to lipids like lithocholic and lauric acids at the 5′-end of the sense strand and showed increased cellular uptake of siRNAs in human liver cells without the use of any transfection reagent. Conjugated cholesterol to the 3′-end of the sense strand of siRNA targeting ApoB, an essential protein for assembly and secretion of very-low-density lipoprotein (VLDL) and low-density lipoprotein (LDL), has been formed that demonstrate significant increase in the levels of this conjugate in the liver.

5.6.5 Nanoparticle Delivery

The generation of nanosized particles is being investigated to enhance the delivery of siRNA based drugs and to circumvent the limitations of difficult to transfect cells such as lymphocytes. Numerous targeted nanoparticles have been used to overcome one or more of the barriers encountered by siRNA when trafficked to the cytosol. Antibody-protamine fusion bodies and antibody-targeted liposomes have been used successfully for primary lymphocytes to target human integrin lymphocyte function associated antigen-1 (LFA-1). Calando Pharmaceuticals Inc. has developed a cationic cyclodextrin-containing polymer that binds siRNAs for systemic delivery. Cyclodextrin containing polycation binds to the anionic backbone of the siRNA and then self-assemble into nanoparticles of approximately 50 nm in diameter.

5.6.6 Hydrodynamic Injection

Quick injection of siRNA in a large volume of physiological buffer effectively localizes duplex siRNA in the liver. In rats, administration of a VEGF (vascular endothelial growth factor)—specific siRNA resulted in more than 75 % inhibition of pathological neovascularization. Due to the invasiveness of the injection technique, hydrodynamics-based transfection is not appropriate for clinical applications at this point. However, recent advances in using a computer-controlled, catheter-guided injection device have greatly improved the precision and reproducibility of this approach. To date, the device has only been used for the delivery of DNA, but siRNA should be equally applicable in this approach.

5.6.7 Direct Administration into Target Organs

For dsRNA delivery several direct methods can be applied. Electroporation is used in simpler organisms, whereas microinjection of dsRNA into germ line or early embryo is preferred in multicellular organisms. In *C. elegans*, injection into the intestine or pseudocoelom is almost as efficient as injection into germ line. Even feeding worms with bacteria that express dsRNA or soaking worms in dsRNA solutions has been applied with some success. However, dsRNA injected into early embryos is diluted upon cell division; as a result early genes are more easily inactivated than late genes, which is especially a problem for higher organisms. In ongoing clinical trials of RNAi therapeutics, direct application of modified or unmodified siRNA into the eye (Allergen Inc. and Opko Health Inc.) and skin (TransDerm Inc.) through injection or by intranasal delivery (Alnylam) are the leading methods of administration. However, these delivery methods may be limited to a few indications in a small number of accessible organs such as the lung, eye and skin.

5.6.8 Delivery into Plants

siRNAs have been delivered into tobacco plants by biolistic pressure to cause silencing of GFP expression. Silencing occasionally was detected as early as a day after bombardment, and it continued to potentiate up to 3–4 days post-bombardment. Systemic spread of silencing occurred 2 weeks later to manifest in the vascular tissues of the non-bombarded leaves that were closest to the bombarded ones. After a month or so, the loss of GFP expression was seen in nonvascular tissues as well. RNA blot hybridization with systemic leaves indicated that the biolistically delivered siRNAs induced the *de novo* formation of siRNAs, which accumulated to cause systemic silencing.

5.7 Applications of RNAi

The new technique was taken up by lots of enthusiasm by scientists. It was soon realized that there has not been a tool this sharp in last many years. Besides being an area of intense, upfront basic research, the RNAi process holds the key to future technological applications.

5.7.1 Gene Knockdown

RNAi is a specific, potent, and highly successful approach for loss-of-function studies in virtually all eukaryotic organisms. As RNAi relies on the sequence-specific interaction between siRNA and mRNA, siRNAs can be tailored to silence almost any gene. Using this mechanism, researchers can cause a drastic decrease in the expression of a targeted gene. Studying the effects of this decrease can show the physiological role of the gene product. Since RNAi may not totally abolish expression of the gene, hence it is more appropriate to call it "knockdown" to differentiate it from "knockout" procedures in which expression of a gene is totally extinguished.

Genome sequencing projects generate a wealth of information. The ultimate goal of such projects is to accelerate the identification of the biological function of genes. The function of genes can be analyzed with an appropriate assay, by examining the phenotype of organisms that contain mutations in the gene, or on the basis of knowledge gained from the study of related genes in other organisms. However, a significant fraction of genes identified by the sequencing projects are new and cannot be rapidly assigned functions by these conventional methods. RNAi technique has come to rescue as it is not only accurate but can also be used on a breathtaking scale. Although RNAi is unlikely to replace the existing knockout technology, it may have a tremendous impact for those organisms that are not amenable to the knockout strategy.

The relative ease with which genes can be silenced using RNAi has caused a minor revolution in molecular biology. siRNAs that have been chemically synthesized or created by *in vitro* transcription systems, can induce silencing in several organisms, including mammalian cells. In the laboratory, this technique seems to be rapid and convenient, and can be used to tackle many genes at the same time. As discussed earlier it can be achieved remarkably easily in *C. elegans* as gene silencing occurs by direct feeding on bacteria which were earlier fed with dsRNA that expressed it or even soaking worms in dsRNA. The effect was found to be transmitted to the next generation.

Extensive efforts in bioinformatics have been directed toward the design of successful dsRNA reagents that maximize gene knockdown but minimize "off-target" effects, which arise when an introduced RNA has a base sequence that can pair with, thus reduce the expression of multiple genes at a time. Such problems occur more frequently when the dsRNA contains repetitive sequences. A multitude of software tools have been developed for the design of general mammal-specific and virus-specific siRNAs that are automatically checked for possible cross-reactivity.

Although long strand of RNA designed to be cleaved by Dicer can also be used but in most mammalian cells, shorter RNAs are preferred so that induction of mammalian interferon response can be avoided. Specialized laboratory techniques have also been developed to improve the utility of RNAi in mammalian systems by avoiding the direct introduction of siRNA, for example, by stable transfection with a plasmid encoding the appropriate sequence from which siRNAs can be transcribed, or by more elaborate lentiviral vector systems allowing the inducible activation or deactivation of transcription, known as *conditional RNAi*. Julie Ahringer's group at the University of Cambridge has created a library of more than 16,000 cloned dsRNAs (around 86 % of the *C. elegans* genome). By feeding these clones to worms, they have determined the function of 1,722 genes, most of which were previously unknown. The research team of Gary Ruvkun at Harvard Medical School has identified many hundreds of genes involved in fat storage. Similarly, Gregory Hannon's group at Cold Spring Harbor is looking to determine the functions of every gene in the human genome. The list of projects in this area is seemingly endless. In another study conducted by Mutti et al. (2006), microinjection of long dsRNA was shown to transiently silence the expression of two marker genes in the pea aphid, *Acyrthosiphon pisum*. They found RNAi knockdown of a salivary transcript leading to lethality.

5.7.2 Therapeutic Possibilities

Given the gene-specific features of RNAi, it is conceivable that this method will play an important role in therapeutic applications. This approach relies on designing the dsRNA from the DNA sequence of a gene that can specifically and effectively silence a disease-related gene. There was an initial problem in developing therapeutic approaches as dsRNAs that are 30 nucleotides in length or longer can trigger non-sequence-specific interferon responses in mammalian cells. But this has been overcome by delivering siRNAs into mammalian cells in culture. Tremendous amount of work is going on in this direction throughout the world on different diseases. Basically every human disease caused by activity from one or a few genes should be amenable to RNAi based intervention. This list includes cancer, autoimmune diseases, genetic disorders, and viral infections.

5.7.2.1 Acquired Immunodeficiency Syndrome

Acquired Immunodeficiency Syndrome (AIDS) causing HIV is first infectious agent targeted by RNAi because the life cycle and pattern of gene expression of HIV is well understood. It was demonstrated by several scientists that siRNAs can inhibit HIV replication effectively in culture. HIV infection can also be blocked by targeting either viral genes such as *gag*, *rev*, *tat* , and *env* or human genes viz. *CD*4, the principal receptor for HIV. This is promising, as antiviral therapies that can attack multiple viral and cellular targets could circumvent genetic resistance of HIV. Synthetic siRNAs have been used to target various early and late HIV-encoded RNAs in cell lines and in primary haematopoietic cells including the TAR element 50, tat 51–53, rev 51, 52, gag 54, vif 50, nef 50 and reverse transcriptase. These results have been achieved by transfecting the siRNAs into cells, but the challenge lays in translating the same effect *in vivo*. However, the work is going on to make better delivery systems to circumvent the problem, many groups have designed promoter systems that can express functional siRNAs when transfected into human cells. Early results have shown that this can decrease replication of HIV considerably.

5.7.2.2 Hepatitis

Hepatitis viral infection is one of the leading causes of liver cirrhosis and hepato cellular carcinoma (HCC) and remains a challenge for modern medicine. Current treatment strategies of Hepatitis B virus (HBV) infection including the use of α-interferon (IFN) and nucleotide analogues have met with only partial success. Therefore, it is necessary to develop more effective antiviral therapies that can clear HBV infection with fewer side effects. Early RNAi studies noted that RNA silencing was prominent in the liver, which made this organ an attractive target for therapeutic approaches. Hepatitis caused by HBV and Hepatitis C virus (HCV) has been an important target for potential RNAi therapy. The first *in vivo* demonstration of RNAi efficacy against a virus implicated a hydrodynamic co-delivery of an HBV replicon and an expression unit encoding an anti-HBV siRNA in mice. This study with siRNA showed a significant 99 % knockdown of HBV core antigen in liver hepatocytes providing an important proof of principle for future antiviral application of RNAi in liver.

Many immune-related liver diseases are characterized by apoptosis, which is mediated by a protein called Fas. Judy Lieberman's group injected siRNA targeting Fas intravenously into two models of autoimmune hepatitis in mice. This decreased Fas mRNA and protein levels in hepatocytes and protected the cells against liver injury from apoptosis, even when siRNA was administered after the induction of injury. Hence, the successful use of chemically synthesized siRNA, endogenous expression of small hairpin RNA (shRNA) or miRNA to silence the target gene make this technology towards a potentially rational therapeutics for HBV infection.

5.7.2.3 Cancer

The potential for using RNAi to treat metastatic cancers depends on finding good cellular targets. With respect to future medical applications, siRNA-based therapy seems to have a great potential to combat carcinomas, myeloma and cancer caused by over expression of an oncoprotein or generation of an oncoprotein by chromosomal translocation and point mutations. Gregory Hannon and colleagues have used RNAi to silence expression of p53, the 'guardian of the genome' which protects against any tumor-associated DNA damage by introducing several p53-targeting short hairpin RNAs (shRNA) into stem cells in mice. The shRNAs produced a wide range of clinical effects, ranging from benign to malignant tumors, the severity and type of which correlated with the extent to which the shRNA had silenced p53. As tumor suppressors such as p53 usually work as part of a complex and finely regulated network, the ability to dampen these networks to varying degrees will be of enormous value when it comes to investigating the early stages of disease. The success of these modified stem cells also gives hope that this could treat diseases in which stem cells can be modified *ex vivo* and then re-introduced into the affected individual.

In chronic myelogenous leukemia (CML), RNAi has been used to target genes expressing oncogenic fusion proteins such as Brc-Abl oncoprotein p210. The feasibility of delivery of siRNAs and expressing siRNAs viral vectors to diseased regions of the brain coupled with selective targeting of SNPs in the mutant transcripts may be promising for using it as an anticancer therapeutic agent. Transfected siRNAs achieve significant gene knockdown for 3–7 days before its natural degradation. This may not be sufficient for therapy but can be useful for studying the immediate effects of inhibiting gene expression for drug targets or drug target validation.

Researchers at the Charity Cancer Research, UK, and the Netherlands Cancer Institute are working to generate a large library of human cells, each containing a silenced gene with a goal of silencing 300–8,000 cancer genes initially, and then eventually covering the entire human genome. This will help to uncover all the genes that become over expressed in human cancers and to find out precisely what needs to be taken away from a cancerous cell in order to make it normal again.

5.7.2.4 Malaria

Malaria, a mosquito-borne infectious disease of humans and animals, is a leading cause of morbidity and mortality worldwide. Recent evidence strongly suggests that RNAi can play a key role in identifying new genetic means of controlling mosquito-borne diseases. Anopheles mosquitoes were transfected with a transgene containing two copies of the target gene arranged in an inverted repeat configuration. Hairpin RNA is expressed *in vivo* whenever the inverted repeat is transcribed from an upstream promoter. By placing dsRNA expression under the control of a tissue-specific and time-specific promoter, dsRNA expression can be tailored to coincide spatially and temporally with the journey of the parasite through the mosquito. Both parasite receptors and immune components protective of the parasite are putative targets for engineering parasite resistance through RNAi and, in principle, mosquito strains that have been rendered refractory to malaria transmission could be released in the field to replace wild type, permissive populations and achieve malaria eradication (Angaji et al. 2010).

5.7.2.5 Ocular Diseases

Several diseases affect the eye, either directly or as part of a system-wide problem. Wallach (2004) illustrated that the enclosed nature of eye makes ocular diseases ideal targets for siRNA-based therapies. siRNA was detected for at least 5 days when injected into the vitreous cavity. The sequence specificity of siRNA resulting in targeting of a single gene combined with local administration in the relatively isolated confines of the eye provides an ideal way to study eye-specific effects of gene disruption.

The first ever clinical trial in man in 2004 by Acuity Pharmaceuticals have been directed at the treatment of age-related macular degeneration (AMD), which causes blindness or limited vision in millions of adults annually in patients. A siRNA targeting the vascular endothelial growth factor (VEGF), a primary cause of overgrowth of blood vessels in the 'wet' form of AMD, was administered by intravitreal injection. The primary goal of this phase 1 study was to evaluate the safety of the siRNA. A single intravitreal dose of Sirna-027, ranging from 100–800 μg, appears to be safe and well tolerated with no systemic or local adverse events related to the drug.

5.7.2.6 Metabolic Diseases

RNAi holds considerable promise as a therapeutic approach to silence disease causing genes, particularly those that encode so-called 'nondrugable' targets. The endogenous genes involved in the cause or pathway of metabolic diseases are silenced. Besides, no toxicity and adverse effects are observed due to high potency, specificity, and chemical structure of RNAi. A few studies utilized RNAi technology to target key genes involved in the regulation of gluconeogenesis and provided *in vivo* proof-of-principle for the development of RNAi-based therapeutics for diabetes. A link between miRNAs and diabetes seems increasingly likely. In 2004, Poy and his group identified several miRNAs expressed selectively in pancreatic endocrine cell lines: among these was miR-375, over expression of which resulted in suppressed glucose-stimulated insulin secretion and whose inhibition enhanced insulin secretion. Taniguchi et al. (2005) developed an adenovirus-mediated RNAi technique that utilizes shRNAs to substantially and stably reduce the expression of insulin receptor substrates IRS-1 and IRS-2 specifically in the liver of mice to better understand the roles of these proteins in hepatic insulin action. By knocking down IRS-1 and IRS-2 in the liver separately and together, they showed that IRS-1 signaling may be more closely linked to the regulation of genes involved in glucose homeostasis, whereas IRS-2 signaling may have specific roles in the regulation of hepatic lipid metabolism.

Vector-based RNAi approach has been used for hyperglycemia through phosphoenol pyruvate carboxykinase (PEPCK), a rate-controlling enzyme in gluconeogenesis. RNAi also holds promise for the development of novel therapeutic strategies for disorders that are yet difficult to treat such as obesity, neuropathic pain and depression. Although development of RNAi-based therapeutics is in its infancy, early clinical studies show the use of this new class of therapeutics to tackle metabolic diseases.

5.7.2.7 Neurodegenerative Disorders

Neurodegenerative disease is an umbrella term for a range of conditions which primarily affect the neurons in the human brain. The diseases such as Alzheimer's, Parkinson's, Huntington's and amyotrophic lateral sclerosis (ALS) are examples of relatively common age-related neurodegenerative disorders that are increasing as average life expectancy increases. Each disorder is characterized by the dysfunction and death of specific populations of neurons. It is believed that specific genetic mutations are responsible for these diseases. Recent studies have shown that cultured neurons can be efficiently transfected with siRNAs to effectively silence the target genes.

5.7.2.8 Cardiovascular and Cerebral Vascular Diseases

Cardiovascular diseases are the leading cause of death in developed countries. siRNA is a recent advancement that provides the possibility of reducing gene expression at the post-transcriptional level in cultured mammalian cells. This technology has been exploited in the process of atherosclerosis (progressive occlusion of arteries) and also to reduce the damage to heart tissue and brain cells that patients suffer following a myocardial infarction or stroke.

5.7.2.9 Genetic Diseases

A genetic disorder is an illness caused by abnormalities in genes or chromosomes, especially a condition that is present from birth. siRNA seems to be a good candidate for the treatment of genetic diseases because the target-specific nature of siRNA may be able to target a single mutation in the genome, which is causing the disease. Once the target has been located, and the siRNA has been effectively delivered, it has the potential to down-regulate the single mutation, which could restore proper health. The challenge in this method is finding the optimal delivery method for specific sites in the genome, while not affecting the normal genes in the cell. Each genetic disease causes an unique mutation, so finding the specific targets for each disease is the limiting factor for the therapeutic potential of siRNA. The use of siRNA for genetic treatment is also very expensive because of the unique sequence target and diversity among individuals, so customized solutions are needed.

5.7.2.10 Other Diseases

Many groups of researchers are involved in studies related to knocking down the diseased genes. Chi et al. (2003) reported that the siRNA-induced gene silencing of transient or stably expressed GFP mRNA was highly specific in the human embryonic kidney (HEK) 293 cell background. Among the first applications to reach clinical trials were the treatment of muscular degeneration and respiratory syncytial virus (RSV). siRNAs were successfully used to silence genes expressed from RSV and parainfluenza virus (PIV), the RNA viruses that cause severe respiratory disease in neonates and infants. Other proposed clinical trials use center on antiviral therapies, including topical microbicide treatments that use RNAi to treat infection by herpes simplex virus type 2, silencing of influenza gene expression and inhibition of measles viral replication.

5.7.3 Functional Genomics

Functional genomics is the field that attempts to describe gene functions and interactions. There has been a lack of powerful tools for systematic analysis of mammalian gene function, but RNAi may now provide such a strategy. Genomes for many organisms including human has already been sequenced and this technique provides a way by which this enormous wealth of information can be translated into functional definitions for every gene. It may be useful to analyze quickly the functions of a number of genes in a wide variety of organisms. Most functional genomics applications of RNAi in animals have used common model organisms such as *C. elegans* and *Drosophila melanogaster*. *C. elegans* is particularly useful for RNAi research because of the heritable effects of the gene silencing and simple delivery of the dsRNA. Chromosomes I and III of *C. elegans* have been screened by RNAi to identify the genes involved in cell division and embryonic development. Recently, a large-scale functional analysis of $\approx$19,427 predicted genes of *C. elegans* was carried out with RNAi. This study identified mutant phenotypes for 1,722 genes. Another study performed a genome-wide screen to identify the genetic components required for RNAi, using an engineered RNAi sensor strain of *C. elegans*. The RNAi screen identified 90 genes including Piwi/PAZ proteins, DEAH helicase, RNA binding/processing factors, chromatin-associated factors, DNA recombination proteins, nuclear import/export factors, and 11 known components of RNAi machinery. Similarly, in *D. melanogaster*, RNAi technology has been successfully applied to identify genes with essential roles in biochemical signaling cascades, embryonic development, and other basic cellular processes. With the recent development of siRNA and shRNA expression libraries, the application of RNAi technology to assign function to cancer genes and to delineate molecular pathways in which these genes affect in normal and transformed cells, are expected to contribute significantly to the knowledge necessary to develop new techniques and also to improve existing cancer therapy.

Stable transcription of RNAi triggered from suitable expression cassettes integrated into the host cell genome by viral gene transfer can induce long-term and heritable gene silencing in mammalian cells. However, the use of RNAi as a genetic tool is limited due to difficulties in identifying efficient RNAi triggers, the problem of effective delivery and off-target effects, as well as potential genotoxic side effects of viral gene transfer strategies.

Functional genomics using RNAi is a particularly attractive technique for genomic mapping and annotation in plants because many plants are polyploid, which presents substantial challenges for more traditional genetic engineering methods. For example, RNAi has been successfully used for functional genomics studies in hexaploid bread wheat as well as *Arabidopsis* and maize, common plant model systems. Plant nematology in the genomics era is now facing the challenge to develop RNAi screens adequate for high-throughput functional analyses. RNAi-mediated suppression of genes essential for nematode development, survival, or parasitism is revealing new targets for nematode control.

5.8 Conclusion

RNAi is a mechanism for controlling normal gene expression. Lately, it has become a powerful tool for gene therapy for many diseases, besides being used for probing gene functions, making transgenic plants, and rationalizing drug design. However, further refinements are required for the development of a delivery system with little or no toxicity, but with high efficiency and selectivity for consistent and stable expression levels before it can be used successfully. Since human subjects are involved in gene therapy, hence it is also important to perform a risk–benefit analysis and considering the risks of generating infection-competent viruses or introducing genetic changes in germ line cells.

References

Agarwal A, Dasaradhi PVN, Mohmmed A, Malhotra P, Bhatnagar RK, Mukherjee SK (2003) RNA interference: biology, mechanism, and applications. Microbiol Mol Biol Rev 67(4):657–685

Angaji SA, Hedayati SS, Poor RH, Madani S, Poor SS, Panahi S (2010) Application of RNA interference in treating human diseases. J Gen 89:527–537

Bernstein E, Caudy A, Hammond S, Hannon G (2001) Role for a bidentate ribonuclease in the initiation step of RNA interference. Nature 409(6818):363–366

Chi JT, Chang HY, Wang NN, Chang DS, Dunthy N, Brown PO (2003) Genomewide view of gene silencing by small interfering RNAs. Proc Natl Acad Sci USA 100:6343–6346

de Jonge J, Holtrop M, Wilschut J, Huckriede A (2006) Rreconstituted influenza virus envelope as an efficient carrier system for the cellular delivery of small interfering RNAs. Gene Ther 13(5):400–411

Fire A, Xu S, Montgomery MK, Kostas SA, Driver SE, Mello CC (1998) Potent and specific genetic interference by double-stranded RNA in *Caenorhabditis elegans*. Nature 391:806–811

Hammond SM, Boettcher S, Caudy AA, Kobayashi R, Hannon GJ (2001) Argonaute2, a link between genetic and biochemical analyses of RNAi. Science 293:1146–1150

Lee RC, Feinbaum RL, Ambros V (1993) The *C. elegans* heterochronic gene lin-4 encodes small RNAs with antisense complementarity to lin-14″. Cell 75(5): 843–854

Mutti NS, Park Y, Reese JC, Reeck GR (2006) RNAi knockdown of a salivary transcript leading to lethality in the pea aphid, *Acyrthosiphon pisum*. J Ins Sci 6:38

Napoli C, Lemieux C, Jorgensen R (1990) Introduction of a chimeric chalcone synthase gene into petunia results in reversible co-suppression of homologous gene in trans. Plant Cell 2:279–289

Nguyen T, Menocal EM, Harborth J, Fruehauf JH (2008) RNAi therapeutics: an update on delivery. Curr Opin Mol Thera 10(2):158–167

Poy MN, Eliasson L, Krutzfeldt J, Kuwajima S, Ma X, Macdonald PE, Pfeffer S, Tuschl T, Rajewsky N, Rorsman P, Stoffel M (2004) A pancreatic islet-specific microRNA regulates insulin secretion. Nature 432:226–230

Taniguchi CM, Ueki K, Kahn CR (2005) Complementary roles of IRS-1 and IRS-2 in the hepatic regulation of metabolism. J Clin Invest 115:718–727

Wallach T (2004) Acuity pharmaceuticals announces completion of financing round. Acuity Pharma 215:966–6181

DNA Microarray

6

Ashwini M. Charpe

Abstract

DNA microarray is a technology that has revolutionized the functional genomics with a wide array of applications. It is an arrangement of a large number of known genes or corresponding cDNA probes, for a given physiological condition of any living being; that are placed precisely as dots on a glass slide or chip or coated on beads. It works on the principle of Southern hybridization wherein DNA is hybridized with DNA to confirm the expression of a gene. The only difference is, in microarrays probes are placed on solid surface and test DNA is in the hybridization solution, which is just opposite to Southern hybridization where DNA to be diagnosed is placed on nylon or nitrocellulose membrane and probe is in the hybridization solution. DNA microarray helps in screening of thousands of genes in one go to understand their expression in a given physiological condition, when hybridized with the test DNA. During hybridization, fluorescent dyes attached to probes produce emissions of specific color based on complete partial and no binding of DNA to the probes. After hybridization these emissions can be observed and recorded under a confocal laser microscope and further analyzed with the help of image analysis software to understand set of genes up or down regulated in the test DNA and to determine fluorescence intensities that allow the quantitative comparison between the two test DNAs for all genes on the array. This technique is useful in gene expression profiling, comparative genomic hybridization, checking GeneID, SNP detection, alternative splicing detection, fusion gene detection and genome tilling to empirically detect expression of transcripts, or alternative splice forms. It has been widely applied in studies related to cancer biology, microbiology, plant science, environmental science, etc.

6.1 Introduction

A microarray is a 2D arrangement of known biochemical or chemical constituent of a biological system on a solid substrate like glass

A. M. Charpe (✉)
Plant Pathology Section, College of Agriculture, Dr.Panjabrao Deshmukh Krishi Vidyapeeth, Akola, Maharashtra 444104, India
e-mail: ashwinicharpe@yahoo.com

I. Ravi et al. (eds.), *Advances in Biotechnology*,
DOI: 10.1007/978-81-322-1554-7_6, © Springer India 2014

slide or silicon chip that is useful in high throughput screening of biological material. Microarray methods were initially developed to study differential gene expression using complex populations of RNA. Refinements of these methods now permit the analysis of copy number imbalances and gene amplification of DNA and have recently been applied to the systematic analysis of expression at the protein level. Many of the guiding principles of global analysis using microarrays are, in principle, applicable at the RNA, DNA or protein level. Till date, various types of microarrays using available biochemical information have been developed that are as following (http://en.wikipedia.org/wiki/Microarray):

- DNA microarrays: cDNA microarrays, oligonucleotide microarrays, and SNP microarrays.
- MMChips: for surveillance of microRNA populations.
- Protein microarrays.
- Tissue microarrays.
- Cellular microarrays or Transfection microarray.
- Chemical compound microarrays.
- Antibody microarrays
- Carbohydrate arrays or glycoarrays.

This chapter deals with the DNA microarrays which are recognized as multiplex technology that either measure DNA or use DNA as part of its detection system. DNA microarrays are prepared by robotic plotting of picomoles (10^{-12} mol) of specific DNA sequences known as probes or reporters on a solid substrate like glass slide, silicon chip, Affymatrix chip (Affy chips in short), and microscopic beads called Illumina. In standard microarrays, the probes are attached to the respective solid surface by surface engineering using epoxy-silane, amino-silane, lysine, polyacrylamide or others that binds the probe to the surface by a covalent bond. With this technique, thousands of probes can be fixed to each cm^2 area on the solid surface appearing like tiny dots known as *features*. An array can contain tens of thousands of probes. These probes might be prepared from-

- Available cDNA sequences from cDNA libraries developed by trapping expressed sequence tags (ESTs), those are usually 200–300 bp long;
- By *in silico* synthesis of 20–30 bp long non-overlapping oligonucleotide sequences; and
- By using the nucleic acid sequences distinguishing the single nucleotide polymorphism (SNPs).

The DNA microarray is hybridized with a cDNA or cRNA sample known as target and is prepared from a biological system under study, which is subjected to a specific condition, under highly stringent situations. For detection and quantification of hybridization of probe with the target, the target is labeled with the signal producing fluorophore, silver, or chemiluminescence labels. These signals help in determining the relative abundance of nucleic acid sequences in the target. Likewise, DNA microarray facilitates the measurement of expression levels, detection of SNPs and genotyping or resequencing of mutant genomes (http://en.wikipedia.org/wiki/DNA_microarray).

6.2 History

DNA microarrays were gradually developed from southern blotting (Maskos and Southern 1992). Hybridization of probe with the blotted nucleic acid fragment results in separation of radioactivity or chromogenic substrate that can be detected by autoradiography or colorimeter.

In the first step DNA microarray was prepared from lysed bacterial colonies having different gene sequences. A total of 378 such colonies were arranged in an array and hybridized with the DNA of normal and tumor tissue in replications (Augenlicht and Kobrin 1982) to study the gene expressions. In the second step, 4,000 human gene sequences were arranged in an array and hybridized with the DNA preparations from human colonic tumors and normal tissue, the results were processed by computer aided scanning and image processing for quantitative analysis of the expressed sequences. This array was also used for comparison of colonic tissues at different genetic risk (Augenlicht et al. 1987, 1991). In the third step, in 1987, a collection of

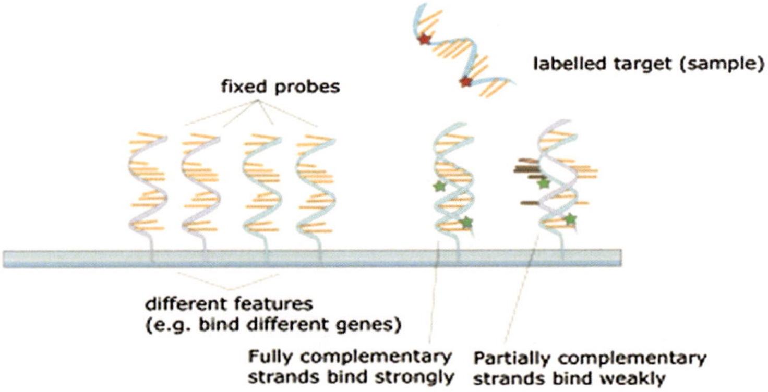

Fig. 6.1 Hybridization of the target to the probe in DNA microarray (*Source* http://en.wikipedia.org/wiki/DNA_microarray)

distinct DNAs was used for preparation of an array and was used for expression profiling leading to identification of genes whose expression was modulated by interferon (Kulesh et al. 1987). Initially, these gene arrays were prepared by spotting cDNAs onto filter paper with a pin-spotting device. In the fourth step, miniaturized microarrays were used for gene expression profiling in the year 1995 (Schena et al. 1995). That led to the preparation of microarray having complete genome of *Saccharomyces cerevisiae,* in 1997 (Lashkari et al. 1997).

6.3 Principle

DNA Microarrays are based on the principle of Southern blotting where fragmented DNA is attached to a solid substrate like nylon or nitrocellulose membrane and then hybridized with the chemiluminiscence or radioactively labeled probe prepared from a known gene or DNA fragment under stringent conditions. The hybridization between two DNA strands takes place by the property of complementary nucleic acid sequences to specifically pair with each other by forming hydrogen bonds between complementary nucleotide base pairs. A high number of complementary base pairs in a nucleotide sequence results in tighter noncovalent bonds between the two strands. After washing-off of nonspecific bonding sequences, only strongly paired strands remain hybridized. In DNA microarrays, this situation is slightly altered wherein, thousands of nonlabeled probes are blotted on the solid surface and these chips are subjected to hybridization with the fluorescently labeled target sequence, i.e., sample DNA (Fig. 6.1). On binding of fluorescently labeled target sequence to a probe sequence, a signal is generated that depends on the strength of the hybridization determined by the number of paired bases, the hybridization conditions (such as temperature), and washing after hybridization. Total strength of the signal, from a spot (called as *feature*), depends upon the amount of target sample binding to the probes present on that spot. Microarrays use relative quantification in which the intensity of a feature is compared to the intensity of the same feature under a different condition, and the identity of the feature is known by its position. The steps involved in microarray experiment are elaborated in Fig. 6.2.

6.4 Types and Fabrication of DNA Microarrays

Broadly DNA arrays are classified as solid phase, beaded and electronic arrays.

6.4.1 Solid-Phase Arrays

The traditional solid-phase array is a collection of orderly microscopic spots (features); each with a specific probe and attached to a solid surface, such as glass, plastic or silicon biochip commonly known as a *genome chip*, *DNA chip,* or *gene array*. Thousands of such features can be

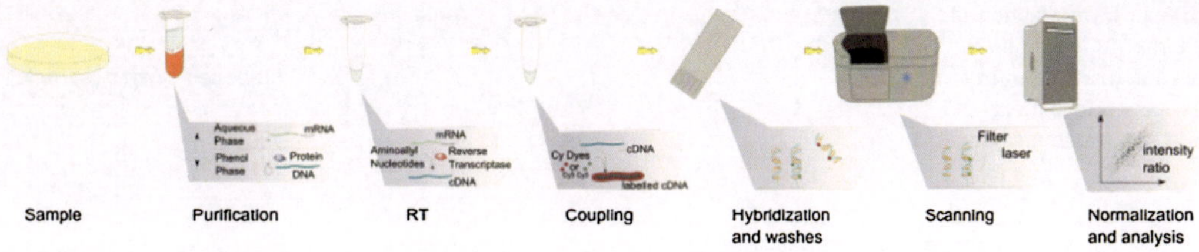

Fig. 6.2 The steps involved in a microarray experiment (*Source* http://en.wikipedia.org/wiki/DNA_microarray)

placed on known locations on a single DNA microarray chip (http://en.wikipedia.org/wiki/DNA_microarray). On the basis of information used for the probes the solid phase DNA microarrays can be of two types, Printed microarrays (oligonucleotide based microarrays) and GeneChips (*in situ* synthesized oligonucleotide microarrays).

6.4.1.1 Oligonucleotide-Based Printed Microarrays

The first DNA arrays used for research work were oligonucleotide-based printed microarrays. These microarrays require either PCR amplified (ds-DNA) or synthesized oligonucleotides (ss-DNA) of known gene sequences as probes. In these microarrays, glass slides are used as solid surface for spotting or printing of the probes. Cheung et al. (1999) observed that glass slides are not only observed to be stable during the varied experimental situations but also produce least background noise during hybridization, thus yielding clearer images. Probes can be printed in two ways by contact printing or by noncontact printing. Whichever the method of printing is used in both the cases an array of 100–150 μm features is created by applying few nanolitres of probe solution per spot. Contact printing means direct application of probe solution on the solid surface with the help of print pins, whereas in noncontact printing a small droplets of probe solution is placed on the solid surface by the print head. Printing is done precisely so as to avoid mixing of adjacent features. In comparison to GeneChip and high density bead arrays, oligonucleotide-based printed arrays have low density of probes ranging from 10,000 to 30,000 features per array due to the relatively large size of the features in it but has more features than suspension bead arrays and electronic microarrays.

When dsDNA is used as probes, it is required to denature them either at the time of printing or after spotting on the solid surface so that they become available for the hybridization (Tomiuk and Hofmann 2001). For getting the probes attached on the glass surface, two methods are used, first, either slides are applied with a positively charged coating that allows the electrostatic interaction with the negative charge of the phosphate backbone of the DNA (Ehrenreich 2006) and second, by UV-cross-linking of amine groups present on treated slides with the covalent bonds present between the thymidine bases in the DNA (Cheung et al. 1999). Each dsDNA probe used in these arrays represents a different gene and is usually 200–800 bp long. These probes should have high specificity and should be free from contaminants so that there is no interference while hybridization. Generally, these probes have a high sensitivity but they suffer in specificity and might be contaminated with nonspecific amplicons (Hager 2006). To select for the probe specificity, amplicons can be tested by sequencing or agarose gel electrophoresis and specificity of the hybridization data can be improved by including multiple gene segments among the probes so as to workout the redundancy. At times decreased specificity is advantageous when the genomic sequence under analysis has high natural polymorphism, although it does not help when we have to differentiate among highly similar target sequences, therefore they are not useful as far as the clinical diagnostics is concerned. At the same

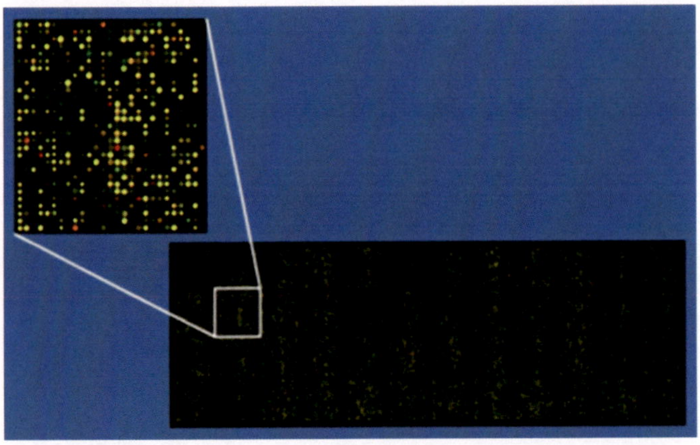

Fig. 6.3 Example of an approximately 40,000 probe spotted oligo microarray with enlarged inset to show details (*Source* http://en.wikipedia.org/wiki/DNA_microarray)

time, use of longer probes results in higher melting temperatures and mismatch tolerance that decreases the specificity of the probes due to random hybridization to nontarget sequences, hence for any microarray experiment probe length should be chosen carefully.

In oligonucleotide microarrays with ssDNA probes, probes are chemically synthesized short sequences usually ranging from 25–80 bp in length, in contrast to gene expression microarrays where they may be up to a length of 150 bp (Chou et al. 2004). In these microarrays due to the shorter length of probes, fewer errors are observed during probe synthesis and they help in studies focusing small genomic regions like polymorphisms. Another benefit due to the small probe length is increased specificity while targeting specific genomic regions although it reduces the sensitivity of detections. It is observed that the sensitivity and the strength of hybridization signal increases with the length of the probe used. In case of very short oligonucleotides, addition of spacers or application of a higher concentration of probe during printing improves the strength of hybridization signal (Chou et al. 2004). Ramdas et al. (2004) observed eightfold increase in sensitivity when 70 mer oligonucleotides were used as probes as compared to 30 mer probes for the genes having low levels of expression. ssDNA oligonucleotide probes are easy to manufacture but care is required that all the probes on the array should have melting temperature lying within a range of 5 °C and should not have palindromic sequences. Hence, it is required to test the probes in advance so that the hybridization data generated through any experiment using ssDNA probes remain unbiased (Chou et al. 2004). Attachment of ssDNA probes to glass surface is facilitated due to covalent linkage between modified 5′ or 3′ ends having amino group with the aldehyde or epoxy functional groups coated on the slides. Oligonucleotide microarrays manufactured by Roche NimbleGen (Madison, WI) and Agilent Technologies (Palo Alto, CA) are good examples. Each NimbleGen microarray contain $>10^6$ features. The available formats are 1 × 2.1 million features, 3 × 720,000 features, 1 × 385,000 features, 4 × 72,000 features, and 12 × 135,000 features per slide. Agilent microarrays are available in the formats having 1 × 244,000 features, 2 × 105,000 features, 4 × 44,000 features, and 8 × 15,000 features per slide (Miller and Tang 2009). An oligo microarray with approximately 40,000 probes is shown in Fig. 6.3 to have an idea of how it looks after hybridization.

6.4.1.2 GeneChips (*in situ* Synthesized Oligonucleotide Microarrays)

GeneChips are produced by *in situ* synthesis of oligonucleotides directly onto quartz or glass wafers having a size of 1.2 cm^2. Tens of thousands of oligonucleotides are placed on one wafer by synthesizing one nucleotide at a time. Each oligonucleotide probe on the array represents one gene. Length of the probes used in these arrays depends on the purpose for which

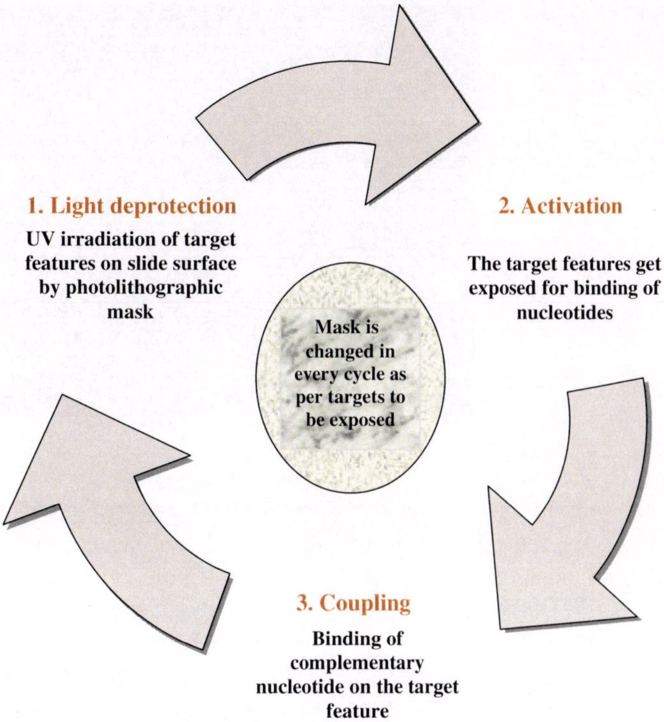

Fig. 6.4 Photolithographic production of oligonucleotide microarrays

these microarrays are designed. GeneChips designed by Agilent technology use longer probes having 60 bp lengths, whereas in Affymatrix GeneChips shorter probes having a length of 25 bp are used. Typically, GeneChips have $>10^6$ features per microarray that actually depends on the distance between the features (Weiszhausz et al. 2006).

For binding of the *in situ* synthesized probes to the solid surface mainly a modification of semiconductor photolithography technology is used. In this technique, quartz surface of the GeneChip is coated with synthetic linkers modified with light-sensitive protecting groups (Fodor et al. 1991). Due to this chemical protection, addition of nucleotide to the microarray surface is restricted till its deprotection by UV light. This means nucleotides modified with a photosensitive protecting group can be added to growing oligonucleotide chains by exposing the array surface to UV light. Further, photolithographic masks are used for directing the specific nucleotides to exact probe sites. Photolithographic mask has a defined pattern of windows; those selectively transmit or block the UV light for the targeted feature on the microarray surface that is chemically protected. Due to this kind of filtration, an area of the microarray surface that does not receive UV light remains protected and does not allow addition of nucleotides, whereas areas exposed to UV light gets deprotected, and specific nucleotides can be added to these sites. Likewise, the order of nucleotide addition is directed by the pattern of windows in each photolithographic mask. Therefore, *in situ* probe synthesis requires cycles of masking, UV exposure, and addition of nucleotide bases (either A, C, T, or G) to the growing oligonucleotide chain on the solid surface (Fig. 6.4) (Weiszhausz et al. 2006). In addition to array synthesis using photolithography mask NimbleGen Systems has developed a technique of maskless array synthesis that provides flexibility of use with large number of probes required in

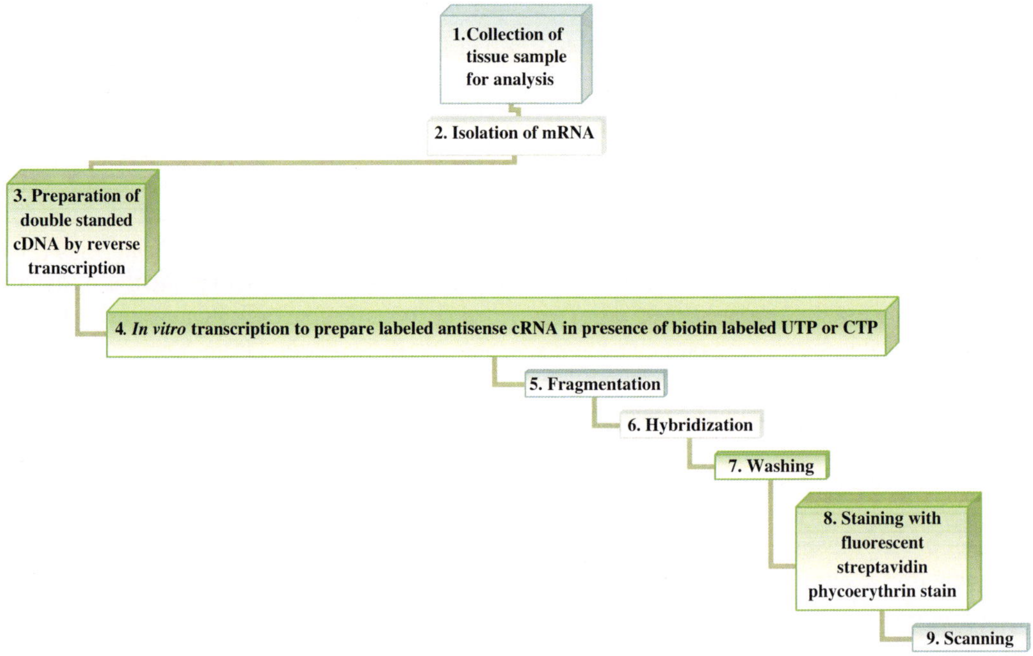

Fig. 6.5 Steps involved in expression analysis using oligonucleotide microarrays hybridization

in situ synthesized oligonucleotide arrays to improve sensitivity, specificity, and statistical accuracy due to shorter probe lengths.

The main advantage of GeneChips is their ability to measure the absolute expression of genes in cells or tissues. Their disadvantages are their higher costs and current inability to simultaneously compare, on the same array, the degree of expression of two related biological samples.

For expression analysis using oligonucleotide microarray hybridization (Fig. 6.5), firstly labeled RNA samples are prepared by converting extracted mRNA to double-stranded cDNA. The cDNA is then copied to antisense RNA (cRNA) by *in vitro* transcription reaction performed in presence of biotin-labeled ribonucleotide triphosphates (UTP or CTP). Fragmented cRNA (50–100 nucleotides) is used for hybridization. After a brief washing step to remove unhybridized cRNA, the microarrays are stained by streptavidin phycoerythrin and scanned (Aharoni and Vorst 2001).

Both types of oligonucleotide microarrays produce cleaner downstream hybridization data. Reproducibility, simpler standardization, and use of controls make these microarrays useful for clinical diagnostics. In comparison to *in situ*-synthesized microarrays, printed microarray seems to be relatively simple and inexpensive with flexibility to quickly adjust spotted probes on the basis of newer informations. However, *in situ*-synthesized microarrays are more robust due to significant control measures included in their design.

Now oligonucleotide arrays have also been developed which combine the flexibilities and qualitative advantages of synthetic probe arrays with the benefits of simultaneous analysis possible by spotted glass array. Further, custom-designed microarrays have also been made available by Nimblegen and Agilent technologies.

6.4.2 Bead Arrays

The bead array is a collection of microscopic silica/polystyrene beads, each with a specific probe and a ratio of two or more dyes, which do not interfere with the fluorescent dyes used on the target sequence. Bead arrays are mainly of two types, high density bead arrays and suspension bead arrays.

6.4.2.1 High-Density Bead Arrays

In high-density bead arrays silica beads of 3 μm size are used as solid support for application of probes. After hybridization, these beads can randomly self-assemble on the substrate. Two substrates available for this purpose are Sentrix Array Matrix (SAM) and the Sentrix BeadChip. These arrays facilitate high density detection of target nucleic acids. One of the examples is Illumina bead arrays, San Diego, CA.

The SAM contains a bundle of 96 fiber-optic cables of 1.4 mm diameter. Each bundle represents an array made up of 50,000 light-conducting fibers of 5 μm diameter; each of this fiber has a microwell for assembling a single bead (Fan et al. 2006). For example, in the Universal Bead array, up to 1,536 bead types can be accommodated. Here, every bead type represents a unique gene sequence and when they assemble on the fiber bundles 30 beads of each type becomes available for analysis in the array (Oliphant et al. 2002). Likewise, each SAM allows the analysis of 96 samples.

In Sentrix BeadChip up to 16 samples can be accommodated on a silicon slide that has microwells prepared for individual beads by microelectromechanical systems technology (Fan et al. 2005). Gunderson (2009) has found BeadChips more appropriate for very-high-density applications like whole-genome genotyping where 10^5–10^6 features are required for identifying genome-wide single nucleotide polymorphisms (SNP).

Since the beads randomly assemble on the substrate, their location is not fixed and it is required to map their position by decoding process after every single application (Gunderson et al. 2004). For the decoding purpose, each bead type is covalently attached with 700,000 copies of a unique capture sequence that acts as the identifier of the bead (Kuhn et al. 2004). For example, in Universal BeadArrays such identifier sequences are called as IllumiCodes and they are designed to specifically avoid homology with human and mouse nucleic acid sequences. The mapping of these Illumina beads requires a series of hybridization and rinse steps, so that the fluorescently labeled complementary oligonucleotides may bind to their specific bead sequence or IllumiCode and ultimately track the location of the corresponding bead type (Fan et al. 2006). This decoding process provides quality control for each feature used in the microarray (Fan et al. 2005).

It is possible to process the SAM by using a standard microtiter plate that facilitates standard automation and high-throughput processing. Similarly, individual arrays on the 16-sample BeadChip are spaced as the standard multichannel pipettor for easy handling while use. 10^5–10^6 features can be supported by Bead arrays. One advantage of bead arrays is their built-in redundancy that provides control for comparing data generated by more than one arrays. Further, due to the uniqueness of each microarray by altering the bead pattern spatial bias can be identified. The work on developing the analysis tools for bead arrays is in progress (Miller and Tang 2009). Bead arrays are found very useful in studies related to DNA methylation (Bibikova and Fan 2009), profiling of gene expressions (Bibikova et al. 2004; Fan et al. 2004; Kuhn et al. 2004), SNP detection, and in the International HapMap Project (Gunderson 2009).

6.4.2.2 Suspension Bead Arrays

In suspension bead arrays, microscopic polystyrene spheres, i.e., microspheres or beads are used as solid support for application of probes and flow cytometry technique is used for detection of bead and target attached to it. In these arrays, beads remain suspended in the hybridization solution and do not require fiber-optic cables or silicon slides to immobilize them before scanning. All the previously discussed microarrays were two-dimensional, or planar arrays whereas, suspension bead arrays are three-dimensional arrays.

These arrays were first described by Horan and Wheeless (1977) for detections involving antigen and antibodies. Further, it was made possible to detect multiple antibodies in one run by using different sized microsphere sets (Scillian et al. 1989; McHugh et al. 1998). In suspension bead arrays if 5.6 μm microspheres are filled with different concentrations of red (658 nm emission) and infrared (712 nm emission) fluorochromes, each bead in the set of 100 microspheres will have a distinct red to infrared ratio providing each bead a unique spectral address. This provision has strengthened multiple detections using suspension bead arrays. When, microspheres with a specific spectral address is coupled with a specific probe, it becomes equivalent to a feature of a two-dimensional microarray but when multiple microspheres with unique spectral addresses are combined with multiple specific probes it is possible to develop up to 100 combinations to study the target nucleic acids. Along with unique spectral address and specific probe suspension beads are also provided with a fluorescent reporter that helps in detecting hybridization of target with the probe. Here, when hybridization takes place reporter separates to emit fluorescence that can be detected by a flow cytometer. During flow cytometry, microsphere suspension is passed through two lasers, i.e., 635 and 532 nm. A 635 nm laser excites red and infrared fluorochromes and helps in detecting the unique spectral address of beads and hence the identity of probe and target can be worked out, whereas, 532 nm laser excites reporter fluorochromes such as *R*-phycoerythrin and Alexa 532 and helps in quantifying the amount of hybridization taken place on individual bead (Fig. 6.6).

Suspension bead arrays are used in following three ways for nucleic acid detections (Dunbar 2006):

1. **Direct DNA hybridization**: In this method, PCR amplicons are directly hybridized to probes hence it is known as direct DNA hybridization. Here, a biotinylated primer is used during PCR amplification, which provides a substrate for binding of the fluorescent streptavidin-*R*-phycoerythrin that result into labeling of hybridized beads (Armstrong et al. 2000; Spiro et al. 2000). High fluorescence is generated due to binding of amplicon with the microspheres.
2. **Competitive DNA hybridization**: In this method, unlabeled PCR amplicons are used for hybridization in combination with biotinylated competitor oligonucleotides. In absence of target DNA competitor DNA binds to the microsphere and produce high fluorescence and when target DNA is present it binds to the competitor DNA which in turn is not able to bind to the microsphere and low fluorescence is generated, which is contrasting to the direct hybridization method.
3. **Solution-based chemistries with microsphere capture**: These chemistries exploit the natural properties of DNA polymerases and ligases and incorporate a capture sequence during the solution-based reactions. In these techniques, universal microspheres with nonspecific capture sequences are used. These universal capture sequences mainly include ZipCode/cZipCode, xTAG of Tm Biosciences, Luminex Molecular Diagnostics, Inc., Toronto, Canada and EraCode of EraGen, Madison, WI (Johnson et al. 2004). ZipCode/cZipCode were the first universal capture sequences those were used to detect SNP's using SBCE (Chen et al. 2000; Iannone et al. 2000; Taylor et al. 2001; Ye et al. 2001). ZipCode sequences are basically 25 bp long random sequences of *Mycobacterium tuberculosis* genome (Chen et al. 2000). They are used by adding a unique ZipCode to the 5′ end of the capture probe and tagging of microspheres with its complementary sequence (cZipCode). In xTAG, only three nucleotides out of four are used to avoid nonspecific hybridization with the naturally occurring sequences. Further, they are thermodynamically matched to avoid variability in their hybridization efficiency. This technology is used in commercial assays conducted by Luminex. In EraCode, synthesized isoguanosine and 5-Me-isocytosine bases are used that are based on the expanded genetic alphabet of MultiCode technology. The isoguanosine and 5-Me-isocytosine bases pairs with each other but they do not

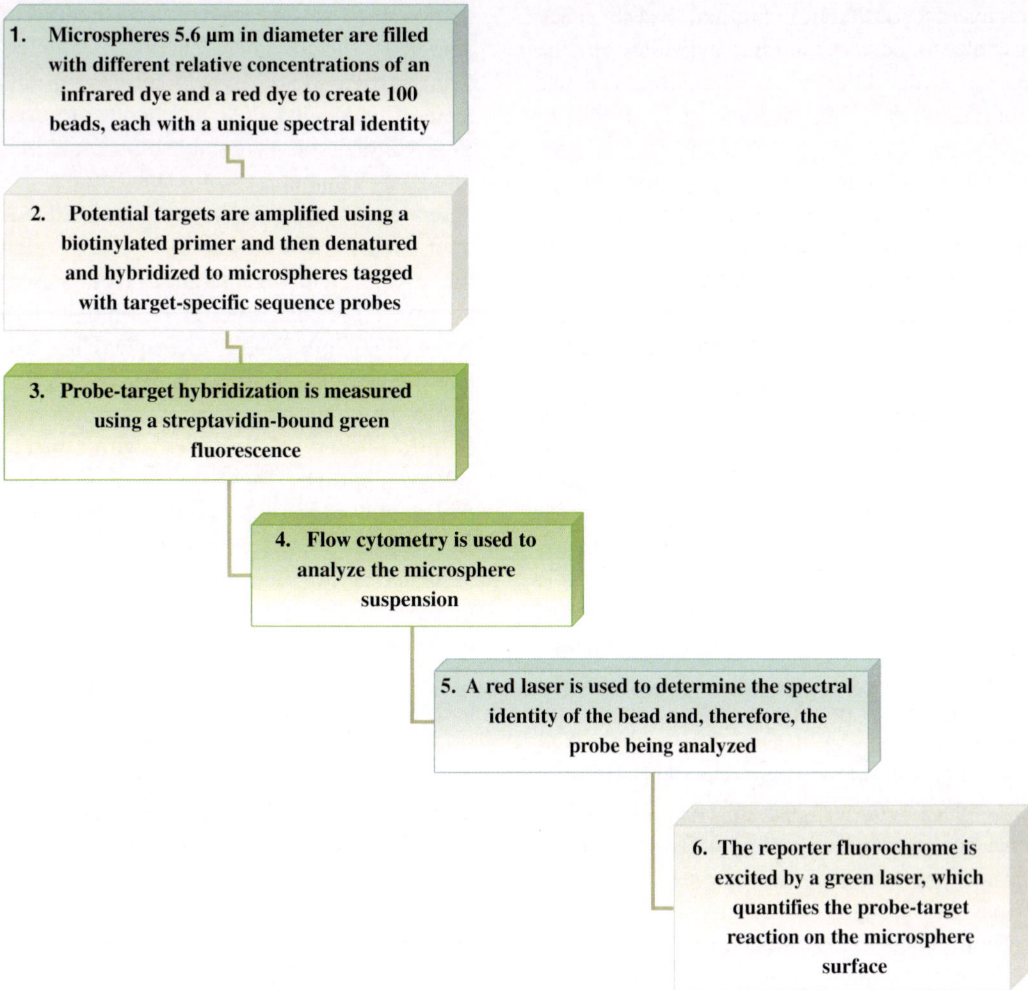

Fig. 6.6 Steps involved in microarray experiment using suspension bead array

pair with natural bases due to which EraCode sequences are highly specific. Solution-based chemistries coupled with subsequent microsphere capture include following procedures:

(a) **Allele-specific primer extension (ASPE) or target-specific primer extension (TSPE)**: In ASPE or TSPE a capture primer with a unique 5′ sequence followed by a target-specific sequence is used. This primer is extended by DNA polymerase if target DNA is present to provide the complementary base for the nucleotide present on 3′ end. Biotinylated deoxynucleotide triphosphate is used in ASPE and/or TSPE to achieve labeling.

(b) **Oligonucleotide ligation assays (OLA)**: Same as in ASPE and/or TSPE a capture primer with a unique 5′ sequence followed by a target-specific sequence is used in OLA. The difference is in the enzymatic reaction which is ligase dependent in OLA. In OLA a biotinylated probe homologous to target DNA is used in addition to the capture primer. Here, if target DNA is present in the sample, capture primer and reporter probe ligates.

(c) **Single-base-chain extension (SBCE)**: This procedure is used for detection of multiple SNP in one run. Here, independent reactions are required for each nucleotide query and for identifying a single SNP one probe with a unique capture sequence is used to identify the possible alleles in separate wells having a different dideoxynucleoside triphosphate in each well. Likewise, in the well in which homologous capture and target sequence is present, a biotinylated dideoxynucleoside triphosphate gets incorporated, hence the further extension gets terminated.

Despite the low feature density due to the availability of universal capture sequences, high ability of multiplexing, relative simplicity, and affordability suspension bead arrays are highly suitable for high-throughput nucleic acid detections including clinical diagnostics if due care is taken to avoid post-amplification intra- and inter-run contaminations.

6.4.3 Electronic Microarrays

Electronic microarray technology is introduced by Nanogen, San Diego, CA. In contrast to all other microarrays that depend on passive movement of complimentary sequences for hybridization, in electronic microarrays their movement is directed by applying electric field. For this purpose in microelectronic cartridges, complementary metal oxide semiconductor technology is used, e.g., NanoChip 400 of Nanogen (Sosnowski et al. 1997). Each NanoChip cartridge is provided with 12 connectors that can control 400 individual test sites. By application of positive current to one or more test sites on the microarray, nucleic acids which are basically negatively charged move towards specific sites, or features. When electronically directed biotinylated probes reaches to the target site it forms streptavidin-biotin bond with the streptavidin present on the microarray surface. Once the probes reaches to their target, positive current is removed from the active features and new test sites are activated. After the probes are attached to the features, fluorescently labeled target DNA is applied to the microarray. Target DNA can hybridize with the probe either passively or by concentrating them electronically. The most common format is the application of probe first, although amplicon may also be applied first and sometimes sandwich assays are also used. Irrespective of the format used, if hybridization takes place, presence of fluorescent reporters can be detected by scanning which is then analyzed (Fig. 6.7) (Miller and Tang 2009).

In electronic microarrays multiplex detection can be achieved at an individual test site because multiple probes, each with a distinct fluorophore, can be sequentially directed to the same feature, e.g., Nanogen's RVA ASR in which P1, parainfluenza virus type 1; P2, parainfluenza virus type 2; P3, parainfluenza virus type 3; FB, influenza B virus; FA, influenza A virus; RSV, respiratory syncytial virus; BKGD, background can be detected simultaneously. On these microarrays single or multiple samples can be used for hybridization to multiple test sites. Further, the commercially available electronic cartridge NanoChip is a universal blank chip, hence the features of the microarray can be directly specified by the user and therefore these microarrays are highly flexible and cost-effective. At present, electronic microarrays facilitate a density of only 400 spots which is sufficient for most of the clinical diagnostic assays.

6.5 Design of DNA Microarrays

Depending on the number of probes under examination, costs, customization requirements, and the type of scientific question to be answered microarrays can be manufactured in different ways. Arrays may have as few as 10 probes or up to 2.1 million µm-scale probes. The designing of microarrays depends on *two channels* (two color microarrays) or *one channel* of detection (one color microarray).

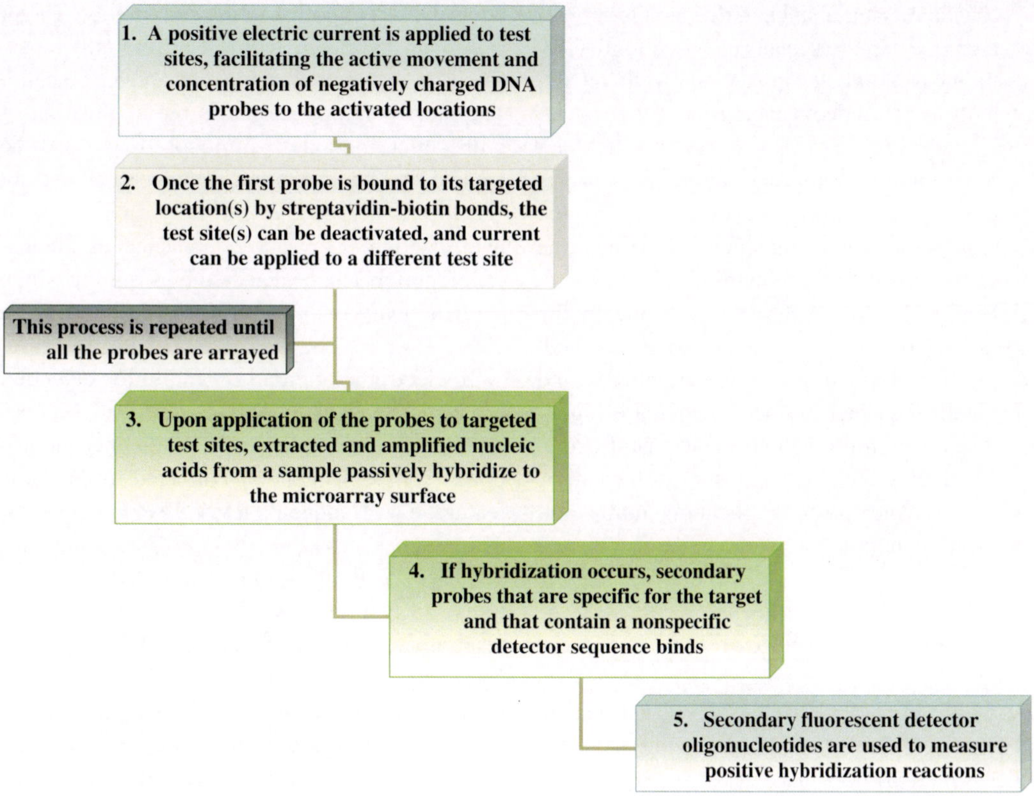

Fig. 6.7 Steps involved in detection using electronic microarray

6.5.1 Two-Color Microarrays or Two-Channel Microarrays

Two-color microarrays or two-channels are typically hybridized with cDNA prepared from two samples to be compared (e.g., diseased tissue versus healthy tissue) and that are labeled with two different fluorophores (Shalon et al. 1996; Fig. 6.8). Fluorescent dyes commonly used for cDNA labeling include Cy3, which has a fluorescence emission wavelength of 570 nm (corresponding to the green part of the light spectrum), and Cy5 with a fluorescence emission wavelength of 670 nm (corresponding to the red part of the light spectrum). The two Cy-labeled cDNA samples are mixed and hybridized to a single microarray that is then scanned in a microarray scanner to visualize fluorescence of the two fluorophores after excitation with a laser beam of a defined wavelength. Relative intensities of each fluorophore may then be used in ratio-based analysis to identify up-regulated and down-regulated genes (Tang et al. 2007a). *In situ* synthesised oligonucleotide microarrays often carry control probes designed to hybridize with RNA spike-ins. The degree of hybridization between the spike-ins and the control probes is used to normalize the hybridization measurements for the target probes. Although absolute levels of gene expression may be determined in the two-color array in rare instances, the relative differences in expression among different spots within a sample and between samples is the preferred method of data analysis for the two-color system. Examples: *Dual-Mode* platform of Agilent, *DualChip* platform for colorimetric Silverquant labeling of Eppendorf, and *Arrayit* of TeleChem International.

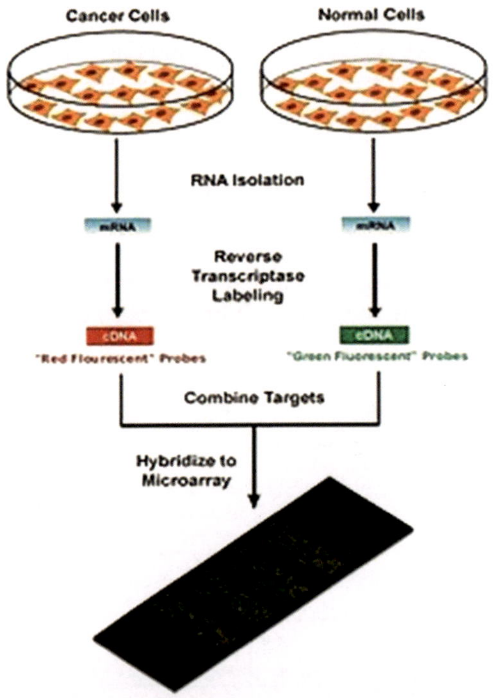

Fig. 6.8 A typical two-color microarray experiment (*Source* http://en.wikipedia.org/wiki/DNA_microarray)

6.5.2 Single-Channel Microarrays or One-Color Microarrays

In single-channel microarrays or one-color microarrays, the arrays provide intensity data for each probe or probe set indicating a relative level of hybridization with the labeled target. However, they do not truly indicate abundance levels of a gene but rather relative abundance when compared to other samples or conditions when processed in the same experiment. Each RNA molecule encounters protocol and batch-specific bias during amplification, labeling, and hybridization phases of the experiment making comparisons between genes for the same microarray uninformative. The comparison of two conditions for the same gene requires two separate single-dye hybridizations. Examples are Affymetrix *GeneChip*, Illumina *BeadChip*, Agilent *single-channel* arrays, the Applied Microarrays *CodeLink* arrays, and the Eppendorf *DualChip and Silverquant* arrays. One strength of the single-dye system is that an aberrant sample cannot affect the raw data derived from other samples, because each array chip is exposed to only one sample (as opposed to a two-color system in which a single low-quality sample may drastically impinge on overall data precision even if the other sample was of high quality). Another benefit is that data are more easily compared to arrays from different experiments so long as batch effects have been accounted for. A drawback to the one-color system is that, when compared to the two-color system, double the number of microarrays are required to compare samples within an experiment.

6.6 Basic Applications of DNA Microarrays

6.6.1 Gene Expression Profiling

DNA microarrays can be used to detect DNA (as in comparative genomic hybridization), or detect RNA (most commonly as cDNA) that may or may not be translated into proteins. The process of measuring gene expression *via* cDNA is called expression analysis or expression profiling. In an mRNA or gene expression profiling experiment, the expression levels of thousands of genes are simultaneously monitored to study the effects of treatments, diseases, and developmental stages on gene expression. For example, microarray-based gene expression profiling can be used to identify genes whose expression is changed in response to pathogens or other stresses as compared to healthy tissues (Adomas et al. 2008).

6.6.2 Comparative Genomic Hybridization

This is meant for assessing genome content in different cells or closely related organisms, for example, tumerogenic cells (Moran et al. 2004).

6.6.3 GeneID

These are small microarrays to check IDs of organisms in food and feed (Mascos and Southern 1992), mostly combining PCR and microarray technology. For example, genetically modified organisms, mycoplasmas in cell culture and pathogens for disease detection.

6.6.4 Chromatin Immunoprecipitation on Chip

DNA sequences bound to a particular protein can be isolated by immunoprecipitating that protein (ChIP), these fragments can be then hybridized to a microarray (such as a tiling array) to determine the protein binding site occupancy throughout the genome. For example, protein to immunoprecipitate are histone modifications (H3K27me3, H3K4me2, H3K9me3, etc.), Polycomb-group protein (PRC2:Suz12, PRC1:YY1) and trithorax-group protein (Ash1) to study the epigenetic landscape and RNA Polymerase II to study the transcription landscape.

6.6.5 DamID

Analogously to ChIP, genomic regions bound by a protein of interest can be isolated and used to probe a microarray to determine binding site occupancy. Unlike ChIP, DamID does not require antibodies but makes use of adenine methylation near the protein's binding sites to selectively amplify those regions, introduced by expressing minute amounts of protein of interest fused to bacterial DNA adenine methyltransferase.

6.6.6 SNP Detection

SNP detection is for identifying single nucleotide polymorphism among alleles within or between populations (Hacia et al. 1999). Several applications of microarrays make use of SNP detection, e.g., genotyping, forensic analysis, measuring predisposition to disease, identifying drug-candidates, evaluating germline mutations in individuals, somatic mutations in cancers, assessing loss of heterozygosity, and genetic linkage analysis.

6.6.7 Alternative Splicing Detection

An *exon junction array* design uses probes specific to the expected or potential splice sites of predicted exons for a gene. It is of intermediate density or coverage, to a typical gene expression array (with 1–3 probes per gene) and a genomic tiling array (with hundreds or thousands of probes per gene). It is used to assay the expression of alternative splice forms of a gene. Exon arrays have a different design; employing probes designed to detect each individual exon for known or predicted genes, and can be used for detecting different splicing isoforms.

6.6.8 Fusion Genes Microarray

A fusion gene microarray can detect fusion transcripts, e.g., from cancer specimens. The principle behind this is building on the alternative splicing microarrays. The oligo design strategy enables combined measurements of chimeric transcript junctions with exon-wise measurements of individual fusion partners.

6.6.9 Tiling Array

Genome tiling arrays consist of overlapping probes designed to densely represent a genomic region of interest, sometimes as large as an entire human chromosome. The purpose is to empirically detect expression of transcripts or alternatively splice forms, which may not have been previously known or predicted.

6.7 Specialized Applications of DNA Microarrays

DNA microarrays have various specific applications to Cancer biology, Microbiology, Plant science, Medicine, Environmental Science, Nephrology, etc. Few of them are explained here.

6.7.1 Applications to Cancer Biology

Cancer is caused by the accumulation of genetic and epigenetic changes resulting from the altered sequence or expression of cancer-related genes, such as oncogenes or tumor suppressor genes, as well as genes involved in cell cycle control, apoptosis, adhesion, DNA repair, and angiogenesis. Because gene expression profiles provide a snapshot of cell functions and processes at the time of sample preparation, comprehensive combinatorial analysis of the gene expression patterns of thousands of genes in tumor cells and comparison to the expression profile obtained with healthy cells should provide insights concerning consistent changes in gene expression that are associated with tumor cellular dysfunction and any concomitant regulatory pathways. Microarray technology has been widely used in the past 10 years to investigate tumor classification, cancer progression, and chemotherapy resistance and sensitivity (Macgregor and Squire 2002). Although the major limiting factors for routine use in a clinical setting at present are cost and access to the microarray technology, it is likely that costs will decrease in the near future and that the technology will become increasingly user friendly and automated. In this section, few examples are discussed to demonstrate that expression arrays can be used to gain a better understanding of the basic biology, diagnosis and treatment of cancer.

6.7.1.1 Identifying Single Nucleotide Polymorphism (SNP)

SNP's are common DNA sequence variations having great significance for biomedical research. In human genome at least one out of every thousand nucleotide is observed to have polymorphism. Most of the SNP's are harmless but few of them that are located in coding regions of the gene alter the composition of amino acids, thus the functions of encoded proteins. Similarly, SNP's located in the non-coding regions are also important if they affect the splicing of the gene, or alter the promoter, enhancer, silencer or any recognition site for DNA binding proteins (Risch 2000). For example, a SNP in 5′ untranslated region of the *RAD51* gene have been identified that may be associated with an increased risk of breast cancer and a lower risk of ovarian cancer among *BRCA2* mutation carriers (Wang et al. 2001).

6.7.1.2 Detection of Mutations

High throughput methods like microarray assays are found useful in rapid detection of mutations among large genes like *BRCA1*, *BRCA2*. For example, a high density oligonucleotide array with about 96,000 oligonucleotides of 20 bp size have been developed by Hacia et al. (1996) that could detect a wide range of heterozygous mutations in the 3.45 Kb exon 11 of the *BRCA1* gene. Wen et al. (2000) found that oligonucleotide based detections are more accurate than DNA sequencing in identifying *p53* mutations in ovarian tumors.

6.7.1.3 Molecular Classification of Tumors

Improvements in tumor classification are central to the development of novel and individualized therapeutic approaches. Histologically indistinguishable tumors often show significant differences in clinical behavior, and sub classification of these tumors based on their molecular profiles may help explain why these tumors respond so differently to treatment (Macgregor and Squire 2002).

For example, (1) microarray technology was applied to develop innovative classifications of leukemias using microarray analysis based on *neighborhood analysis* and the utilization of tumor class predictors. This strategy was able to distinguish between acute myeloid leukemia and acute lymphocytic leukemia without supervisory analysis (Golub et al. 1999); (2) gene expression pattern analysis has been used to classify, at the molecular level, breast tumors (Sorlie et al. 2001), B cell lymphoma (Alizadeh et al. 2000), cutaneous melanoma (Bittner et al. 2000), and lung adenocarcinoma (Garber et al. 2001; Bhattacharjee et al. 2001); (3) by analyzing molecular profiles of 50 nonneoplastic and neoplastic prostate samples, signature expression profiles of healthy prostate, benign prostatic neoplasia, localized prostate cancer, and metastatic prostate cancer has been established (Dhanasekaran et al. 2001). These studies established the feasibility of combining large-scale molecular analysis of expression profiles with classic morphologic and clinical methods of staging and grading cancer for better diagnosis and outcome prediction.

6.7.1.4 Discoveries of Novel Genes

Many tumor suppressor genes and oncogenes are regulated by the expression of other genes at transcription level. Even some of them like *p53* and *Myc* are transcription factors. Upon activation, *p53* induces growth arrest or apoptosis by transcriptionally activating its target genes. With the help of microarray-based expression profiling target genes for several gene products have been identified which directly or indirectly regulates transcription. For example, oligonucleotide array having 6,000 genes is used to identify *p53* target genes (Zhao et al. 2000). Similarly, microarray-based expression profiling have been used to identify *Gadd45* as one of the targets of *BRCA1* tumor suppressor gene (Harkin et al. 1999).

6.7.1.5 Detection of Sensitivity to Drugs

Despite considerable advances in cancer treatment, acquired resistance to chemotherapeutic drugs continues to be a major obstacle in patient treatment and overall outcome. Anticancer drug resistance is thought to occur through numerous mechanisms, and microarrays offer a new approach to studying the cellular pathways implicated in these mechanisms and in predicting drug sensitivity and unexpected side effects. Most array studies have been carried out using cancer cell lines that are rendered resistant to commonly used anticancer drugs. Obtaining further insights into the mechanism of action of anticancer drugs and the diverse pathways involved in drug resistance may eventually be invaluable for design of more strategic treatments that are most appropriate for an individual tumor (Macgregor and Squire 2002).

For example, (1) in an attempt to obtain molecular fingerprinting of anticancer drugs in cancer cells the expression profiles of doxorubicin-induced and -resistant cancer cells were monitored (Kudoh et al. 2000); (2) similarly, in another study, a subset of 1,400 genes were analyzed to study the correlation between expression profiles and drug mechanism of action of a panel of 118 anticancer drugs (Ross et al. 2000; Scherf et al. 2000).

6.7.1.6 Tumor Metastasis Related Studies

Conversion of noninvasive tumors into invasive tumors is known as metastasis when cancer cells start spreading from one organ to another organ or tissue with blood stream or lymphatic system. DNA microarrays have been used to identify the genes involved in metastasis.

For example, clonal relationship of 22 liver tumor foci from six patients have been investigated by Cheung et al. (2002) with the help of cDNA microarray having 23,000 genes. With an aim to identify metastasis-related genes, they could identify 63 up-regulated genes, of which 39 were known genes and 24 were expressed sequence tags and 27 down regulated genes, of which 14 were known genes and 13 were expressed sequence tags. Similar studies have been conducted with osteosarcoma, colorectal tumor and brain metastasis.

6.7.1.7 Identification of Molecular Markers Specific to Tumor

Several research groups have focused on identifying subsets of genes that show differential expression between healthy tissues or cell lines and their tumor counterparts to identify biomarkers for several solid tumors, including ovarian carcinomas, oral cancer, melanoma, colorectal cancer, and prostate cancer.

In a recent study, a subset of genes showing differential expression between healthy ovaries and ovarian tumors has been reported. Some of these genes, such as metallothionein 1G, which was found to be up-regulated in tumor samples, are implicated in resistance to the anticancer drug cisplatin and might be an indicator of pretreatment resistance of these tumors to cisplatin (Bayani et al. 2002). Other gene identified in this study is the osteopontin gene, which was strongly up-regulated in some tumor samples that have been shown to be secreted in the serum of patients with metastatic cancer, might be an excellent candidate for biomarkers of tumor progression in EOC (Singhal et al. 1997).

One of the most important challenges investigators are facing while using microarray analysis is determining which of the plethora of new differentially expressed genes is biologically relevant to the tumor system being studied. Even when rigorous efforts are made to minimize the number of variables in a microarray study, there may be an unmanageable number of differentially expressed genes that will contribute excessive background values. Therefore, combining expression microarray analysis with other approaches, particularly cytogenetics techniques, such as spectral karyotyping and comparative genomic hybridization array (CGH) (Pollack et al. 1999), offers the possibility to focus on significantly smaller subsets of genes of direct relevance to tumor biology (Bayani et al. 2002). A combination of expression arrays and CGH array techniques was used on breast cancer cell lines and has identified a limited number of genes that are both amplified and over expressed (Monni et al. 2001).

Finally, validation of the relative expression obtained from genome-wide microarray analysis is critical. Several approaches can be chosen from basic Northern analysis or semiquantitative reverse transcription-PCR to *in situ* hybridization (ISH) using tissue microarrays. For example, expression of several candidate genes associated with prostate cancer has been analyzed (Mousses et al. 2002) those were previously identified by cDNA microarray analysis. Tissue microarrays constructed from 544 histological biopsies were analyzed by ISH using RNA probes and/or by immunohistochemistry (IHC) using antibodies. There was excellent correlation between the cDNA microarray results and the results obtained with ISH and Northern blot analysis. In addition, protein expression assessed by IHC was also consistent with RNA expression. Similarly, comparable technologies to confirm over expression of hepsin and *PIM-1* in prostate cancer have been used (Dhanasekaran et al. 2001).

6.7.2 Applications to Microbiology

DNA microarrays can be applied to various aspects of microbial studies like microbial detection and identification, determination of antimicrobial drug, drug resistance, microbial typing, microbial gene expression profiling, host gene expression profiling during microbial infections, and determination of host genomic polymorphism.

6.7.2.1 Detection and Identification of Microbes

DNA microarrays could potentially be an assay method to address multiple questions for species identification in both clinical and environmental settings. It identifies their phylogenetic status based on unique 16S rRNA sequences and provides information related to the presence of antibiotic markers and pathogenicity regions (Ye et al. 2001).

For example, (1) a DNA probe array have been described for species identification and detection of rifampin resistance in *M. tuberculosis* (Troesch et al. 1999). In this 70 mycobacterial isolates from

27 different species and 15 rifampin-resistant strains were tested. A total of 26 of the 27 species as well as all of the rifampin-resistant mutants were correctly identified. (2) For field applications, a portable system for microbial sample preparation and oligonucleotide microarray analysis has also been reported (Bavykin et al. 2001). This portable system contained three components: (a) a universal silica mini-column for successive DNA and RNA isolation, fractionation, fragmentation, fluorescent labeling, and removal of excess free label and short oligonucleotides; (b) microarrays of immobilized oligonucleotide probes for 16S rRNA identification; and (c) a portable battery-powered device for imaging the hybridization of fluorescently labeled RNA fragments on the arrays. Beginning with whole cells, it takes approximately 25 min to obtain labeled DNA and RNA samples and an additional 25 min to hybridize and acquire the microarray image using a stationary image analysis system or the portable imager. (3) Respiratory viral pathogen has been detected in connection with multiplex PCR amplification. (4) Simultaneous detection and typing of human papillomaviruses and (5) rapid detection and characterization of methicillin resistant *Staphylococcus aureus*, have also become possible through DNA microarrays.

6.7.2.2 Virulence Factor Determination of Pathogenic Microbes

Many genes associated with virulence are regulated by specific conditions. One way to determine the candidate virulence factors is to investigate the genome-wide gene expression profiles under relevant conditions, such as physiological changes during interaction with the host. A second approach would rely on comparative genomics (Ye et al. 2001).

For example, in a genome comparison study among *Helicobacter pylori* strains, a class of candidate virulence genes was identified by their coinheritance with a pathogenicity island using DNA microarray technique. The whole genome microarray of *H. pylori* was also shown to be an effective method to identify differences in gene content between two *H. pylori* strains that induce distinct pathological outcomes. It is demonstrated that the ability of *H. pylori* to regulate epithelial cell responses related to inflammation depends on the presence of an intact *cag* pathogenicity island (Salama et al. 2000).

6.7.2.3 Responses of Host to Pathogenic or Resident Microbes

DNA microarray is useful in study of the host transcriptional profiles and identifying the differentially expressed genes in host-microbe interaction. Such experiments demonstrate that host genomic transcriptional profiling, in combination with functional assays to evaluate subsequent biological events, provides insight into the complex interaction of host and human pathogens. At the same time, such studies help in developing our understanding about the essential nature of the interactions between resident microorganisms and their hosts.

For example, (1) the host transcriptional profiles during the interaction of *Bordetella pertussis* with a human bronchial epithelial cell line BEAS-2B were investigated using high-density DNA microarrays (Belcher et al. 2000). The early transcriptional response to this pathogen is dominated by the altered expression of cytokines, DNA-binding proteins, and NFkB-regulated genes. It was found that *B. pertussis* induces mucin gene transcription by BEAS-2B cells and then counters this defense by using mucin as a binding substrate. This result indicates the host defensive and pathogen counter-defensive strategies. (2) A DNA microarray was also used to identify the host genes that were differentially expressed upon infection by *Psuedomonas aeruginosa* to the A549 lung pneumocyte cell line (Ichikawa et al. 2000). Differential expression of genes involved in various cellular functions was found, and one of those genes encodes the transcription factor interferon regulatory factor-1. (3) Further, monitoring of cellular responses to *Listeria monocytogenes* and *Chlamydia pneumoniae* has also been reported (Cohen et al. 2000; Coombes and Mahony 2001).

(4) DNA microarrays have been used to investigate the global intestinal transcriptional responses to the residential colonization of *Bacteroides thetaiotaomicron*, a prominent component of the normal mouse and human intestinal microflora (Hooper et al. 2001). The results reveal that this commensal bacterium modulates the expression of genes involved in several important intestinal functions, including nutrient absorption, mucosal barrier fortification, xenobiotic metabolism, angiogenesis, and postnatal intestinal maturation.

6.7.2.4 Study of Mode of Action of Drugs, Inhibitors and Toxic Compounds

Inhibition of a particular cellular process may result in a regulatory feedback mechanism, leading to changes in gene expression patterns. Exploring the gene expression profiles with DNA microarrays may reveal information on the mode of action for drugs and inhibitors or toxic compounds. Insights gained from this approach may define new drug targets and suggest new methods for identifying compounds that inhibit those targets. In addition to the alternation in gene expression patterns related to the drug's mode of action, drugs can induce changes in genes related to stress responses that are linked to the toxic consequences of the drug. The secondary effects of a drug may reveal information on the potential resistance mechanism, which may help to design the drugs that have fewer side effects but have high efficacy by reducing the ability of bacteria to neutralize the drug. Each type of compound often generates a signature pattern of gene expression. A database populated with these signature profiles can serve as a guide to elucidate the potential mode of action as well as side effects of uncharacterized compounds.

For example, DNA microarray hybridization experiments have been conducted in *M. tuberculosis* to explore the changes in gene expression induced by the antituberculous drug isoniazid (Wilson et al. 1999). Isoniazid selectively interrupts the synthesis of mycolic acids, which are branched β-hydroxy fatty acids. Microarray experiments showed that isoniazid induced several genes that encode proteins that are physiologically relevant to the drug's mode of action, including an operonic cluster of five genes encoding type II fatty acid synthase enzymes and *fbpC*, which encodes trehalose dimycolyl transferase.

6.7.2.5 Studies Related to Evolution and Epidemiology of Microbes

DNA microarrays can be used to explore the variability in genetic content and in gene expression profiles within a natural population of the same or related species and between the ancestor and the descendents. As a result, it provides very rich information on the molecular basis of microbial diversity, evolution, and epidemiology (Ye et al. 2001).

For example, genomes within the species of *M. tuberculosis* have been compared with a high density oligonucleotide microarray to detect small-scale genomic deletions among 19 clinically and epidemiologically well-characterized isolates (Kato-Maeda et al. 2001). This study reveals that deletions are likely to contain ancestral genes whose functions are no longer essential for the survival of organism, whereas genes that are never deleted constitute the minimal mycobacterial genome. As the amount of genomic deletion increased, the likelihood that the bacteria will cause pulmonary cavitation decreased, suggesting that the accumulation of mutations tends to diminish their pathogenicity.

6.7.2.6 Process Optimization and Pathway Engineering

Traditional approaches of biocatalysis optimization use random screening, mutagenesis, and engineering improvement. While these methods are still very effective, a better understanding of the underlying physiology using genomic tools can accelerate these efforts. Information obtained by the DNA microarrays can help pathway engineering and process optimization in following ways.

1. Regulatory circuitry and coordination of gene expression among different pathways under different growth conditions can be measured by DNA microarray
2. The physiological state of the cells during fermentation can be assessed by the genome-wide transcriptional patterns
3. DNA microarray can help in identifying genes involved in a production process if they are coregulated
4. The differences in genetic contents and expression profiles between wild-type and improved strains can be compared
5. The actual array data can be incorporated into the mathematical models to describe a cellular process
6. General applications of DNA microarray technology to understand microbial physiology will continue to generate very large amounts of information that will ultimately benefit the pathway engineering and fermentation optimization effort.

Current research in using array information in pathway engineering and bioprocessing is at its early stage. Arrays containing genes involved central metabolism, key biosyntheses, some regulatory functions, and stress response have been used to investigate the metabolic responses to protein overproduction and metabolic fluxes in *E. coli* (Oh and Liao 2000a, b; Gill et al. 2001). Gene array analysis was also used as a tool to investigate the differences in the expression levels for 30 genes involved in xylose catabolism in the parent, strain B, and the engineered strain, KO11 (Tao et al. 2001). Increased expression of genes involved in xylose catabolism is proposed as the basis for the increase in growth rate and glycolytic flux in ethanologenic KO11 strain. Refer Table 6.1 for information on different types of micro arrays used in microbial systems and useful links to various websites with information on microbial genome.

6.7.3 Applications to Plant Science

DNA microarrays are playing significant role in understanding the expressions of genes in situations like effect of biotic stresses like disease, insect and herbivore attacks; effect of abiotic stresses like cold, heat, soil salinity, water deficiency, etc.; effect of circadian rhythm on diurnal cycle; and various physiological functions and genomic composition of plants. Some of the applications are explained here.

6.7.3.1 Studies on Biotic Stresses

In plants, biotic stresses are responsible for reducing the yield potential. Studies using DNA microarrays has provided some understanding of differential gene expressions and defense responses to biotic stresses like diseases, insect pests, and herbivore attack on plants. The genomic expressions in biotic stresses could be understood largely due to the applications of cDNA microarrays (Ahroni and Vorst 2001). Some of them are given below based on the work mostly done on *Arabidopsis*.

For example, (1) the expression of two *Arabidopsis* accessions having rosette leaves has been studied with the help of cDNA microarray having 673 clones (Kehoe et al. 1999); (2) the trait for defense response to pathogen in *Arabidopsis* has been mapped by oligonucleotide microarray having 412 polymorphisms (Cho et al. 1999); (3) cDNA microarrays having 150 clones have been used to study the response of *Arabidopsis* to mechanical wounding and insect feeding (Reymond et al. 2000); (4) cDNA microarray having 2375 clones has been employed to understand the response to treatments with defense-related signaling molecules and fungal pathogen in *Arabidopsis* (Schenk et al. 2000); (5) in *Arabidopsis* genomic expressions associated with systemic acquired resistance (SAR) have been studied by using cDNA microarray having 10,000 clones (Maleck et al. 2000); (6) expression of downstream genes in MAP kinase 4 signaling pathways has also been identified

Table 6.1 Links of sources with information on microbial genomes and DNA microarray construction and applications (Ye et al. 2001)

S.N.	Web link	Reference and available information
1	http://www.affymetrix.com/	It provides commercial high-density oligonucleotide arrays and services
2	http://www.operon.com/	Operon. It provides 70 mer oligonucleotide array sets and primer synthesis service
3	http://www.genosys.com/	SigmaGenosys. It provides membrane arrays and whole genome array sets. Other services include primer synthesis
4	http://www.eurogentec.com/	Eurogentec. It provides oligonucleotides and whole genome microbial arrays. It also performs custom array experiments
5	http://www.nanogen.com/	Nanogen. Microelectronics 10–400 bp fragments
6	http://www.cmgm.stanford.edu/pbrown/	The Brown laboratory home page. It includes a complete guide to microarraying for the molecular biologist and other links
7	http://www.jgi.doe.gov/tempweb/JGI_microbial/html	Joint Genome Institute. This site contains information on the current microbial genome-sequencing project
8	http://www.tigr.org/tigr-scripts/CMR2/CMR HomePage.spl	Tigr's. Comprehensive Microbial Resource (CMR) home page
9	http://www.genome.wi.mit.edu/genome_software/other/primer3.html	Software Primer3 for primer design
10	http://www.microarrays.org/index.html	Protocols and other information
11	http://www.arrayit.com/index.html	TeleChem International. It provides slides, software and other services
12	http://www.corning.com/	Corning. It provides slides and commercial DNA arrays
13	http://www.genemachines.com/	GeneMachines. It provides spotter and other robotics
14	http://www.axon.com/	Axon Instruments. Scanner
15	http://www.mdyn.com/	Molecular Dynamics. It provides spotter and scanners
16	http://www.stat.berkeley.edu/users/	Terry Speed's microarray data analysis group page
17	http://www.rana.lbl.gov/EisenResearch.html	Michael Eisen's software Cluster and TreeView
18	http://www.lionbioscience.com/	Lion Bioscience. It provides arraySCOUT software and an integrated database platform based on SRS system

(continued)

Table 6.1 (continued)

S.N.	Web link	Reference and available information
19	http://www.sigenetics.com/cgi/SiG.cgi/index.smf	Silicon Genetics. It provides GeneSpring Suite for data analysis, presentation and storage
20	http://www.genome-www4.stanford.edu/MicroArray/SMD/	Stanford Microarray Database
21	http://www.genomics.lbl.gov/ ~ ecoreg/	The *E. coli* regulation consortium. A site for primary and secondary data on pathway regulation
22	http://www.sghms.ac.uk/depts/medmicro/bugs/	Bacterial microarrays at St. George's Hospital Medical School
23	http://www.web.wi.mit.edu/young/location/	Methods for genome-wide protein–DNA binding studies
24	christoph.dehio@unibas.ch	PrimeArray software for genome scale primer design

with the help of cDNA microarray having 9,861 clones (Peterson et al. 2000); (7) cDNA microarray having 11,521 clones have been used to solve large-set of biological questions, e.g., amino acid metabolism, apoptosis, development, effect of environmental conditions, effect of hormonal treatment, effect of metals, functions of mitochondria, attack by pathogen, stress, RNA stability, comparative genomics and viruses in *Arabidopsis* (Schaffer et al. 2000; Wisman and Ohlroggge 2000); and (8) in lima bean responses to herbivory and herbivore-induced volatiles have been studied with the help of cDNA microarray having 2,032 clones.

6.7.3.2 Studies on Abiotic Stresses

Like biotic stresses abiotic stresses also play significant role in healthy growth of plants. Abiotic stresses like unfavorable temperature, humidity, light, soil pH, soil water, etc., adversely affect plant physiology, thereby affecting its proper growth and yield potential. Therefore, understanding of biochemical changes taking place in plants during stressed condition may help in minimizing the adverse effect on yield by adopting appropriate management practices or by developing tolerant transgenic varieties (Ahroni and Vorst 2001). Most of such studies have been conducted using cDNA, DNA and oligonucleotide microarrays.

For example, (1) Genomic expressions of *Arabidopsis* under drought and cold stresses are explained with cDNA microarray having 1,300 clones (Seki et al. 2001). (2) Response of *Synechocystis* sp to high light has been studied with the help of DNA microarray having 1.0 kb PCR fragments of 3079 PCC6803 clones (Hihara et al. 2001). (3) Similarly, expression of cold-regulated genes in hik33 mutant of *Synechocystis* sp have been understood with the use of DNA microarray prepared with 1.0 kb PCR fragments of 3079 PCC6803 clones (Suzuki et al. 2001). (4) Gene expressions due to salt stress have been studied in rice with the help of cDNA microarray having 1,728 clones (Kawasaki et al. 2001). (v) At Stanford university gene expression following exposure to high salinity in ice plant and *Arabidopsis* have been studied by Arabidopsis Functional Genomics Consortium using DNA, cDNA and oligonucleotide arrays having 2,600 clones (Bohnert et al. 2001, http://afgc.stanford.edu).

6.7.3.3 Studies on Diurnal Changes

Diurnal changes regulated by circadian clock are well understood in human beings. Better understanding of circadian rhythm in plants may help in selecting area specific crop, responsive variety and to device efficient crop management practices for sustainable food security. Efforts have been made to understand the existence of such circadian rhythm in plants by using oligonucleotide microarray representing 8,200 genes (Harmer et al. 2000) and cDNA microarray having 11,521 clones (Schaffer et al. 2001).

6.7.3.4 Studies on Physiology of Plants

Understanding biochemical details of normal physiology of plants may help in improving the management practices of crops in terms of date of sowing, irrigation, and fertilizer application, etc., due to clearer understanding of various requirements of plants during different growth stages. This may further help in developing transgenics with desirable traits. Efforts have been made to understand the gene expressions during various physiological activities in different plants with the help of DNA, cDNA and oligonucleotide microarrays (Ahroni and Vorst 2001).

For example, (1) expression in roots and leaves (cDNA; 48 clones) (Schena et al. 1995); in major plant organs (cDNA; 1,443 clones) (Ruan et al. 1998); response to nitrate treatments (cDNA; 5,524 clones) (Wang et al. 2000); phytochrome A mediated response (Oligo; 412 polymorphisms) (Spiegelman et al. 2000); expression in developing seeds (cDNA; 2715 clones) (Girke et al. 2000); expression in different tissues, organs, genetic conditions and growth environments (Oligo; 8,200 genes represented) (Zhu and Wang 2000) have been studied in *Arabidopsis*. (2) In strawberry, genes expressed for ripening and flavor has been identified by employing cDNA microarray having 1,701 clones (Ahroni and Vorst 2001). (3) In Petunia, genes expressed during flower development have been studied using cDNA microarray having 480 petunia clones (Aharoni et al. 2000). (4) In Maize, expression analysis of the glutathione-*S*-transferase gene family has been done with cDNA microarray of 42 maize clones (McGonigle et al. 2000). (5) Similarly, gene expressions in different tissues of rice have been studied by cDNA microarray having 1,265 clones.

6.7.3.5 Genetic Studies

Analysis of gene expressions using DNA microarrays may help in ascertaining the phylogenetic relationship of plant species, races of plant pathogens and insect pests. Further, it may help in developing molecular markers like ESTs and SNPs that are helpful in tracing the movement of the gene in breeding populations while developing new varieties bearing specific traits. There are few studies where cDNA microarrays have been used.

For example, cDNA microarray having 1,152 clones has helped in identifying repetitive genomic fragments in 17 *Vicia* species that helped in their phylogenetic resconstruction (Nouzov'a et al. 2001).

6.7.3.6 Multiple Pathogen Detection and Quantification of Population Densities

Crop plants are known to be infected with various pathogens in a crop season. It makes it a difficult task to differentiate and identify the multiple pathogens in a single assay at a given time. DNA microarrays were originally developed for gene expression analysis (Schena et al. 1996) but may also be useful in detecting multiple pathogens in a single assay (Lievens and Thomma 2005; Lievens and Thomma 2007). For pathogen detection using DNA microarray, the target DNA is amplified using universal primers those anneal to conserved sequences flanking the diagnostic domains. The amplicons are labeled and then hybridized with the array. Likewise by using universal primers it is possible to amplify multiple targets which can be then detected by hybridization to the array. Using this strategy various microorganisms like fungi (Lievens et al. 2003; Nicolaisen et al. 2005), oomycetes (Lévesque et al. 1998; Tambong et al. 2006),

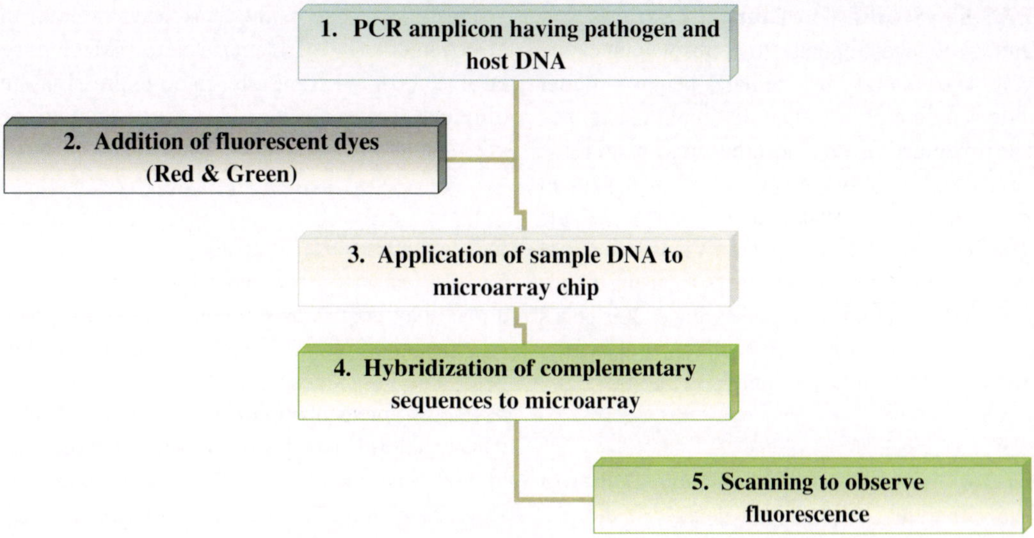

Fig. 6.9 Diagnosis of plant pathogens by DNA microarrays

bacteria (Fessehaie et al. 2003), nematodes (Uehara et al. 1999) and viruses (Boonham et al. 2003) have been successfully detected. Further, to increase the number of multiple detections in one run, combination of sequence-nonspecific amplification techniques with DNA microarrays is found to be useful in detecting the viruses for which appropriate universal primers are not available. For multiple detections, the discrimination power of DNA microarrays is very high because they can discriminate the targets differing even for a SNP if proper hybridization conditions are maintained (Lievens et al. 2006).

For example, (a) arrays developed to identify and detect more than 100 species of a fungal genus *Pythium* (Tambong et al. 2006). (b) The DNA array developed for detecting a wide range of fungal and bacterial pathogens occurring on horticultural crops and turf grasses (Lievens and Thomma 2005).

Other than multiple pathogen detections, DNA microarray technology is also been found useful in quantitative determination of microbial population densities because under some situations hybridization signals on the array are found proportional to the quantity of target DNA in the sample (Lievens et al. 2005). Likewise, it is possible to detect and quantify multiple pathogens in a single assay.

For diagnosis of plant pathogens by DNA microarray technique, firstly, RNA is extracted from infected plant sample which is then converted to complementary DNA (cDNA) and second-strand synthesis is performed using random pentamers with an anchor primer sequence at the 5′ end. Likewise due to the use of anchor primers unbiased amplification of plant and viral RNA takes place, which is then used as a target for hybridization of array as shown in Fig. 6.9.

6.8 Designing DNA Microarray Experiment

Careful planning of design is crucial to the success of microarray experiments if statistically and biologically valid conclusions are to be drawn from the data. There are three main elements to consider when designing a microarray experiment. First, replication of the biological samples is essential for drawing conclusions from the experiment. Second, technical replicates (two RNA samples obtained from each experimental unit) help to ensure precision and allow for testing differences within treatment groups. The biological replicates include independent RNA extractions and technical replicates may be two aliquots of the same extraction. Third, spots of each cDNA clone or oligonucleotide are present as replicates (at least duplicates) on the microarray slide, to provide a measure of technical precision while hybridization. At the same time, importance of replicates cannot be overemphasized because variability can be very high in microarray experiments. Many groups also choose to carry out *dye reversals*, in which one replicate array is hybridized with the experimental sample labeled with one fluorophore and the reference sample with the other dye. The corresponding duplicate array is then hybridized with experimental samples and reference samples labeled with the opposite fluorophores. This strategy generates replicate data while balancing the possible differential efficiency of dye incorporation among RNA samples. It is critical that information about the sample preparation and handling is discussed before planing an experiment, in order to help identify the independent units in the experiment and to avoid inflated estimates of statistical significance (Churchill 2002).

Microarray experiments are conducted using three basic designs (Macgregor and Squire 2002):

6.8.1 Case control design

In a case control study, two samples from a single individual, e.g., tumor and healthy tissues are compared directly. Because patient variability and genetic heterogeneity are key issues in microarray data analysis, the case control design is an excellent solution to avoid their effect on results, when feasible.

6.8.2 Blocked Design

Blocked designs are typically used to study the effect of a treatment or growth condition on a sample such as a cell line. They have been successfully used to examine cell lines grown under different conditions (e.g., cultured in the presence or absence of an anticancer drug) or different related cell lines (e.g., wild type v/s mutant, nontransfected cells v/s transfected cell).

6.8.3 Random Profile Design

Random profile designs are widely used in microarray experiments when cell lines or patient samples are selected and profiled. Most of the *profiling papers* have used this design, which offers the ability to use data from many different individuals but offers no intrinsic control for bias in the patient populations or cell populations used.

In both the blocked and randomized profile designs, the sample is typically compared with a common or universal reference, which should have adequate representation of the majority of genes on the array being profiled and be easily available. Commercially available reference RNA is often a good choice because of wide gene representation (e.g., Stratagene and Clontech). The use of a common reference also offers the advantage of allowing longitudinal comparative analysis among several microarray projects between different research groups interested in a common aspect of cancer research, such as tumor progression or resistance to anticancer drugs (Lashkari et al. 1997)

6.9 Preparation of Target and Hybridization

Total RNA and mRNA are used for microarray experiments that help in generating high-quality data with high degree of confidence. For successful microarray experiments, isolation of high-quality RNA from sample is required. Different standard RNA extraction methodologies have been used successfully. Experimenters can choose protocol for RNA extraction by their personal experience. Quantitative and qualitative evaluation of the RNA obtained can be carried out by standard techniques, such as agarose gel electrophoresis, but it requires relatively large amounts of RNA samples. More recently, assessment of RNA quality and quantity has been greatly facilitated by the use of microcapillary-based devices such as the Agilent Bioanalyzer (Agilent Technologies), which can be used with as little as 5 ng of total RNA.

Insufficient availability of RNA is one of the major limitations in the routine application of microarray technology to disease detection in patient samples. Thus, there has been considerable interest in the development of RNA amplification strategies that facilitate RNA extraction from laser capture microdissected (LCM) samples, such as fine-needle biopsies. For standard microarray experiments, the isolated RNA is reverse-transcribed into target cDNA in the presence of fluorescent (generally Cy3-dNTP or Cy5-dNTP) or radio labeled deoxynucleotides ([^{33}P]- or [^{32}P]-α-dCTP) to develop labeled targets. After purification and denaturation, these labeled targets are hybridized to the microarrays at a temperature determined by the hybridization buffer used. After hybridization, the arrays are washed under stringent conditions to remove nonspecific target binding and are air-dried before image acquisition and quantification (Macgregor and Squire 2002).

6.10 Imaging and Quantification

Processing of microarray image requires differential excitation and emission wavelengths of the two fluors to obtain a scan of the array for each emission wavelength, typically as two 16-bit grayscale TIFF images. These images are then analyzed to identify the spots, calculate their associated signal intensities, and assess local background noise. Most image acquisition software packages also contain basic filtering tools to flag spots such as extremely low-intensity spots, ghosts spots (where background is higher than spot intensity), or damaged spots (e.g., dust artifacts). These results allow an initial ratio of the evaluated channel/reference channel intensity to be calculated for every spot on the chip. The products of the image acquisition are the TIFF image pairing and a quantified data file that has not yet been normalized (Macgregor and Squire 2002). Assessment of different image analysis methods is available on website http://oz.berkeley.edu/tech-reports/.

6.11 Warehousing and Normalization of Data

The quantity of data generated in a microarray experiment typically requires a dedicated database system to store and organize the microarray data and images. The first role of a local microarray database is the storage and annotation (description of experimental parameters) of microarray experiments by the investigator who designed and carried out the microarray experiments. In addition, there is currently an increasing global interest in making microarray data sets publicly available in a standardized format. This would allow other investigators to reproduce published microarray experiments, thereby independently verify them, to compare data sets across different microarray platforms, and importantly, to interrogate published microarray data sets by use of various bioinformatics tools to explore different biological problems. To answer this need, the Minimal Information About a Microarray Experiment, or MIAME standard, has been proposed by the MGED (http://www.mged.org/) organization as a series of criteria that should be used when defining microarray experiment parameters. One of such database is *GeneTraffic* of Iobion Informatics, which holds all of the microarray data files and TIFF images, as well as a MIAME supportive annotation of experiments (Macgregor and Squire 2002).

Once data have been loaded into the database, they are normalized, and aggregate statistics are calculated. Normalization is a process that scales spot intensities such that the normalized ratios provide an approximation of the ratio of gene expression between the two samples. Different strategies are adopted for normalization of microarray data. The choice of a robust and adequate normalization method is as crucial for the quality of the data obtained as the experimental design of the microarray experiment itself. A discussion of normalization methods is available on website www.utoronto.ca/cancyto/CLINCHEM.

6.12 Statistical Analysis and Data Mining

Microarray data sets are commonly very large, and analytical precision is influenced by a number of variables. Statistical challenges include taking into account effects of background noise and appropriate normalization of the data. Normalization methods may be suited to specific platforms and, in the case of commercial platforms, the analysis may be proprietary. Algorithms that affect statistical analysis include:

(a) Image analysis: gridding, spot recognition of the scanned image (segmentation algorithm), removal or marking of poor-quality and low-intensity features (called *flagging*).
(b) Data processing: background subtraction (based on global or local background), determination of spot intensities and intensity ratios, visualization of data (e.g., MA plot), and log-transformation of ratios, global or local normalization of intensity ratios.
(c) Identification of statistically significant changes: *t*-test, ANOVA, Bayesian method (Ben-Gal et al. 2005), Mann–Whitney test methods tailored to microarray data sets, which take into account multiple comparisons or cluster analysis (Priness et al. 2007). These methods assess statistical power based on the variation present in the data and the number of experimental replicates, and can help minimize Type I and Type II errors in the analyses (Wei et al. 2004).
(d) Network-based methods: Statistical methods that take the underlying structure of gene networks into account, representing either associative or causative interactions or dependencies among gene products (Streib and Dehmer 2008).

Microarray data may require further processing aimed at reducing the dimensionality of the data to aid comprehension and more focused analysis (Wouters et al. 2003). Other methods permit analysis of data consisting of a low number of biological or technical replicates; such as the Local Pooled Error (LPE) test pools standard deviations of genes with similar expression levels in an effort to compensate for insufficient replication (Jain et al. 2003). Gene expression values from microarray experiments can be represented as heat maps to visualize the result of data analysis.

Data mining methods typically fall into one of two classes: supervised and unsupervised. In unsupervised analysis, the data are organized without the benefit of external classification information. *Hierarchical clustering* (Adomas et al. 2008), *K-means clustering* (Moran et al. 2004), or *self-organizing maps* (Hacia et al. 1999) are examples of unsupervised clustering approaches that have been widely used in microarray analysis (Nuwaysir et al. 2002; Bammler et al. 2005; Adomas et al. 2008). Supervised analysis uses some external information, such as the disease status of the samples studied. Supervised analysis involves choosing from the entire data set a training set and a testing set and also involves construction of classifiers, which assign predefined classes to expression profiles. Once the classifier has been trained on the training set and tested on the testing set, it can then be applied to data with unknown classification. Supervised methods include *k-nearest neighbor classification*, *support vector machines*, and *neural nets*. For example, Shalon et al. (1996) used a *k-nearest neighbor strategy* to classify the expression profiles of leukemia samples into two classes: acute myeloid leukemia and acute lymphocytic leukemia. Tang et al. (2007b) used Large-scale RNA profiling and supervised machine learning algorithms to construct a molecular classification for 10 carcinomas i.e. prostate, lung, ovary, colorectum, kidney, liver, pancreas, bladder/

ureter, and gastroesophagus. Similarly, *neural network analysis* has been used by Churchill (2002) to delineate consistent patterns of gene expression in cancer.

Recently, a strategy called *Significance Analysis of Microarrays* has been proposed by Micro Array Quality Control (MAQC) project of NCTR centre for toxicoinformatics, which allows the determination of significantly differentially expressed genes between groups of samples analyzed by expression arrays. This approach has been used to narrow down the analysis to a subset of genes that were also shown to be differentially expressed when analyzed by conventional two-dimensional hierarchical clustering. The approach has helped in identifying the genes that show differential expression between early-stage epithelial ovarian cancer (EOC), late-stage EOC, and healthy ovary.

6.13 Redundancy and Challenges

DNA microarray technology is a new developing era in gene expression analysis that has its own limitations and challenges. Few of them are as follows:

1. The relation between a probe and the mRNA that it is expected to detect is not trivial. Some mRNAs may cross-hybridize probes in the array that are supposed to detect another mRNA.
2. mRNAs may experience amplification bias that is sequence or molecule-specific.
3. Probes that are designed to detect the mRNA of a particular gene may be relying on genomic EST information that is incorrectly associated with that gene.
4. Bioinformatics challenges:

(a) The multiple levels of replication in experimental design create difficulties in data handling.

(b) The number of platforms and independent groups and data format is a challenge for standardization. Microarray data is difficult to exchange due to the lack of standardization in platform fabrication, assay protocols, and analysis methods. This presents an interoperability problem in bioinformatics. Various grass-roots open-source projects are trying to ease the exchange and analysis of data produced with nonproprietary chips. For example,

The MIAME checklist helps to define the level of detail that should exist and is being adopted by many journals as a requirement for the submission of papers incorporating microarray results. But MIAME does not describe the format for the information, so while many formats can support the MIAME requirements, as of 2007, no format permits verification of complete semantic compliance.

The MAQC *Project* is being conducted by the US Food and Drug Administration (FDA) to develop standards and quality control metrics which will eventually allow the use of micro array data in drug discovery, clinical practice and regulatory decision-making.

The MGED Society has developed standards for the representation of gene expression experiment results and relevant annotations.

(c) The statistical analysis of the data is complicated.

(d) Accuracy and precision solely depends on relation between probe and gene.

(e) To share the sheer volume (in bytes) of data dedicated data storage system is required.

6.14 Pitfalls of DNA Microarray Technology

Like any other technology, DNA microarray technology has some lackings and drawbacks (Vacha 2003). To list a few:

1. Focusing on image quality over data quality.
2. Paying more attention to absolute signal than signal-to-noise.
3. Failing to interpret replicate results within the context of the 'level' of replication.
4. Assuming that statistical significance is equivalent to biological significance.
5. Ignoring experimental design considerations.
6. Using experimental conditions that are different from the error model conditions.

7. Paying more attention to the magnitude of the log ratio than the significance of the log ratio.
8. Automatically assuming that q-pcr, rpa, or northern blot analysis is needed to confirm every microarray result.
9. Cutting upfront costs at the expense of downstream results.
10. Pursuing one path in data interpretation.

6.15 Conclusion

DNA microarray is a very useful technology to understand the gene functions from sequences identified through small or large genome sequencing projects as it helps in handling multiple genes in a go, therefore speeding the functional genomics e.g., human genome sequence, rice genome sequence, pigeon pea genome sequence, etc. It is also helpful in understanding the biochemical details of any physiological state of an individual, i.e., diseased, stressed, mutated, or constitutive. This technique has tremendous possibilities in diagnostics encompassing human, animals, plants, birds, etc. Proper experimental planning, data warehousing and logical interpretation of data may help in developing molecular medicines for the diseases that are not curable by conventional medicines. It may also be useful to identify valuable genes and gene combinations that may help in developing crop varieties with improved qualitative and quantitative yield traits, to understand genetics, evolution and epidemiology of races of any living being, host responses to biotic and abiotic stresses, and drug sensitivity of host and pathogen. Likewise, an in-depth knowledge of many other biological phenomenons with practical applicability may also be obtained.

References

Adomas A, Heller G, Olson A, Osborne J, Karlsson M, Nahalkova J, Van Zyl L, Sederoff R, Stenlid J, Finlay R, Asiegbu FO (2008) Comparative analysis of transcript abundance in *Pinus sylvestris* after challenge with a saprotrophic, pathogenic or mutualistic fungus. Tree Physiol 28(6):885–897

Aharoni A, Keizer LCP, Bouwmeester HJ, Sun ZK, Alvarez Huerta M, Verhoeven HA, Blaas J, van Houwelingen A, de Vos RCH, van der Voet H, Jansen RC, Guis M, Mol J, Davis RW, Schena M, van Tunen AJ, O'Connell AP (2000) Identification of the *SAAT* gene involved in strawberry flavor biogenesis by use of DNA microarrays. Plant Cell 12:647–661

Aharoni A, Vorst O (2001) Reports on the use of DNA microarrays in plants. Plant Mol Biol 48:99–118

Alizadeh AA, Eisen MB, Davis RE, Ma C, Lossos IS, Rosenwald A, Boldrick JG, Sabet H, Tran T, Yu X, Powell JI, Yang LM, Marti GE, Moore T, Hudson J, Lu LS, Lewis DB, Tibshirani R, Sherlock G, Chan WC, Greiner TC, Weisenburger DD, Armitage JO, Warnke R, Levy R, Wilson W, Grever MR, Byrd JC, Botstein D, Brown PO, Staudt LM (2000) Distinct types of diffuse large B-cell lymphoma identified by gene expression profiling. Nature 403:503–511

Arimura G, Tashiro K, Kuhara S, Nishioka T, Ozawa R, Takabayashi J (2000) Gene responses in bean leaves induced by herbivory and by herbivore-induced volatiles. Biochem Biophys Res Commun 277:305–310

Armstrong B, Stewart M, Mazumder A (2000) Suspension arrays for high throughput, multiplexed single nucleotide polymorphism genotyping. Cytometry 40:102–108

Augenlicht et al (1987) Cancer Res 47:6017–6021

Azugenlicht et al (1991) Proc Nat Acad Sci 88: 3286–3289

Augenlicht, Kobrin (1982) Cancer Research 42: 1088–109

Bammler T, Beyer RP, Consortium, Members of the Toxicogenomics Research, Kerr X, Jing LX, Lapidus S, Lasarev DA, Paules RS, Li JL et al (2005) Standardizing global gene expression analysis between laboratories and across platforms. Nat Methods 2(5):351–356

Bavykin SG, Akowski JP, Zakhariev VM, Barsky VE, Perov AN, Mirzabekov AD (2001) Portable system for microbial sample preparation and oligonucleotide microarray analysis. Appl Environ Microbiol 67:922–928

Bayani J, Brenton JD, Macgregor PF, Beheshti B, Albert M, Nallainathan D et al (2002) Parallel analysis of sporadic primary ovarian carcinomas by spectral karyotyping, comparative genomic hybridization and expression microarrays. Cancer Res 62:3466–3476

Belcher CE, Drenkow J, Kehoe B, Gingeras TR, McNamara N, Lemjabbar H, Basbaum C, Relman DA (2000) The transcriptional responses of respiratory epithelial cells to *Bordetella pertussis* reveal host defensive and pathogen counter-defensive strategies. Proc Natl Acad Sci USA 97:13847–13852

Ben-Gal I, Shani A, Gohr A, Grau J, Arviv S, Shmilovici A, Posch S, Grosse I (2005) Identification of transcription factor binding sites with variable-order bayesian networks. Bioinformatics 21(11):2657–2666

Bhattacharjee A, Richards WG, Staunton J, Li C, Monti S, Vasa P et al (2001) Classification of human lung carcinomas by mRNA expression profiling reveals distinct adenocarcinoma subclasses. Proc Natl Acad Sci USA 98:13790–13795

Bibikova M, Fan JB (2009) GoldenGate assay for DNA methylation profiling. Methods Mol Biol 507:149–163

Bibikova M, Talantov D, Chudin E, Yeakley JM, Chen J, Doucet D, Wickham E, Atkins D, Barker D, Chee M, Wang Y, Fan JB (2004) Quantitative gene expression profiling in formalin-fixed, paraffin-embedded tissues using universal bead arrays. Am J Pathol 165:1799–1807

Bittner M, Meltzer P, Chen Y, Jiang Y, Seftor E, Hendrix M et al (2000) Molecular classification of cutaneous malignant melanoma by gene expression profiling. Nature 406:536–540

Bohnert HJ, Ayoubi P, Borchert C, Bressan RA, Burnap RL, Cushman JC, Cushman MA, Deyholes M, Fischer R, Galbraith DW, Hasegawa PM, Jenks M, Kawasaki S, Koiwa H, Kore-eda S, Lee BH, Michalowski CB, Misawa E, Nomura M, Ozturk N, Postier B, Prade R, Song CP, Tanaka Y, Wang H, Zhu JK (2001) A genomics approach towards salt stress tolerance. Plant Physiol Biochem 39:295–311

Boonham N, Walsh K, Smith P, Madagan K, Graham I, Barker I (2003) Detection of potato viruses using microarray technology: towards a generic method for plant viral disease diagnosis. J Virol Methods 108:181–187

Chen J, Iannone MA, Li MS, Taylor JD, Rivers P, Nelsen AJ, Slentz KKA, Roses A, Weiner MP (2000) A microsphere-based assay for multiplexed single nucleotide polymorphism analysis using single base chain extension. Genome Res 10:549–557

Cheung ST, Chen X, Guan XY, Wong SY, Tai LS, Ng IO, So S, Fan ST (2002) Identify metastasis-associated genes in hepatocellular carcinoma through clonality delineation for multinodular tumor. Cancer Res 62:4711–4721

Cheung VG, Morley M, Aguilar F, Massimi A, Kucherlapati R, Childs G (1999) Making and reading microarrays. Nat Genet 21:15–19

Cho RJ, Mindrinos M, Richards DR, Sapolsky RJ, Anderson M, Drenkard E, Dewdney J, Reuber TL, Stammers M, Federspiel N, Theologis A, Wei HY, Hubbell E, Au M, Chung EY, Lashkari D, Lemieux B, Dean C, Lipshutz RJ, Ausubel FM, Davis RW, Oefner PJ (1999) Genomewide mapping with biallelic markers in *Arabidopsis thaliana*. Nat Genet 23:203–207

Chou CC, Chen CH, Lee TT, Peck K (2004) Optimization of probe length and the number of probes per gene for optimal microarray analysis of gene expression. Nucleic Acids Res 32:e99

Churchill GA (2002) Fundamentals of experimental design for cDNA microarrays. Nature genetics supplement 32: 490. doi:10.1038/ng1031. PMID12454643

Cohen P, Bouaboula M, Bellis M, Baron V, Jbilo O, Chazel CP, Galiegue S, Hadibi EH, Casellas P (2000) Monitoring cellular responses to *Listeria monocytogenes* with oligonucleotide arrays. J Biol Chem 275:11181–11190

Coombes BK, Mahony JB (2001) cDNA array analysis of altered gene expression in human endothelial cells in response to *Chlamydia pneumoniae* infection. Infect Immun 69:1420–1427

Dhanasekaran SM, Barrette TR, Ghosh D, Shah R, Varambally S, Kurachi K et al (2001) Delineation of prognostic biomarkers in prostate cancer. Nature 412:822–826

Dunbar SA (2006) Applications of Luminex xMAP technology for rapid, high-throughput multiplexed nucleic acid detection. Clin Chim Acta 363:71–82

Ehrenreich A (2006) DNA microarray technology for the microbiologist: an overview. Appl Microbiol Biotechnol 73:255–273

Fan JB, Gunderson KL, Bibikova M, Yeakley JM, Chen J, Garcia EW, Lebruska LL, Laurent M, Shen R, Barker D (2006) Illumina universal bead arrays. Methods Enzymol 410:57–73

Fan JB, Hu SX, Craumer WC, Barker DL (2005) BeadArray-based solutions for enabling the promise of pharmacogenomics. Biotechniques 39:583–588

Fan JB, Yeakley JM, Bibikova M, Chudin E, Wickham E, Chen J, Doucet D, Rigault P, Zhang B, Shen R, McBride C, Li HR, Fu XD, Oliphant A, Barker DL, Chee MS (2004) A versatile assay for high-throughput gene expression profiling on universal array matrices. Genome Res 14:878–885

Fessehaie A, de Boer SH, Lévesque CA (2003) An oligonucleotide array for the identification and differentiation of bacteria pathogenic on potato. Phytopathology 93:262–269

Fodor SP, Read JL, Pirrung MC, Stryer L, Lu AT, Solas D (1991) Light-directed, spatially addressable parallel chemical synthesis. Science 251:767–773

Garber ME, Troyanskaya OG, Schluens K, Petersen S, Thaesler Z, Gengelbach MP et al (2001) Diversity of gene expression in adenocarcinoma of the lung. Proc Natl Acad Sci USA 98:13784–13789

Gill RT, DeLisa MP, Valdes JJ, Bentley WE (2001) Genomic analysis of high-cell density recombinant *Escherichia coli* fermentation and A cell conditioning B for improved recombinant protein yield. Biotechnol Bioeng 72:85–95

Girke T, Todd J, Ruuska S, White J, Benning C, Ohlrogge J (2000) Microarray analysis of developing *Arabidopsis* seeds. Plant Physiol 124:1570–1581

Golub TR, Slonim DK, Tamayo P, Huard C, Gaasenbeek M, Mesirov JP et al (1999) Molecular classification of cancer: class discovery and class prediction by gene expression monitoring. Science 286:531–537

Gunderson KL (2009) Whole-genome genotyping on bead arrays. Methods Mol Biol 529:197–213

Gunderson KL, Kruglyak S, Graige MS, Garcia F, Kermani BG, Zhao C, Che D, Dickinson T, Wickham E, Bierle J, Doucet D, Milewski M, Yang R, Siegmund C, Haas J, Zhou L, Oliphant A, Fan JB, Barnard S, Chee MS (2004) Decoding randomly ordered DNA arrays. Genome Res 14:870–877

Hacia JG, Brody LC, Chee MS, Fodor SP, Collins FS (1996) Detection of heterozygous mutations in *BRCA1* using high density oligonucleotide arrays and two color fluorescence analysis. Nat Genet 14:441–447

Hacia JG, Fan JB, Ryder O, Jin L, Edgemon K, Ghandour G, Mayer RA, Sun B, Hsie L, Robbins CM, Brody LC, Wang D, Lander ES, Lipshutz R, Fodor SP, Collins FS (1999) Determination of ancestral alleles for human single-nucleotide polymorphisms using high-density oligonucleotide arrays. Nat Genet 22(2):164–167

Hager J (2006) Making and using spotted DNA microarrays in an academic core laboratory. Methods Enzymol 410:135–168

Harkin DP, Bean JM, Miklos D, Song YH, Truong VB et al (1999) Induction of *GADD 45* and *JNK/SAPK*-dependent apoptosis following inducible expression of *BRCA1*. Cell 97:575–586

Harmer SL, Hogenesch LB, Straume M, Chang HS, Han B, Zhu T, Wang X, Kreps JA, Kay SA (2000) Orchestrated transcription of key pathways in Arabidopsis by the circadian clock. Science 290:2110–2113

Hihara Y, Kamei A, Kanehisa M, Kaplan A, Ikeuchi M (2001) DNA microarray analysis of cyanobacterial gene expression during acclimation to high light. Plant Cell 13:793–806

Hooper LV, Wong MH, Thelin A, Hansson L, Falk PG, Gordon JI (2001) Molecular analysis of commensal host–microbial relationships in the intestine. Science 291:881–884

Horan PK, Wheeless LL Jr (1977) Quantitative single cell analysis and sorting. Science 198:149–157

Iannone MA, Taylor JD, Chen J, Li MS, Rivers P, Kesler KAS, Weiner MP (2000) Multiplexed single nucleotide polymorphism genotyping by oligonucleotide ligation and flow cytometry. Cytometry 39:131–140

Ichikawa JK, Norris A, Bangera MG, Geiss GK, van't Wout AB, Bumgarner RE, Lory S (2000) Interaction of *Pseudomonas aeruginosa* with epithelial cells: identification of differentially regulated genes by expression microarray analysis of human cDNAs. Proc Natl Acad Sci USA 97:9659–9664

Jain N, Thatte J, Braciale T, Ley K, O'Connell M, Lee JK (2003) Local-pooled-error test for identifying differentially expressed genes with a small number of replicated microarrays. Bioinformatics 19(15): 1945–1951

Johnson SC, Marshall DJ, Harms G, Miller CM, Sherrill CB, Beaty EL, Lederer SA, Roesch EB, Madsen G, Hoffman GL, Laessig RH, Kopish GJ, Baker MW, Benner SA, Farrell PM, Prudent JR (2004) Multiplexed genetic analysis using an expanded genetic alphabet. Clin Chem 50:2019–2027

Kato-Maeda M, Rhee JT, Gingeras TR, Salamon H, Drenkow J, Smittipat N, Small PM (2001) Comparing genomes within the species *Mycobacterium tuberculosis*. Genome Res 11:547–554

Kawasaki S, Borchert C, Deyholos M, Wang H, Brazille S, Kawai K, Galbraith D, Bohnert HJ (2001) Gene expression profiles during the initial phase of salt stress in rice. Plant Cell 13:889–906

Kehoe DM, Villand P, Somerville SC (1999) DNA microarrays for studies of higher plants and other photosynthetic organisms. Trends Plant Sci 4:38–41

Kudoh K, Ramanna M, Ravatn R, Elkahloun AG, Bittner ML, Meltzer PS et al (2000) Monitoring the expression profiles of doxorubicin-induced and doxorubicin-resistant cancer cells by cDNA microarray. Cancer Res 60:4161–4166

Kuhn K, Baker SC, Chudin E, Lieu MH, Oeser S, Bennett H, Rigault P, Barker D, McDaniel TK, Chee MS (2004) A novel, high-performance random array platform for quantitative gene expression profiling. Genome Res 14:2347–2356

Kulesh DA, Clive DR, Zarlenga DS, Greene JJ (1987) Identification of interferon-modulated proliferation-related cDNA sequences. Proc Natl Acad Sci USA 84(23):8453–8457

Lashkari DA, DeRisi JL, McCusker JH, Namath AF, Gentile C, Hwang SY, Brown PO, Davis RW (1997) Yeast microarrays for genome wide parallel genetic and gene expression analysis. Proc Natl Acad Sci USA 94(24):13057–13062. doi:10.1073/pnas.94.24.13057. PMC 24262. PMID 9371799

Lévesque CA, Harlton CE, de Cock AWAM (1998) Identification of some oomycetes by reverse dot blot hybridization. Phytopathology 88:213–222

Lievens B, Brouwer M, Vanachter ACRC, Lévesque CA, Cammue BPA, Thomma BPHJ (2003) Design and development of a DNA array for rapid detection and identification of multiple tomato vascular wilt pathogens. FEMS Microbiol Lett 223:113–122

Lievens B, Brouwer M, Vanachter ACRC, Lévesque CA, Cammue BPA, Thomma BPHJ (2005) Quantitative assessment of phytopathogenic fungi in various substrates using a DNA macroarray. Environ Microbiol 7:1698–1710

Lievens B, Claes L, Vanachter ACRC, Cammue BPA, Thomma BPHJ (2006) Detecting single nucleotide polymorphisms using DNA arrays for plant pathogen diagnosis. FEMS Microbiol Lett 255:129–139

Lievens B, Thomma BPHJ (2005) Recent developments in pathogen detection arrays: implications for fungal plant pathogens and use in practice. Phytopathology 95:1374–1380

Lievens B, Thomma BPHJ (2007) Quantification in multiplex format as a challenging goal for plant pathogen molecular diagnostic assays. Pest Technol (Global Science Books) 17–25

Macgregor PF, Squire JA (2002) Application of microarrays to the analysis of gene expression in cancer. Clin Chem 48:1170–1177

Maleck K, Levine A, Eulgem T, Morgan A, Schmid J, Lawton KA, Dangl JL, Dietrich RA (2000) The transcriptome of *Arabidopsis thaliana* during systemic acquired resistance. Nat Genet 26:403–410

Maskos U, Southern EM (1992) Oligonucleotide hybridizations on glass supports: a novel linker for oligonucleotide synthesis and hybridization properties of oligonucleotides synthesized *in situ*. Nucleic Acids Res 20(7):1679–1684

McGonigle B, Keeler SJ, Lan SMC, Koeppe MK, O'Keefe DP (2000) A genomics approach to the comprehensive analysis of the glutathione *S*-transferase gene family in soybean and maize. Plant Physiol 124:1105–1120

McHugh RS, Ratnoff WD, Gilmartin R, Sell KW, Selvaraj P (1998) Detection of a soluble form of *B7–1 (CD80)* in synovial fluid from patients with arthritis using monoclonal antibodies against distinct epitopes of human *B7–1*. Clin Immunol Immunopathol 87:50–59

Miller MB, Tang YW (2009) Basic concepts of microarrays and potential applications in clinical microbiology. Clin Microbiol Rev 22(4):611–633

Monni O, Barlund M, Mousses S, Kononen J, Sauter G, Heiskanen M et al (2001) Comprehensive copy number and gene expression profiling of the *17q23* amplicon in human breast cancer. Proc Natl Acad Sci USA 98:5711–5716

Moran G, Stokes C, Thewes S, Hube B, Coleman DC, Sullivan D (2004) Comparative genomics using *Candida albicans* DNA microarrays reveals absence and divergence of virulence-associated genes in *Candida dubliniensis*. Microbiology 150(10):3363–3382

Mousses S, Bubendorf L, Wagner U, Hostetter G, Kononen J, Conelison R et al (2002) Clinical validation of candidate genes associated with prostate cancer progression in the *CWR22* model system using tissue microarrays. Cancer Res 62:1256–1260

Nicolaisen M, Justesen AF, Thrane U, Skouboe P, Holmstrom K (2005) An oligonucleotide microarray for the identification and differentiation of trichothecene producing and non-producing *Fusarium* species occurring on cereal grain. J Microbiol Methods 62:57–69

Nouzová M, Neumann P, Navrátilová A, Galbraith DW, Macas J (2001) Microarray-based survey of repetitive genomic sequences in *Vicia* spp. Plant Mol Biol 45:229–244

Nuwaysir EF, Huang W, Albert TJ, Singh J, Nuwaysir K, Pitas A, Richmond T, Gorski T, Berg JP, Ballin J, McCormick M, Norton J, Pollock T, Sumwalt T, Butcher L, Porter D, Molla M, Hall C, Blattner F, Sussman MR, Wallace RL, Cerrina F, Green RD (2002) Gene expression analysis using oligonucleotide arrays produced by maskless photolithography. Genome Res 12(11):1749–1755

Oh MK, Liao JC (2000a) DNA microarray detection of metabolic responses to protein overproduction in *Escherichia coli*. Metab Eng 2:201–209

Oh MK, Liao JC (2000b) Gene expression profiling by DNA microarrays and metabolic fluxes in *Escherichia coli*. Biotechnol Prog 16:278–286

Oliphant A, Barker DL, Stuelpnagel JR, Chee MS (2002) BeadArray technology: enabling an accurate, cost-effective approach to high-throughput genotyping. BioTechniques (Suppl) 56–58:60–61

Pease AC, Solas D, Sullivan EJ, Cronin MT, Holmes CP, Fodor SP (1994) http://www.pubmedcentral.nih.gov/articlerender.fcgi?tool=pmcentrez&artid=43922 Light-generated oligonucleotide arrays for rapid DNA sequence analysis. PNAS 91(11):5022–5026.

Perou CM, Sorlie T, Eisen MB, van de Rijn M, Jeffrey SS, Rees CA et al (2000) Molecular portraits of human breast tumors. Nature 406:747–752.

Petersen M, Brodersen P, Naested H, Andreasson E, Lindhart U, Johansen B, Nielsen HB, Lacy M, Austin MJ, Parker JE, Sharma SB, Klessig DF, Martienssen R, Mattsson O, Jensen AB, Mundy J (2000) *Arabidopsis* MAP kinase 4 negatively regulates systemic acquired resistance. Cell 103:1111–1120

Pollack JR, Perou CM, Alizadeh AA, Eisen MB, Pergamenschikov A, Williams CF, Jeffrey SS, Botstein D, Brown PO (1999) Genome-wide analysis of DNA copy-number changes using cDNA microarrays. Nat Genet 23(1):41–46

Priness I, Maimon O, Ben-Gal I (2007) Evaluation of gene-expression clustering by mutual information distance measures. BMC Bioinform 8(1):111

Ramdas L, Cogdell DE, Jia JY, Taylor EE, Dunmire VR, Hu L, Hamilton SR, Zhang W (2004) Improving signal intensities for genes with low-expression on oligonucleotide microarrays. BMC Genomics 5:35

Reymond P, Weber H, Damond M, Farmer EE (2000) Differential gene expression in response to mechanical wounding and insect feeding in *Arabidopsis*. Plant Cell 12:707–719

Risch NJ (2000) Searching for genetic determinants in new millennium. Nature 405:847–856

Ross DT, Scherf U, Eisen MB, Perou CM, Rees C, Spellman P et al (2000) Systematic variation in gene expression patterns in human cancer cell lines. Nat Genet 24:227–235

Ruan Y, Gilmore J, Conner T (1998) Towards *Arabidopsis* genome analysis: Monitoring expression profiles of 1,400 genes using cDNA microarrays. Plant J 15:821–833

Salama N, Guillemin K, McDaniel TK, Sherlock G, Tompkins L, Falkow S (2000) A whole-genome microarray reveals genetic diversity among *Helicobacter pylori* strains. Proc Natl Acad Sci USA 97:14668–14673

Schaffer R, Landgraf J, Amador MP, Wisman E (2000) Monitoring genome-wide expression in plants. Curr Opin Biotechnol 11:162–167

Schaffer R, Landgraf J, Accerbi M, Simon V, Larson M, Wisman E (2001) Microarray analysis of diurnal and circadian regulated genes in *Arabidopsis*. Plant Cell 13:113–123

Schena M, Shalon D, Davis RW, Brown PO (1995) Quantitative monitoring of gene expression patterns with a complementary DNA microarray. Science 270(5235):467–470

Schena M, Shalon D, Heller R, Chai A, Brown PO, Davis RW (1996) Parallel human genome analysis: microarray-based expression monitoring of 1000 genes. Proc Nat Acad Sci USA 93:10614–10619

Schenk PM, Kazan K, Wilson I, Anderson JP, Richmond T, Somerville SC, Manners JM (2000) Coordinated plant defense responses in *Arabidopsis* revealed by microarray analysis. Proc Natl Acad Sci USA 97:11655–11660

Scherf U, Ross DT, Waltham M, Smith LH, Lee JK, Tanabe L et al (2000) A gene expression database for the molecular pharmacology of cancer. Nat Genet 24:236–244

Scillian JJ, McHugh TM, Busch MP, Tam M, Fulwyler MJ, Chien DY, Vyas GN (1989) Early detection of antibodies against rDNA-produced HIV proteins with a flow cytometric assay. Blood 73:2041–2048

Seki M, Narusaka M, Abe H, Kasuga M, Shinozaki KY, Carninci P, Hayashizaki Y, Shinozaki K (2001) Monitoring the expression pattern of 1,300 *Arabidopsis* genes under drought and cold stresses by using a full-length cDNA microarray. Plant Cell 13:61–72

Shalon D, Smith SJ, Brown PO (1996) A DNA microarray system for analyzing complex DNA samples using two-color fluorescent probe hybridization. Genome Res 6(7):639–645

Singhal H, Bautista DS, Tonkin KS, O'Malley FP, Tuck AB, Chambers AF et al (1997) Elevated plasma osteopontin in metastatic breast cancer associated with increased tumor burden and decreased survival. Clin Cancer Res 3:605–611

Sorlie T, Perou CM, Tibshirani R, Aas T, Geisler S, Johnsen H et al (2001) Gene expression patterns of breast carcinomas distinguish tumor subclasses with clinical implications. Proc Natl Acad Sci USA 98:10869–10874

Sosnowski RG, Tu E, Butler WF, O'Connell JP, Heller MJ (1997) Rapid determination of single base mismatch mutations in DNA hybrids by direct electric field control. Proc Natl Acad Sci USA 94:1119–1123

Spiegelman JI, Mindrinos MN, Fankhauser C, Richards D, Lutes J, Chory J, Oefner PJ (2000) Cloning of the arabidopsis *RSF1* gene by using a mapping strategy based on high-density DNA arrays and denaturing high-performance liquid chromatography. Plant Cell 12:2485–2498

Spiro A, Lowe M, Brown D (2000) A bead-based method for multiplexed identification and quantitation of DNA sequences using flow cytometry. Appl Environ Microbiol 66:4258–4265

Streib FE, Dehmer M (2008) Analysis of microarray data a network-based approach. Wiley, New York. ISBN 3-527-31822-4

Suzuki I, Kanesaki Y, Mikami K, Kanehisa M, Murata N (2001) Cold-regulated genes under control of the cold sensor *Hik33* in *Synechocystis*. Mol Microbiol 40:235–244

Tambong JT, de Cock AWAM, Tinker NA, Lévesque CA (2006) Oligonucleotide array for identification and detection of *Pythium* species. Appl Environ Microbiol 72:2691–2706

Tang T, François N, Glatigny A, Agier N, Mucchielli MH, Aggerbeck L, Delacroix H (2007a) Expression ratio evaluation in two-colour microarray experiments is significantly improved by correcting image misalignment. Bioinformatics 23(20):2686–2691

Tang BM, McLean AS, Dawes IW, Huang SJ, Lin RC (2007b) The use of gene-expression profiling to identify candidate genes in human sepsis. Am J Respir Crit Care Med 176:676–684

Tao H, Gonzalez R, Martinez A, Rodriguez M, Ingram LO, Preston JF, Shanmugam KT (2001) Engineering a homoethanol pathway in *Escherichia coli*: increased glycolytic flux and levels of expression of glycolytic genes during xylose fermentation. J Bacteriol 183:2979–2988

Taylor JD, Briley D, Nguyen Q, Long K, Iannone MA, Li MS, Ye F, Afshari A, Lai E, Wagner M, Chen J, Weiner MP (2001) Flow cytometric platform for high-throughput single nucleotide polymorphism analysis. Biotechniques 30(661–666):668–669

Tomiuk S, Hofmann K (2001) Microarray probe selection strategies. Brief Bioinform 2:329–340

Troesch A, Nguyen H, Miyada CG, Desvarenne S, Gingeras TR, Kaplan PM, Cros P, Mabilat C (1999) *Mycobacterium* species identification and rifampin resistance testing with high-density DNA probe arrays. J Clin Microbiol 37:49–55

Uehara T, Kushida A, Momota Y (1999) Rapid and sensitive identification of *Pratylenchus* spp. using reverse dot blot hybridization. Nematology 1:549–555

Vacha SJ (2003) Ten pitfalls of microarray analysis. Technical note 72 dna. Agilent Technologies, Inc. http://www.chem.agilent.com/scripts/LiteraturePDF.asp?iWHID=32461

Wang R, Guegler K, LaBrie ST, Crawford NM (2000) Genomic analysis of a nutrient response in Arabidopsis reveals diverse expression patterns and novel metabolic and potential regulatory genes induced by nitrate. Plant Cell 12:1491–1509

Wang WW, Spurdle AB, Kolachana P, Bove B, Modan B, Ebbers SM, Suthers G, Tucker MA, Kaufman DJ, Doody MM et al (2001) A single nucleotide polymorphism in the 5′ translated region of *RAD 51* and risk of cancer among *BRCA1/2* mutation carriers. Cancer Epidemiol Biomarkers Prev 10:955–960

Wei C, Li J, Bumgarner RE (2004) Sample size for detecting differentially expressed genes in microarray experiments. BMC Genomics 5(1):87

Weiszhausz DDD, Warrington J, Tanimoto EY, Miyada CG (2006) The Affymetrix GeneChip platform: an overview. Methods Enzymol 410:3–28

Wen WH, Bernstein L, Lescallett J, Beazer-Barclay Y, Sullivan-Halley J, White M, Press MF (2000) Comparison of *TP53* mutations identified by oligonucleotide microarray and conventional DNA sequence analysis. Cancer Res 60:2716–2722

Wilson M, DeRisi J, Kristensen HH, Imboden P, Rane S, Brown PO, Schoolnik GK (1999) Exploring drug-induced alterations in gene expression in *Mycobacterium tuberculosis* by microarray hybridization. Proc Natl Acad Sci USA 96:12833–12838

Wisman E, Ohlrogge J (2000) Arabidopsis microarray service facilities. Plant Physiol 124:1468–1471

Wouters L, Gõhlmann HW, Bijnens L, Kass SU, Molenberghs G, Lewi PJ (2003) Graphical exploration of gene expression data: a comparative study of three multivariate methods. Biometrics 59(4): 1131–1139

Ye RW, Wang T, Bedzyk L, Croker KM (2001) Applications of DNA microarrays in microbial systems. J Microbiol Methods 47:257–272

Zhao R, Gish K, Murphy M, Yin Y, Notterman D, Hoffman WH, Tom E, Mack DH, Levine AJ (2000) Analysis of *p53*-regulated gene expression patterns using oligonucleotide arrays. Genes Dev 14:981–993

Zhu T, Wang X (2000) Large-scale profiling of the *Arabidopsis* transcriptome. Plant Physiol 124: 1472–1476

7 Metagenomics: The Exploration of Unculturable Microbial World

G. K. Joshi, J. Jugran, and J. P. Bhatt

Abstract

Metagenomics refers to the direct extraction and analysis of the DNA from an assemblage of microorganisms for accessing genomic wealth of microorganisms which cannot be cultivated in the laboratory by conventional methods. Given the fact that only 1 % of total microorganisms of an environmental sample are actually culturable, advancement in metagenomic studies has far increased our access to the genes of microbial community that remained uncultured so far. In today's world where products of microbial origin have proved their utility in almost every sphere of life, metagenomic studies have become highly important as it gives a clue to the hidden wealth of microbial world. This chapter deals with various aspects of metagenomics.

7.1 Introduction

Studies in microbiology could never be possible without a microscope. So, the event of a human being (Antonie van Leeuwenhoek) seeing first time a bacterium in the year 1663 is highly significant. Equally appreciable was the work of botanist Ferdinand Cohn, who classified many bacteria and described the life cycle of *Bacillus subtilis* based on his microscopic observations. Mycologist Franz Unger had understood the concept of pure culture as early as in 1850s. But it was in large part the emphasis on disease causality that solidified pure culture as the standard bacteriological technique for laboratory microbiology. Robert Koch's postulates and his own innovations in developing culture media were instrumental in this shift. Microbiologists were attracted to the power and precision of studies of microorganisms in pure culture and as a result, most of the knowledge that fills modern microbiology textbooks is derived from the organisms maintained in pure culture. For a long time microbiologists focused on discoveries based on readily cultured model organisms leading to eruption of knowledge in microbial physiology and genetics in the last century. Over the years microorganisms and their products

G. K. Joshi (✉) · J. Jugran · J. P. Bhatt
Department of Zoology and Biotechnology, HNB Garhwal University, Srinagar (Garhwal), Uttarakhand, India
e-mail: gkjoshi@rediffmail.com

J. Jugran
e-mail: jyotijugran28@gmail.com

J. P. Bhatt
e-mail: profjpbhatt@gmail.com

I. Ravi et al. (eds.), *Advances in Biotechnology*,
DOI: 10.1007/978-81-322-1554-7_7, © Springer India 2014

proved to be useful for mankind in a variety of ways. As a result, a number of industries were sprung up in different parts of the world to harness the benefits of extraordinary versatile metabolic capacities of these tiny creatures. In an attempt to isolate and maintain novel microorganisms, microbiologists across the world explored nearly every available ecological niche/climatic zone. However, their success was based on their ability to develop suitable media to provide growth support to the desired organisms and in this course many of actual inhabitants remained uncultured owing to a variety of causes like lack of necessary symbionts, nutrients, surfaces, excess inhibitory compounds, incorrect combinations of temperature, pressure, or atmospheric gas composition, accumulation of toxic waste products from their own metabolism, etc. There was paradigm shift in this approach as direct analysis of 5S and 16S rRNA gene sequences in the environment were used to describe the diversity of microorganisms in an environmental sample bypassing culturing steps (Pace et al. 1986). The technical difficulties in this approach were largely overcome with the development of PCR technology and primer designing methods that can be used to amplify almost the entire gene. The study of soil bacterial diversity, demonstrated with the help of DNA–DNA re-association techniques revealed that the complexity of bacterial DNA in the soil was at least 100-fold greater than that could be accounted for by traditional culturing approach. This work led to the suggestion that the diversity of uncultured world could far exceed the previous estimates.

It is now estimated that less than 1 % of the total microbes in an environment are actually cultivable through traditional culturing approach (Kimura 2006). Advances in the area of metagenomics have enabled us to have access to a far greater number of microorganisms in a habitat than any time before. Metagenomics is a combination of two separate words—*meta* and *genomics*—derived from the concept of meta-analysis (statistically combining separate analyses) and genomics (the wide ranging analysis of an organism's genetic material) as defined by Schloss and Handelsman in 2003. It can be referred to as the culture-independent analysis of microbial communities. It involves the direct isolation of genomic DNA from an environment followed by its cloning in easily cultivable microorganisms (Fig.7.1). The approach thus obviates the need for culturing the microorganism in order to have access to its genomic wealth. There are other terms used for the same approach as community genomics, environmental genomics, and population genomics. Pace et al. (1985) first proposed the thought of cloning DNA directly from environmental samples and the report for such cloning in phage vector appeared 6 years later (Schmidt et al. 1991). Over the years, following the initial report, culture-independent analysis have made it apparent that a large number of 'yet to be cultivated' bacteria belong to phylogentically distinct genotypic classes.

7.2 Steps in Metagenomic Analysis

7.2.1 DNA Isolation

In general, metagenomic DNA isolation strategy should result in representation from a broad range of organisms. DNA isolation is, therefore, the most important step for the subsequent downstream genome analysis. Shearing of DNA should be avoided during extraction as long intact fragments are required for community analysis. Further, the DNA should be free from contaminating elements for easy downstream studies. It is difficult to develop general protocol for DNA isolation as the environmental sample may contain enormous group of microorganisms belonging to different genotypes, classes, and divisions having marked difference in their cell wall structure. In addition to that, problem for recovery of DNA straight from environmental sample lies in the fact that cells are in a physiological state that is hard to lyse. Designing of a unique ideal DNA isolation protocol for recovery from every microbial cell in the sample is, therefore, extremely difficult. Despite every effort certain groups are represented more in terms of the isolated genomic material compared to others. However, in certain cases a genomic biasness is deliberately introduced where the aim is to collect DNA from certain group of bacteria with some desirable properties. There are two main strategies for extraction of metagenomic DNA viz. 'direct method' involving *in situ* cell lysis and 'indirect method' where cells are first separated from the sample and then lysed. Both methods are equally good as far as the DNA analysis for determination of diversity spectrum is concerned. However, the former method results in more amount of DNA in an impure form. In case of soil the problem of contaminating agents such as humic acid may interfere with the process of DNA isolation. On the other hand, a large amount of water has to be processed in order to have a significant yield of DNA from aquatic samples.

Vigorous cell lysis methods are generally employed considering the equal representation of each community member from the diverse population in an environmental sample. This may, however, lead to the shearing of genomic fragments and thus making the intact gene recovery difficult. Also, there is an increased risk of chimera formation from small template DNA during subsequent PCR. Extraction of total metagenomic DNA is, therefore, a compromise between the vigorous extraction and quality of DNA required for the representation of all microbial genomes. Other physical methods for DNA isolation include freeze-thawing, ultrasonication, etc. Chemical methods, on the other hand, are preferred when the aim is to select certain taxa by exploiting their unique biochemical properties. SDS-based cell lysis is the most common type of method for release of intracellular DNA. Mechanical method is shown to recover more diversity than the chemical method. A combination of different physical and chemical methods can also be used depending on the soil characteristics and associated microbial diversity.

Since not all members of a community are evenly represented in the extracted genome, experimental normalization is sometimes necessary as a partial solution to it. One such approach involves density gradient separation of genotypes in the presence of an intercalating agent. This leads to the separation of genomes based on their % G and C content as distinct bands in centrifugation tube. A normalized metagenome is then obtained by combining equal amount of each band. Another method for normalization for rare sequences involve denaturing metagenomic DNA followed by re-annealing under stringent conditions. Compared to rare sequences, abundant single-stranded DNA will anneal rapidly and acquire double-stranded form. Single-stranded sequences can then be separated from the double-stranded nucleic acids.

A precultivation step improves the quality of the community DNA. This, of course, reduces the chance of representation of maximum diversity of the original environmental sample and enhances the possibilities of recovering genomes from microorganisms with desired trait. The important strategies that have been employed for enriching the metagenome for sequences of interest before cloning, are summarized in Table 7.1. Other latest methods for enrichment are based on microarray, suppressive subtractive hybridization, differential expression analysis, and multiple displacement amplification.

7.2.2 Library Construction and Choice of Vector

Metagenomic library has been created in almost all types of commonly used gene cloning vectors such as plasmid, phage, cosmid, fosmid, bacterial artificial chromosome (BAC), etc. During the early studies plasmid-based vectors were primarily used for metagenomic library with less than 10 kb insert size. Large capacity vectors are now fast replacing the plasmid-based vectors. The choice, however, depends on the overall aim of the study. Small insert library is suitable for cloning of individual gene. These libraries are useful for functional screening of clones expressing various enzymes like chitinase, lipase, esterase, amylase, nitrilase, etc. It also provides large amount of sequencing

Table 7.1 Metagenomic library enrichment methods

Enrichment method	Principle	References
Based on GC content	Enrichment for the sequences of interest by its separation from the bulk metagenomic DNA based on the variation in GC contents through ultracentrifugation.	Schloss and Handelsman (2005)
BrdU enrichment	Incorporation of bromodeoxyuridine (BrdU) in the soil leading to its incorporation into the DNA of metabolically active subset of the community. The labeled DNA can then be separated by immunocapture technique	Urbach et al. (1999)
Stable isotope probing (SIP)	C^{13} labeled substrate is provided to the soil bacteria and those capable of metabolizing the substrate incorporate the isotope into their DNA. The heavier DNA due to C^{13} is separated from the bulk by density gradient centrifugation	Radajewski and Murrell (2002)
Culture enrichment	The sample can be enriched for desired microbes by varying nutritional, physical, and chemical conditions during precultivation step	Singh et al. (2009)

information. The main drawback with small capacity vectors is the huge number of clones in the library which are sometimes very difficult to handle. On the other hand, large insert library is required when gene cluster or operons are to be isolated and studied. Cosmid or fosmid vectors have been used to create library with an insert size of about 40 kb, whereas BAC vectors can markedly reduce the size of the metgenomic library by cloning up to 150 kb. The recovery of high molecular weight DNA is, however, a prerequisite for using the high capacity vectors, which is sometimes very difficult to achieve as the soil compositions demand vigorous isolation strategies leading to extensive fragmentation of DNA. *Escherichia coli* is the most preferred choice of host. Other hosts such as *Streptomyces lividans*, *Pseudomonas putida*, and *Rhizobium* sp. have also been successfully used.

7.2.3 Screening Strategies

7.2.3.1 Function Based

In a metagenomic library clones with useful traits can be traced by mainly two methods viz. function-based screening and sequence-based screening. The functional screening depends on the faithful expression of the cloned gene in a heterologous host. Since no pre-sequence analysis of the gene of interest is required, there is very high chance to identify entirely new classes of genes for new or known functions. The expression of the gene may confer some selective advantage to the desired clone making its visual identification possible in a pool of clones. For example, antibiotic resistance genes or enzyme producing genes can be selected on specific plates (Fig. 7.2). However, very few clones in a metagenomic library express any given activity. Many thousand clones need to be screened before getting one with desirable trait. In the recent years highly efficient screening and selection systems have been developed for selecting positive clones. Uchiyama et al. (2005) have developed Substrate-induced gene expression screening (SIGEX) for rapid identification of clones showing catabolic gene expression. The technique is based on the fact that catabolite-gene expression requires certain substrates as inducers which, in many cases, is controlled by regulatable elements situated near to catabolic genes. A plasmid-based operon-trap expression vector (p18GFP) containing gene for green fluorescent protein (GFP) immediately after the insert site, is constructed and used for creating metagenomic library in this technique. The clones are exposed to an inducer which easily diffuses to the cytoplasm of all transformed cells in the library. In positive clones (in which the entire operon responding to that

Fig. 7.1 Steps in metagenomic analysis

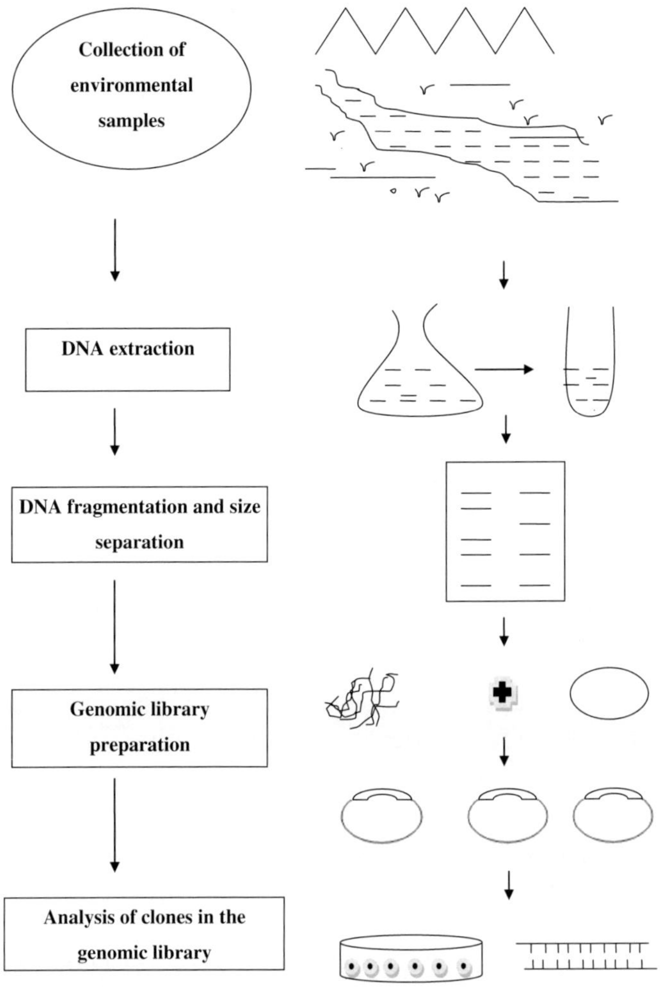

substrate/inducer is cloned), the inducer binds to the cloned regulatory elements and causes expression of downstream sequences for GFP. The GFP positive cells can then be separated by fluorescence activated cell sorter (FACS). Another similar approach is METREX in which product of metagenomic DNA induces bacterial quorum sensing. The metagenomic library is created in host cells which already contain a reporter plasmid. The reporter plasmid contains a reporter gene such as *gfp* under the control of a regulatory promoter. The promoter can be upregulated in the presence of the product whose gene is present only in the positive clones in the library. The positive clones thus express *gfp* and can be identified and separated from other clones in the library.

7.2.3.2 Sequence Based

This strategy is based on the prior information of target gene sequence. The classical approach of hybridization with labeled probe can be used to identify the positive clones in the metagenomic library (Fig. 7.3). Complete sequencing of clones that contains phylogenetic anchors indicates the taxonomic group of the source of DNA. Random sequencing can also be done to identify a gene of interest followed by assessment of flanking DNA to contain any phylogenetic anchor. The powerful approach of sequence analysis guided by the identification of phylogenetic markers produced the first genomic sequence linked to a 16S rRNA gene that affiliated with γ–*Proteobacteria* (Handelsman 2004). Sequencing random clones is an alternative to phylogenetic marker-based

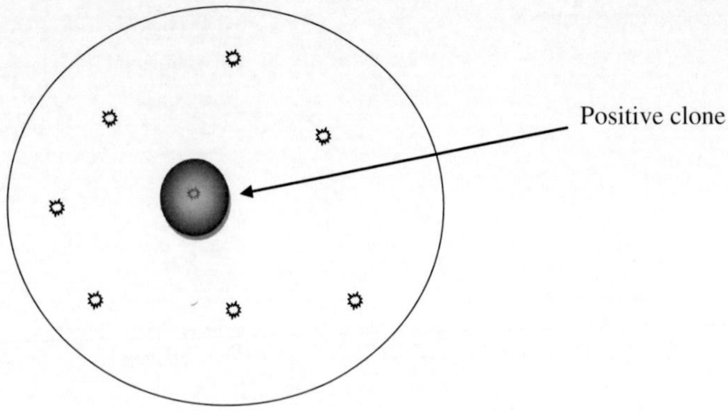

Fig. 7.2 Function-based visual identification of positive clone on solid medium

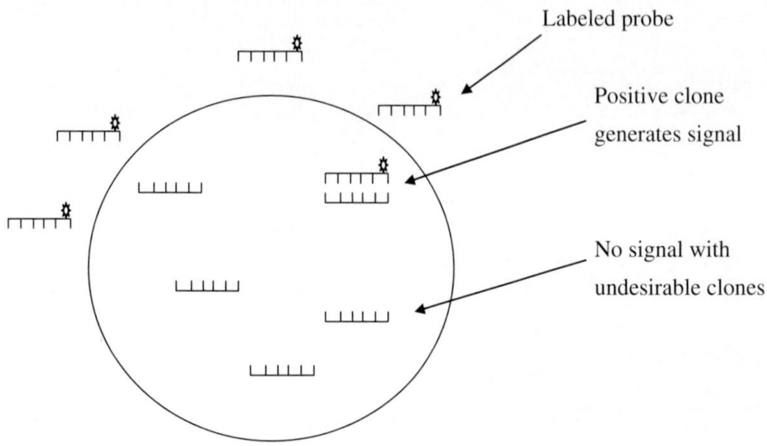

Fig. 7.3 Sequence-based identification of positive clone

approach. The field of metagenomics has been transformed by the application of a whole genome shotgun sequencing approach during the last 6–7 years. Next-generation ultra high-throughput sequencing techniques have revolutionized the field with heavy price cut allowing sequencing at scale larger then anytime before. Sequencing projects have been carried out to assess the actual microbial diversity and their ecological inference in different environment such as sea and acid mine drainage, etc. In the past few years, techniques of DNA microarray has also been widely used for screening clones in metagenomic library.

The technique of PCR has been applied to the isolation and detection of novel genes from community genome. However, the major limitation with PCR is to rely on the flanking DNA sequence information for isolation of an unknown gene from an uncultured organism. Furthermore, when PCR is directed at conserved sequence motifs, only partial genes are recovered. In an attempt to solve this problem, a novel strategy for recovering complete open reading frame from environmental DNA samples was proposed by Stokes in 2001. His team designed PCR assays targeted toward the 59-base element family of recombination sites that flank integrons associated gene cassettes. This approach has resulted in the amplification of diverse gene cassettes containing complete open reading frame from the environmental DNA.

7.3 Applications of Metagenomics

Metagenomic studies are important and have a range of applications for both academia and industries. The recent findings in metagenomics suggest the role of microorganism in influencing human life far beyond the previous expectations. The important applications and outcome from these studies are summarized below.

7.3.1 Ecological Inferences from Microbial Diversity Estimation

In microbial ecology it is important to know how microorganisms acquire nutrients and produce energy, form symbioses and compete, and communicate with other members of the community. Metagenomic investigations have resulted in the identification of various novel life forms in geographically distinct region and attempts are still underway to describe their possible role in that environment. As defined below metagenomic findings have resulted in restructuring of our understanding of the ecological balance in most of the environments.

7.3.1.1 Community Structure

Genomic information of the 'unculturable' bacteria can help in understanding their physiology and also their role in ecosystems. Metagenomic investigation has revealed that bacteriorhodopsins that function as light- driven proton pump for energy generation in halophilic archaea are also present in eubacteria. Presence of rhodopsin in marine proteobacteria suggested the possibility of a new phototrophic pathway that may influence the flux of carbon and energy in the ocean's photozone worldwide. By utilizing shotgun sequencing approach complete genomes of *Leptospirillum* group II and *Ferroplasma* type II were assembled from a natural acidophilic biofilm. Pathways of carbon and nitrogen fixation and energy generation for every organism of the simple biofilm community were established. The almost complete assembly of the genome of an uncultured bacterium, *Kuenenia stuttgartiensis*, has revealed unique metabolic adaptations associated with anaerobic ammonium oxidation. PCR-DGGE (Denaturing gradient gel electrophoresis) is a sophisticated method that enables us to visualize a complete microbial community as a simple banding pattern and also allows to easily monitor community dynamics within the environment. In addition to usual 16S rRNA, various functional genes have also been targeted to investigate the diversity and community structure in this approach. Sargasso Sea genome project carried out by Venter et al. (2004) has resulted in the identification of genes for transport of phosphonates and utilization of polyphosphates and pyrophosphates. The finding indicates the microbial community's capability to survive in the extremely phosphate-limited environment of Sargasoo Sea.

7.3.1.2 Diversity Analysis

The present barriers in estimation of microbial biodiversity are largely overcome by the developments in the area of metagenomics. A large proportion of uncultivated bacteria belonging to new genotypes, classes, and divisions in the domains eubacteria and archaea have been discovered so far through this approach. Bacteria those were previously unknown in most well-characterized samples like dental plaques, sea water, and garden soil have been identified. Unlike previous assumptions presence of archaea in habitats like soil, seawater, etc., has been noticed. Sargasso Sea genome project has led to identification of 1.2 million previously unknown genes and 148 previously unknown phylotypes (Venter et al. 2004). Metagenomic studies have also revealed important information about the viral diversity of the human and marine environments. Breitbart et al. (2002) after studying genomic analysis of two uncultured marine viral communities reported that over 65 % of the sequences were not significantly similar to previously reported sequences. The most common significant hits among the known sequences were found belonging to viruses. In the recent years, phage diversity has also been revealed by metagenomic technology.

7.3.1.3 Symbiosis

Many of the bacteria which have intimate association with their eukaryotic hosts particularly in the marine environment cannot be cultivated in the laboratory. Metagenomics provides a way to reconstruct their genome, thus leading insight into the evolution of the symbiosis between the host and the bacterium. The first genome reconstruction of an uncultured organism was that of *Buchnera aphidicola*, the endosymbiont of aphids (Handelsman 2004). Since then various other bacterial partners of different eukaryotic hosts have been studied through metagenomic approach. The study becomes important in view of the fact that a number of biomedically relevant natural products like anticancer agents, anti-infective agents of other bioactivities are obtained from invertebrate animals, e.g., sponges, tunicates, and bryozoans and there is increasing evidence that many of these metabolites are produced by these animals in association with their bacterial symbionts.

7.3.1.4 Paleogenomics

Another application of metagenomics has been suggested in the area of paleogenomics where it has been useful for resolving phylogenetic relationships between extinct and modern animals. PCR-based amplification of mitochondrial DNA strategies were also used to set the missing link. It is assumed that over the years the ancient animal DNA got mixed with the genomes of the abundant opportunistic microbes. A metagenomic approach has been defined to obtain cave bear sequence and homology search with PCR amplified genes from black, brown, and polar bears verified the origin and a phylogenetic tree was constructed (Singh et al. 2009). The results of such novel strategies suggest the active need for initiatives to take up genome projects targeting extinct species to revolutionize the area of paleobiology.

7.3.2 Biotechnological Prospects of Metagenomics

The usefulness of the microbes has been harnessed by mankind in the form of billion dollar industries utilizing these tiny organisms for the production of a variety of products. The advances in metagenomics have revealed that what we know today is just a tip of the iceberg. Many more novel gene(s) and their products are yet to be discovered in near future. Metagenomics provides a way to peep into this hidden wealth of microbial world. Following are some of the areas important from biotechnological point of view, where metagenomics can revolutionize the existing trends for industrial production of metabolites.

7.3.2.1 Bioactive Compounds

Natural products have been a productive source of lead structures for designing and developing new antimicrobial substances. Metagenomics provides a way for cloning the biosynthetic gene clusters of bioactive natural products, which may in turn provide a renewable source of compounds that have often been difficult to isolate in sufficient quantities to permit extensive biological testing. Pederin, an anticancer agent originally isolated from the beetle *Paederus fuscipes* is synthesized by cooperative activity of many genes. The biosynthetic gene cluster for it was recovered from a beetle derived metagenomic DNA library that was later found to originate from an uncultured symbiotic Pseudomonad. Clones that produce proteins with potential anti-infective properties have also been identified through metagenomics. A range of novel antibiotics have been detected in metagenomic libraries. Brady et al. (2001) have reported the production of deoxyviolacein and a broad spectrum antibiotic violacein from a soil metagenomic library. In an investigation two-colored triaryl cation antibiotics, i.e., Turbomycin A and Turbomycin B and their genes have been isolated from a soil metageomic library (Gillespie et al. 2002). Genes for vitamins like biotin and vitamin C have also been detected in metagenomic library.

7.3.2.2 Antibiotic Resistance Gene

Metagenomic libraries have also been explored for isolation of antibiotic resistance genes from environmental samples. This may give a major insight into the mechanism of microbial drug resistance. In today's world where development of antibiotic resistance by pathogens has become a major concern for health practitioners and pharmaceutical industries, metagenomic studies have given a vital clue in terms of the discovery and identification of novel antibiotic resistance genes. Identification of 9 clones expressing resistance to aminoglycoside antibiotics and one expressing resistance to tetracycline from a soil metagenomic library has been reported by Riesenfeld et al. (2004). Environment like human oral cavity has been found to be a source of tetracycline, amoxicillin, and gentamycin resistance genes. The search for novel anti-microbial genes and compounds is probably the field that is advancing most rapidly in metagenomics.

7.3.2.3 Novel Enzymes

A wide range and a remarkable number of biocatalysts have been discovered from unculturable microbes. This may be attributed to the fact that gene expressed in nature represents proteins (here enzymes) which presumably, through the evolutionary process has been undergoing hard and long selection pressure. The pressure thus exerted drives the gene evolution toward enzyme production which is most fit for solving interaction with the substrate needed for a given organism in an ecological niche it is adapted to inhabit. The single most biotechnological application of metagenomic investigations having the largest number of publications worldwide so far can be seen in the area of biocatalysts. There are more than 100 publications for a single class of enzyme like lipase production through metagenomic clones. In a similar manner, a vast number of reports worldwide indicate the possibility of identification and isolation of novel enzymes and their genes through metagenomic approach. There are several reports of isolation of novel variants for majority of known industrial enzymes like lipase, protease, amylase, chitinase, cellulase, xylanase, agarase, nitrilase, etc., from metagenomic library created from different types of soil or aquatic samples. Table 7.2 enlists some of the latest publications in just last 3 years reporting novel enzyme identification through metagenomic library.

7.4 Role of Bioinformatics in Development of Metagenomics

Developments in the area of bioinformatics have a marked role in establishing the approaches of metagenomics to the current level. Many promising advances in terms of correct analysis and interpretation of vast data generated by metagenomic studies across the globe have come from the bioinformatics people. Specific computational and statistical tools have been designed for metagenomic data analysis and comparison. Programs like MetaGene and Orphelia are available for gene prediction for short reads of around 700 bp. MEGAN, MLTreeMap, AMPHORA, and CARMA are similarity-based and phylogeny-based phylotyping tools that utilize similarity searches of the metagenomic sequence against a database of known genes/proteins. ARB is an interactive software tool for sequence database (RNA/DNA/amino acids). MG-RAST is an automated server that facilitates automatic phylogenetic and functional analysis of genome. PHACCS is an online tool for estimating the structure and diversity of uncultured viral communities using metagenomic information. AMOVA and HOMOVA are statistical tools for genetic diversity estimation among members of different communities in a metagenome. A tool like DOTUR has been developed for analysis of community membership and structure comparison. It is used to determine whether a metagenomic library contains enough number of genes for it to be considered representatives of the diversity in the original microbial community.

Table 7.2 Metagenomic discovery of novel enzymatic genes in recent years

S.No.	Enzyme name	Source	References
1.	Esterase	Sheep rumen	Bayer et al. (2010)
2.	Alginate lyase	Gut microflora of abalone	Sim et al. (2012)
3.	Cellulase	Soil	Liu et al. (2011)
4.	Amylase	Soil	Sharma et al. (2010)
5.	α-amylase	Soil	Vidya et al. (2011)
6.	Protease	Desert land	Neveu et al. (2011)
7.	Monooxygenase	Industrial effluent sludge	Singh et al. (2010)
8.	Laccase	Soil	Ye et al. (2011)
9.	Chitinase	Marine sediment	Bhuiyan et al. (2011)
10.	Nitrilase	Oil contaminated soil	Bayer et al. (2011)
11.	Dichlorohydroxylase	Pesticide contaminated soil	Lu et al. (2011)
12.	Lipase	Mangrove sediment	Couto et al. (2010)

7.5 Conclusion and Future Prospects

Metagenomics is a study of uncultivated microorganisms in the environment. It is expected that the number of novel genes identified using this approach will soon exceed the number identified through sequencing of novel genes from isolated individual microbes. It has potential to provide insight into the functional dimensions of environmental genomic datasets. Although the outcomes of metagenomic studies worldwide have refined some of our current understanding of microbial diversity and its possible interaction with the nature, it is too early to describe it as 'come of age' technology. Finding a useful gene through functional screening is still like searching a needle in a haystack as thousands of clones are to be screened in order to find a positive one. Every year new approaches and technical innovations in this area are reported but many of the technical difficulties are yet to be fully resolved. Better designing of vectors and perfect host are required to overcome this problem. Similarly, steps of gene enrichment, DNA extraction, library construction, and phylogenetic affiliation of isolated genes need to be improved to yield optimum results. The upgradation and sophistication of sequencing techniques are bound to revolutionize this area by enabling scientists to take up Gb-scale metagenomic projects for sequencing complex communities to (nearly) saturation. However, much deeper sampling methods along with special assembly, analysis tools, and data storage infrastructure are required before such projects are initiated. Metatranscriptomics and metaproteomics are the complementary approaches that are seen with great expectation to provide promising results in near future.

References

Bayer S, Kunert A, Ballschmiter M, Greiner-Stoeffele T (2010) Indication for a new lipolytic enzyme family: isolation and characterization of Two esterases from a metagenomic library. J Mol Micobiol Biotechnol 18(3):181–187

Bayer S, Birkemeyer C, Ballschmiter M (2011) A nitrilase from a metagenomic library acts regioselectively on aliphatic dinitriles. Appl Microbiol Biotechnol 89(1):91–98

Bhuiyan FA, Nagata S, Ohnishi K (2011) Novel chitinase genes from metagenomic DNA prepared from marine sediments in southwest Japan. Pak J Biol Sci 14(3):204–211

Brady SF, Chao CJ, Handelsman J, Clardy J (2001) Cloning and heterologous expression of a natural product biosynthetic gene cluster from cDNA. Org Lett 3:1981–1984

Breitbart M, Salmon P, Andresen B, Mahaffy JM, Segall AM, Mead D, Azam F, Rohwer F (2002) Genomic analysis of uncultured marine viral communities. Proc Natl Acad Sci USA 99:14250–14255

Couto GH, Glogauer A, Faoro H, Chubatsu LS, Souza EM, Pedros FO (2010) Isolation of a novel lipase from a metagenomic library derived from mangrove sediment from the south Brazilian coast. Genet Mol Res 9(1):514–523

Gillespie DE, Brady SF, Bettermann AD, Cianciotto NP, Liles MR, Rondon MR, Clardy J, Goodman RM, Handelsman J (2002) Isolation of antibiotics Turbomycin A and B from a metagenomic library of soil microbial DNA. Appl Envion Micorobol 68(9): 4301–4306

Handelsman J (2004) Metagenomics: application of genomics to uncultured microorganisms. Microbiol Mol Bio Rev 68(4):669–685

Kimura N (2006) Metagenomics: access to unculturable microbes in the environment. Microbes Environ 21(4):201–215

Liu W, Shen W, Cao H, Cui Z (2011) Cloning and functional characterization of a novel endo-β-1,4-glucanase gene from a soil-derived metagenomic library. Appl Mirobiol Biotechnol 89(4):1083–1092

Lu Y, Yu Y, Zhou R, Sun W, Dai C, Wan P, Zhang L, Hao D, Ren H (2011) Cloning and characterization of a novel 2,4-dichlorophenol hydroxylase from a metagenomic library derived from polychlorinated biphenyl-contaminated soil. Biotechnol Lett 33:1159–1167

Neveu J, Regard C, Dubow MS (2011) Isolation and characterization of two serine proteases from metagenomic libraries of the Gobi and Death Valley deserts. Appl Microbiol Biotechnol 91(3):635–644

Pace NR, Stahl DA, Lane DJ, Olsen GJ (1985) Analyzing natural microbial populations by rRNA sequences. ASM News 51:4–12

Pace NR, Stahl DA, Lane DJ, Olsen GJ (1986) The analysis of natural microbial populations by ribosomal RNA sequences. Adv Microb Ecol 9:1–55

Radajewski S, Murrell JC (2002) Stable isotope probing for detection of methanotrophs after enrichment with $^{13}CH_4$. Methods Mol Biol 179:149–157

Riesenfeld CS, Goodman RM, Handlesman J (2004) Uncultured soil bacteria are a new reservoir of new antibiotic resistance gene. Environ Microbiol 6:981–989

Schloss PD, Handelsman J (2003) Biotechnological prospects from metagenomics. Curr Opn Biotechnol 14:303–310

Schloss PD, Handelsman J (2005) Introducing DOTUR, a computer program for defining operational taxonomic units and estimating species richness. Appl Environ Microbiol 71(3):1501–1506

Schmidt T, DeLong EF, Pace NR (1991) Analysis of a marine picoplankton community by 16S rRNA gene cloning and sequencing. J Bacteriol 173:4371–4378

Sharma S, Khan FG, Qazi GN (2010) Molecular cloning and characterization of amylase from soil metagenomic library derived from Northwestern Himalayas. Appl Microbiol Biotechnol 86:1821–1886

Sim SJ, Baik KS, Park SC, Choe HN, Seong CN, Shin TS, Woo HC, Cho JY, Kim D (2012) Characterization of alginate lyase gene using a metagenomic library constructed from the gut microflora of abalone. J Industrial Microbial Biotechnol 39(4):585–593. doi:10.1007/s10295-011-1054-0

Singh J, Behal A, Singla N, Joshi A, Birbian N, Singh S, Bali V, Batra N (2009) Metagenomics: concept, methodology, ecological inference and recent advances. Biotechnol J 4:480–494

Singh A, Chauhan N, Thulasiram HV, Taneja V, Sharma R (2010) Identification of two flavin monooxygenases from an effluent treatment plant sludge metagenomic library. Biores Technol 101(21):8481–8484

Stokes HW, Holmes AJ, Nield BS, Holley MP, Nevalainen KMH, Mabbutt BC, Gillings MR (2001) Gene: cassette PCR: sequence independent recovery of entire genes from environmental DNA. Appl Environ Microbiol 67:5240–5246

Uchiyama T, Abe T, Ikemura T, Watanabe K (2005) Substrate induced gene expression screening of environmental metagenomic libraries for isolation of catabolic genes. Nature Biotechnol 23(1):88–93

Urbach E, Vergin KL, Giovannoni SJ (1999) Immunochemical detection and isolation of DNA from metabolically active bacteria. Appl Environ Microbiol 65:1207–1213

Venter JC, Remington K, Heidelberg JF, Halpern AL, Rusch D, Eisen JA, Wu D, Paulsen I, Nelson KE, Nelson W, Fouts DE, Levy S, Knap AH, Lomas MW, Nealson K, White O, Peterson J, Hoffman J, Parsons R, Tillson HB, Pfannkoch C, Rogers YH, Smith HO (2004) Environmental genome shotgun sequencing of the Sargasso sea. Science 304:66–74

Vidya J, Swaroop S, Singh SK, Alex D, Sukumaran RK, Pandey A (2011) Isolation and characterization of a novel α-amylase from a metagenomic library of Western Ghats of Kerala, India. Biologia 66(6): 939–944

Ye M, Li G, Qu W, Liu YH (2011) Molecular cloning and characterization of a novel metagenome-derived multicopper oxidase with alkaline laccase activity and highly soluble expression. Appl Microbiol Biotechnol 87(3):1023–1031

8 Proteomics

Indu Ravi

Abstract

Proteins are macromolecules and the main components of the physiological metabolic pathways of cells. The post-genomic era focuses on the complete understanding of the structure and functions of proteins. Proteome, the protein complement of a given cell at a specific time is the main focus of proteomics. Proteomics has recently been of interest to scientists because it gives a much better understanding of an organism than genomics. A number of advanced techniques such as 2D-gel electrophoresis, HPLC, Mass spectrometry, MALDI-TOF, NMR, DNA microarray, etc., are used for the study of proteomics. The most important applications of proteomics are in the discovery of biomarkers and drug targets. It will also be useful in diseases which involve multigenes, e.g., cancer, diabetes II, asthma, hypertension, cardiovascular and neurological diseases. This chapter focuses on the basic steps involved, techniques used, and applications of proteomics.

8.1 Introduction

Within a given human proteome, the number of proteins can be as large as 2 million. Proteins themselves are macromolecules, i.e., long chains of amino acids which are constructed when the cellular machinery of the ribosome translates RNA transcripts from DNA in the cell's nucleus. The transfer of information within cells is commonly from DNA to RNA to protein (Fig. 8.1).

Proteins perform vital functions in the body such as:

- Catalyzing various biochemical reactions through enzymes.
- Acting as messengers through neuro transmitters.
- Acting as control elements which regulate cell reproduction.
- Influencing growth and development of various tissues through trophic factors.
- Transporting oxygen in the blood through hemoglobin.
- Defending the body against diseases through antibodies.

The proteome changes constantly in response to tens of thousands of intra- and extracellular environmental signals. The proteome varies with

I. Ravi (✉)
Indira Gandhi National Open University (IGNOU), Regional Centre Jaipur, 70/79, Patel Marg, Mansarovar, Jaipur, Rajasthan 302020, India
e-mail: induravi11@yahoo.com

I. Ravi et al. (eds.), *Advances in Biotechnology*,
DOI: 10.1007/978-81-322-1554-7_8, © Springer India 2014

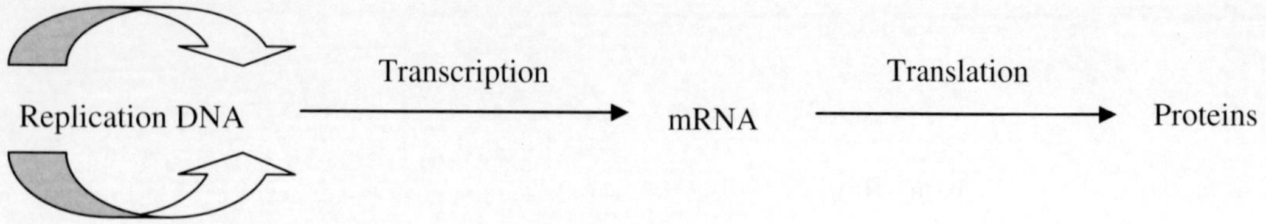

Fig. 8.1 Information transfer in the central dogma of biology

health or disease, the nature of each tissue, the stage of cell development and effects of drug treatments. As such, the proteome often is defined as "the proteins present in one sample (tissue, organism, cell culture) at a certain point in time."

The sequencing of the human genome has provided comprehensive resources for genomic data. In addition, the numerous fields such as transcriptomics, proteomics, and metabolomics, etc., and systems biology are providing new ways to study living processes. Genomics provides an overview of the complete set of genetic instructions provided by the DNA, while transcriptomics looks into gene expression patterns. Proteomics studies dynamic protein products and their interactions, while metabolomics is also an intermediate step in understanding organism's entire metabolism (Fig. 8.2).

Proteomics attempts to study the structure, function, and control of biological systems and processes by systematic and quantitative analysis of proteins. The term "proteomics" was first coined in 1997 to make an analogy with genomics, the study of the genes. The word "proteome" is a blend of "protein" and "genome", and was coined by Marc Wilkins in 1994 while working on the concept as a Ph.D. student. Different definitions of proteomics have been given by different workers from time to time as, the study of full set of proteins encoded by the genome or the study of all the proteins expressed in a cell.

As the genome describes the genetic content of an organism, a proteome defines the protein complement of the genome. The proteome is dynamic, and is the set of proteins expressed in a specific cell, given a particular set of conditions. Proteomics begins with the functionally modified protein and works back to the gene responsible for its production.

8.1.1 Structure of Proteins

Proteins can be organized in four structural levels:

1. Primary—The amino acid sequence, containing members of 20 amino acids.
2. Secondary—Local folding of the amino acid sequence into α helices and β sheets.
3. Tertiary—3D conformation of the entire amino acid sequence.
4. Quaternary—Interaction between multiple small peptides or protein subunits to create a large unit.

Each level of protein structure is essential to the biological function of the protein. The primary

Fig. 8.2 Relationship between genomics, transcriptomics, proteomics, and metabolomics

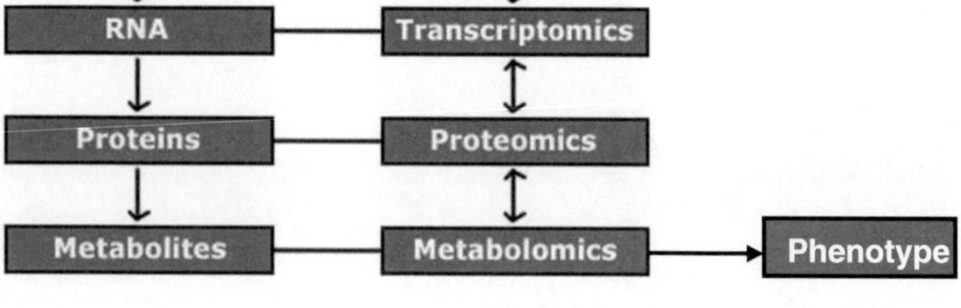

Table 8.1 Broad-based proteomics approach versus traditional focused approach

	Broad-based approach	Focussed approach
Goal	Understand the proteome as a whole	Understand specific protein function
Basic steps	1. Identify organism	1. Identify protein
	2. Understand sample type and preparation	2. Understand sample type and preparation
	3. Utilize analytical technology compatible with sample type	3. Isolate protein
	4. Bioinformatic analysis of the proteome sample	4. Utilize analytical technology that is compatible
	5. Build a proteomic model	1. Bioinformatic analysis of the protein sample
		2. Model protein's function and/or structure
Pros	1. Proteomic information about a specific tissue under certain conditions can be gained	1. Inexpensive and results can be generated much quicker than a large proteomic study
	2. Relationships between many proteins can be understood	1. Functional and structural information about a protein can be determined
Cons	1. Extensive upfront planning	1. Protein information may not be valuable without a global proteomic understanding
	2. The study will cost more and last longer than a focused study	2. Hard to develop global relationships from many focused experiments taken together
	3. No guarantee that quality proteomic data will be generated in the end	
Common technologies implemented	1. SDS-PAGE	1. SDS-PAGE
	2. 2DE-DIGE	2. HPLC
	3. HPLC	3. Mass spectroscopy (MS)
	4. Mass spectroscopy (MS)	4. Molecular modeling tools (bioinformatics)
	5. Proteomic modeling tools (bioinformatics)	

sequence of the amino acid chain determines where secondary structures will form, as well as the overall shape of the final 3D conformation. The 3D conformation of each small peptide or subunit determines the final structure and function of the protein (Garret and Grisham 1995).

The proteome is the entire complement of proteins. It is now known that mRNA is not always translated into protein, and the amount of protein produced for a given amount of mRNA depends on the gene it is transcribed from and on the current physiological state of the cell. Proteomics confirms the presence of the protein and provides a direct measure of the quantity present. The level of transcription of a gene gives only a rough estimate of its level of expression into a protein. A mRNA produced in abundance may be degraded rapidly or translated inefficiently, resulting in a small amount of protein. Many proteins experience post-translational modifications that profoundly affect their activities and many transcripts give rise to more than one protein, through alternative splicing or alternative post-translational modifications. Further, many proteins form complexes with other proteins or RNA molecules, and only function in the presence of these other molecules and finally, protein degradation rate plays an important role in protein content.

8.1.2 Broad-Based Proteomics

The first step when utilizing broad-based proteomics is to develop a hypothesis specific to the proteome being studied. For this an organism that already has a great deal of genomic information available is taken, since the genome is always a useful supplement to proteomic information. Then the technologies are chosen which should be compatible with the sample. Some proteomic methods include High-performance liquid chromatography (HPLC), Mass Spectrometry, SDS-PAGE, two-dimensional gel electrophoresis, and *in silico* protein modeling. There are many sample type, sample preparation, and analytical technology combinations which can be used for study in proteomics. Table 8.1 compares the broad-based approach with traditional focused approach in proteomics.

8.1.3 Human Proteome Organization

Human Proteome Organization (HUPO) is an international scientific organization representing and promoting proteomics through international cooperation and collaborations by fostering the development of new technologies, techniques, and training. It is an international consortium of national proteomics research associations, government researchers, academic institutions, and industry partners. HUPO, founded in June 2001, promotes the development and awareness of proteomics research, advocates proteomics researchers throughout the world, and facilitates scientific collaborations between HUPO members and initiatives, organized to gain better and complete understanding of the human proteome. The Human Proteome Organization is currently working on establishing a defined standard for data submission and annotation for the many different proteomics techniques currently used to identify and annotate proteins. The proteomics standards initiative (PSI) is a working group of HUPO. It aims to define data standards for proteomics in order to facilitate data comparison, exchange, and verification. PSI focuses on the following:

- Minimum Information about a Proteomics Experiment (MIAPE) defines the metadata that should be provided along with a proteomics experiment.
- Data Markup Languages for encoding the data and metadata.
- Ontologies for consistent annotation and representation.

8.2 Classification of Proteomics

Proteomics is mainly concerned with determining the structure, expression, localization, biochemical activity, interactions, and cellular roles of proteins. According to Graves and Haystead (2002) proteomics can be broadly classified into three types:

1. **Structural proteomics**—It is in-depth large-scale analysis of protein structure. It deals with determination of the 3D-structure of protein complexes or the proteins present in a specific cellular organelle. Protein structure comparisons can help to identify the functions of newly discovered genes. Structural analysis can also show where drugs bind to proteins and where proteins interact with each other. This is achieved using technologies such as X-ray crystallography and NMR spectroscopy.
2. **Expression proteomics**—Large-scale analysis of expression and differential expression of proteins is called as expression proteomics. This can help to identify the main proteins found in a particular sample and proteins differentially expressed in related samples, such as diseased versus healthy tissue. A protein found only in a diseased sample may represent a useful drug target or diagnostic marker. Proteins with similar expression profiles may also be functionally related. Technologies such as 2D-PAGE and mass spectrometry are used for this.
3. **Interaction proteomics**—It deals with analysis of interactions between proteins to characterize complexes and determine function. The characterization of protein–protein interactions helps to determine protein

functions and can also show how proteins assemble in larger complexes. Technologies such as affinity purification, mass spectrometry, and the yeast two-hybrid system are particularly useful for this.

8.3 Basic Steps in Proteomics

The following steps are involved in a proteomics experiment (Fig. 8.3):

1. protein isolation from a biological sample (e.g., a cell extract) following some experimental treatment.
2. fractionation of the resulting proteins (or peptides, the products of proteome digestion) by methods such as two-dimensional polyacrylamide gel electrophoresis (2D-PAGE) or liquid chromatography (LC).
3. protein or peptide detection by MS.

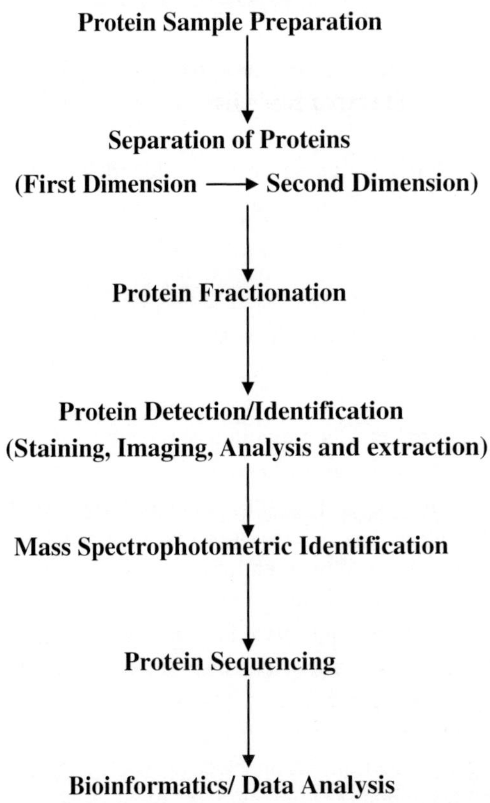

Fig. 8.3 Flow diagram of the steps involved in proteomics

4. protein identification through manual interpretation or database correlation of mass spectra.

8.3.1 Protein Sample Preparation

With technological advances in proteomics, procedures for preparation of protein samples prior to any particular procedure have also advanced. A number of issues arise in this respect, including sample clean-up, fractionation, enrichment, and also sample condition optimization. This facet of proteomics is becoming particularly critical in case of high-throughput protocols where the necessary conditions of a sample in one stage may directly conflict with the efficacy of a second stage. For example, during the initial step in 2D electrophoresis and isoelectric focusing, all proteins in a sample are given a net charge of zero; while in the second step, gel electrophoresis requires a negative charge on all products in the sample in order to induce movement through the gel matrix. Many companies, e.g., Millipore, Bio-Rad, Gelifesciences, Invitrogen, Aligent, Beckmancoulter, Bioproximity, etc., offer pre-packaged kits that allow to prepare samples for many different techniques. They also offer many protein samples and other protein technologies.

8.3.2 Determining the Existence of Proteins in Complex Mixtures

Classically, antibodies to particular proteins or to their modified forms have been used in biochemistry and cell biology studies. For more quantitative determinations of protein amounts, techniques such as ELISAs can be used. For proteomic study, more recent techniques such as Matrix-assisted laser desorption or ionization has been employed for rapid determination of proteins in particular mixtures.

8.3.3 Determining Post-translationally Modified Proteins

A particular protein can be studied by developing an antibody which is specific to that modification. For example, there are antibodies which only recognize certain proteins when they are tyrosine-phosphorylated, also, there are antibodies specific to other modifications. These can then be used to determine the set of proteins that have undergone the modification of interest. For sugar modifications, such as glycosylation of proteins, certain lectins have been discovered which bind sugars. A more common way to determine post-translational modification of interest is by 2D gel electrophoresis. Recently, another approach called PROTOMAP has been developed which combines SDS-PAGE with shotgun proteomics to enable detection of changes in gel-migration such as those caused by proteolysis or post-translational modification.

8.4 Techniques Used in Proteomics

Some of the techniques used in proteomics study are given below:

1. **One- and two-dimensional gel electrophoresis** is used to identify the relative mass of a protein and its isoelectric point.
2. **HPLC** is used for separation of proteins.
3. **X-ray crystallography** and **nuclear magnetic resonance** are used to characterize the 3D structure of peptides and proteins. However, low-resolution techniques such as circular dichroism, Fourier transform infrared spectroscopy, and small angle X-ray scattering can be used to study the secondary structure of proteins.
4. **Tandem mass spectrometry (MS/MS)** combined with reverse phase chromatography or 2D electrophoresis is used to identify (by *de novo* peptide sequencing) and quantify all the levels of proteins found in cells.
5. **Mass spectrometry**, often **MALDI-TOF**, is used to identify proteins by peptide mass fingerprinting (PMF). Less commonly this approach is used with chromatography and/or high-resolution mass spectrometry.
6. **Affinity chromatography, Yeast two-hybrid techniques, Fluorescence resonance energy transfer (FRET), and Surface plasmon resonance (SPR)** are used to identify protein–protein and protein-DNA binding reactions.
7. **X-ray tomography** is used to determine the location of labeled proteins or protein complexes in an intact cell. It is frequently correlated with images of cells from light-based microscopes.
8. **Software-based image analysis** is utilized to automate the quantification and detection of spots within and among gel samples.

Some of these techniques have been discussed in detail.

8.4.1 One- and Two-Dimensional Gel Electrophoresis

One- and two-dimensional gel electrophoresis is used to identify the relative mass and isoelectric point of a protein.

8.4.1.1 Electrophoresis

The migration of charged colloidal particles or molecules through a solution under the influence of an applied electric field is usually provided by immersed electrodes. It can also be described as a method of separating substances, especially proteins and analyzing molecular structure based on the rate of movement of each component in a colloidal suspension under the influence of an electric field. An analyte is a chemical substance that is the subject of chemical analysis. Separation by electrophoresis depends on differences in the migration velocity of ions or solutes through a given medium in an electric field. The electrophoretic migration velocity of an analyte is

$v_p = \mu_p E$ where E is the electric field strength and μ_p is the electrophoretic mobility.

The electrophoretic mobility is inversely proportional to frictional forces in the buffer, and directly proportional to the ionic charge of the analyte. The forces of friction against an analyte are dependent on the analyte's size and the viscosity (η) of the medium. Analytes with different frictional forces or different charges will separate from one another when they move through a buffer. At a given pH, the electrophoretic mobility of an analyte is:

$$\mu_p = \frac{z}{6\pi r}$$

where, r is the radius of the analyte and z is the net charge of the analyte.

Differences in the charge to size ratio of analyte cause differences in electrophoretic mobility. Small, highly charged analytes have greater mobility, whereas large, less charged analytes have lower mobility. Electrophoretic mobility is an indication of an analyte's migration velocity in a given medium. The net force acting on an analyte is the balance of two forces: the electrical force acting in favor of motion, and the frictional force acting against motion. The two forces remain steady during electrophoresis, thus the electrophoretic mobility is a constant for a given analyte under a given set of conditions.

Electrophoresis has a wide variety of applications in proteomics, forensics, molecular biology, genetics, biochemistry, and microbiology. One of the most common uses of electrophoresis is to analyze differential expression of genes. Healthy and diseased cells can be identified by differences in the electrophoretic patterns of their proteins. Proteins can also be characterized similarly, and some information about their structure can be derived from the masses of fragments in the gel.

8.4.1.2 Two-Dimentional Gel Electrophoresis

Two-dimensional polyacrylamide gel electrophoresis (2DE) was first described by O'Farrell in 1975 and has evolved markedly as one of the core technologies for the analysis of complex protein mixtures extracted from biological samples since then. The proteins are separated in 2 steps according to 2 independent properties [isoelectric point (pI) and molecular weight (MW)].

The proteins are made up of amino acids which may have positive, negative, or no charges. In addition, amino acids can be hydrophobic or hydrophilic. These amino acid properties are combined in the molecule to determine the electrostatic and amphiphilic properties of the proteins. When a protein of charge $\boldsymbol{q}$ is placed in an electric field, $\boldsymbol{E}$, it experiences an electrical force given by: $\mathbf{F} = \boldsymbol{q} \times \boldsymbol{E}$. Under the influence of this force, the protein moves until the force becomes zero ($\mathbf{F} = 0$). This principle is used to make proteins move in a liquid or solid media. A common medium used is Polyacrylamide (PAA). PAA is a flexible, elastic polymer that allows particles to move inversely proportional to their sizes; i.e., smaller particles move faster than larger particles. Thus a mixture of proteins in a PAA gel under the influence of a force F will separate in individual molecules depending on their sizes and charges. Two-dimensional polyacrylamide gel electrophoresis (2D-PAGE) is a form of gel electrophoresis in which proteins are separated and identified in two dimensions oriented at right angles to each other. Small changes in charge and mass can easily be detected by this method, because it is rare that two different proteins will resolve to the same place in both dimensions. A 2D gel can resolve one thousand to two thousand proteins, which appear, after staining, as dots in the gel.

The three main advantages of this technique are its robustness, its parallelism, and its unique ability to analyze complete proteins at high resolution. This technique is useful when comparing two similar samples to find specific protein differences. Using 2D-PAGE, hundreds to thousands of polypeptides can be analyzed in a single run. The proteins can be separated in pure form from the resultant spots which can be quantified and further analyzed by mass spectrometry, depending on their resolution. Polypeptides can also be probed with antibodies and tested for post-translational modifications. 2D-PAGE is also used to study differential expression of proteins between cell types. The two main drawbacks are its very low efficiency in the analysis of hydrophobic proteins, and high sensitivity toward the dynamic range and quantitative distribution issues. It requires a large amount of sample handling, limited reproducibility, and a smaller dynamic range than other separation methods. It is also not automated for high-throughput analysis. Certain proteins are difficult for 2D-PAGE to separate such as those that are in low abundance, acidic, basic, hydrophobic, very large, or very small.

8.4.2 Centrifugation

Centrifugation is one of the most important and widely applied research techniques in biochemistry, cellular and molecular biology, and in medicine. In proteomics it plays a vital role in the fundamental and necessary process of isolating proteins. This process begins with intact cells or tissues. Before the proteins can be obtained, the cells must be broken open by processes such as snap freezing, sonication, homogenization by high pressure, or grinding with liquid nitrogen. Once the cells have been opened up, all of their contents including cell membranes, RNA, DNA, and organelles will be mixed in the solvent with the proteins. Centrifugation is used for separating out all the non-protein material. Within the centrifuge samples are spun at high speeds and the resulting force causes particles to separate based on their density. Centrifugation is also used for removing cells or other suspended particles from their surroundings, isolating viruses and macromolecules, including DNA, RNA, proteins, and lipids or establishing physical parameters of these particles from their observed behavior during centrifugation, separating from dispersed tissue the various subcellular organelles including nuclei, mitochondria, chloroplasts, golgi bodies, lysosomes, peroxisomes, glyoxysomes, plasma membranes, endoplasmic reticulum, polysomes, and ribosomal subunits. Once the mixture of proteins has been isolated using centrifugation, one of several methods to separate out individual proteins can be used for further study.

8.4.3 Protein Separation-Chromatography

To obtain a pure protein sample, a protein has to be isolated from all other proteins and cellular components. This can prove to be difficult because a single protein often makes up only 1 % of the total protein concentration of a cell. Therefore, 99 % of the protein components of a sample must be removed before it can be classified as pure. Protein separations can be done by chromatography. There are several properties of proteins that can be used to separate them. Different types of chromatography take advantage of different properties. Proteins can be separated on the basis of size, shape, hydrophobicity, affinity to molecules or charge. All methods utilize an insoluble stationary phase and a mobile phase that passes over it. The mobile phase is commonly a liquid solution which contains the protein that has to be isolated. The stationary phase on the other hand is made up of a group of beads, usually based on a carbohydrate or acrylamide derivative, that are bound to ionically charged species, hydrophobic characters, or affinity ligands.

In column chromatography, when a protein sample is applied to the column, it equilibrates between the stationary phase and the mobile phase. Depending on the type of chromatography, proteins with certain characteristics will bind to the stationary phase while those lacking the sought characteristics will remain in the mobile phase and pass through the column. For example in ion exchange chromatography, a positively charged protein binds to a negatively charged stationary phase, while the negatively charge protein will be eluted from the column with the mobile phase. The final step involves

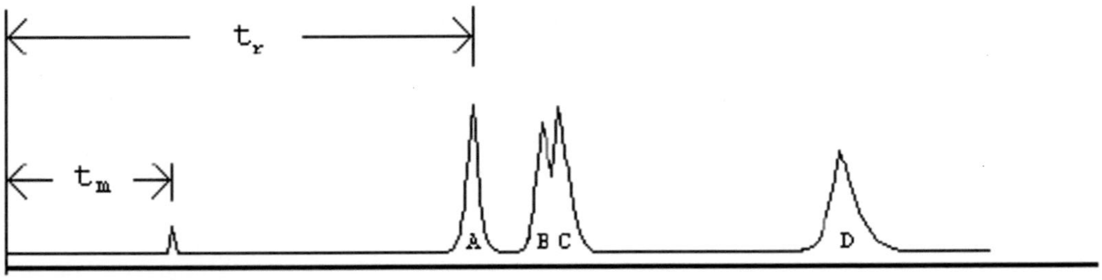

Fig. 8.4 Chromatogram showing separation based on signals given by a detector. X axis: Time in Min or volume in ml. Y axis: Signal. t_m — the time required for the mobile phase to travel the entire length of the column, t_r—the time required for a specific protein to elute from the column

displacing the protein from the stationary phase, also known as elution. This is done by introducing a particle which will compete with the protein binding site on the stationary phase. Various commercial columns are available; specifically Bio-Rad, Sigma-Aldrich, GE Healthcare, etc., offer a variety of chromatography columns (Fig. 8.4).

8.4.4 Emerging and Miscellaneous Technologies in Proteomics

8.4.4.1 Multi-dimentional Protein Identification Technology or Orthogonal Separations

Multi-dimentional protein identification technology (MudPIT) is a chromatography-based proteomic technique in which a complex peptide mixture is prepared from a protein sample and loaded directly onto a triphasic microcapillary column packed with reversed phase, strong cation exchange, and reversed phase HPLC grade materials. Once the complex peptide mixture is loaded the column is placed directly in-line with a tandem mass spectrometry (MS/MS). The MS/MS data generated from a MudPIT run is then searched to determine the protein content of the original sample.

MudPIT is a robust and widely accepted method for protein identification from a wide variety of samples. It is an excellent tool for both qualitative and quantitative proteomic analyses. Through on-line 2D HPLC, complex peptides mixtures can be well separated. For a relatively simple sample, better results are obtained from MudPIT, compared to single dimension liquid chromatography method. It was developed as a method to analyze the highly complex samples necessary for large-scale proteome analysis by electrospray ionization, MS/MS, and database searching. This method couples a 2D liquid chromatography separation of peptides on a microcapillary column with detection in a tandem mass spectrometer. In the MudPIT, a protein or mixture of proteins is first reduced (to break cysteine disulfide bonds), alkylated (to prevent reformation of disulfide bonds), and digested into a complex mixture of peptides.

MudPIT has been used in a wide range of proteomics experiments, including large-scale catalogs of proteins in cells and organisms, profiling of organelle and membrane proteins, identification of protein complexes, determination of post-translational modifications and quantitative analysis of protein expression. In MudPIT biochemical fractions containing many proteins are directly proteolyzed and the enormous number of peptides generated, are separated by 2D liquid chromatography before entering the mass spectrometer. Instead of MALDI-TOF, MS/MS is employed so that, after the mass of a peptide is measured, the peptide is fragmented using a collision-induced dissociation cell, and the masses of the fragmentation products are determined.

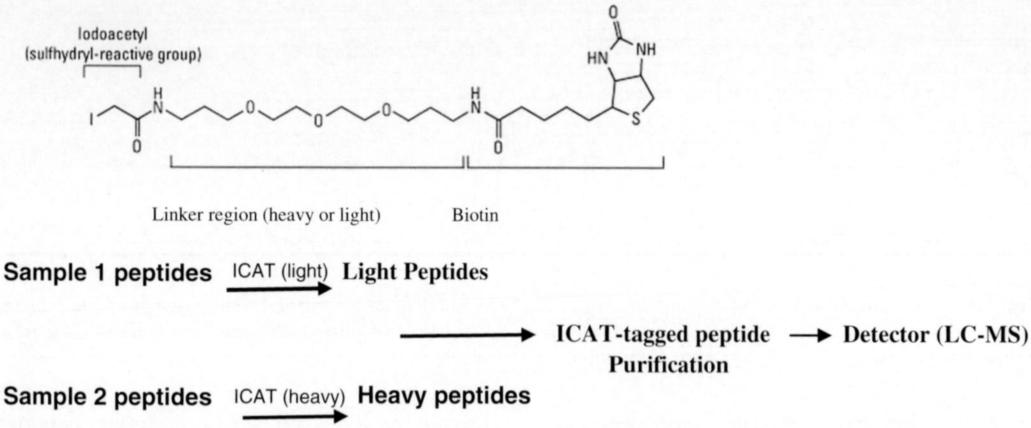

Fig. 8.5 Process of ICAT used in proteomics

8.4.4.2 Isotope-Coded Affinity Tag

Isotope-coded affinity tags (ICATs) are gel-free method for quantitative proteomics that relies on chemical labeling reagents. These chemical probes consist of three general elements: a reactive group capable of labeling a defined amino acid side chain (e.g., iodoacetamide to modify cysteine residues), an isotopically coded linker and a tag (e.g., biotin) for the affinity isolation of labeled proteins/peptides. For the quantitative comparison of two proteomes, one sample is labeled with the isotopically light (d0) probe and the other with the isotopically heavy (d8) version (Fig. 8.5). To minimize error, both samples are then combined, digested with a protease (i.e., trypsin), and subjected to avidin affinity chromatography to isolate peptides labeled with isotope-coded tagging reagents. These peptides are then analyzed by liquid chromatography–mass spectrometry (LC–MS). The ratios of signal intensities of differentially mass-tagged peptide pairs are quantified to determine the relative levels of proteins in the two samples. The original tags were developed using deuterium, but later tags using ^{13}C were used instead to circumvent issues of peak separation during liquid chromatography due to the interaction of deuterium with the stationary phase of the column.

For samples that are not amenable to metabolic labeling, such as when analyzing clinical samples (e.g., biological fluids, tissue samples) or when experimental time is limited, chemical or enzymatic stable isotopic labeling methods are available for quantitative proteomic analyses. These include strategies to add isotopic atoms or isotope-coded tags to peptides or proteins. A rapid and relatively inexpensive method of chemical labeling is stable isotope dimethylation which uses formaldehyde in deuterated water to label primary amines with deuterated methyl groups. This approach also does not change the ionic state of the labeled peptides because of the reductive amination that occurs, so their chemical properties remain the same as those of unlabeled peptides. Benefit of this approach is that many samples are amenable to formaldehyde fixation, which is fast and cheap compared to other labeling reagents. This requires using pure samples or sample preparation to reduce the complexity of biological samples to minimize the number of peaks detected by MS.

Protein labeling with ICAT followed by MS/MS allows sequence identification and accurate quantification of proteins in complex mixtures, and has been applied to the analysis of global protein expression changes, protein changes in subcellular fractions, components of protein complexes, protein secretion, and body fluids.

8.4.4.3 Label-Free Tags

Label-free methods for both relative and absolute quantitation have been developed as a rapid and low-cost alternative to other quantitative proteomic approaches. These strategies are ideal for large-sample analyses in clinical screening or biomarker discovery experiments but are less reliable for measuring small changes. Unlike other quantitation methods, label-free samples are separately collected, prepared, and analyzed by liquid chromatography–mass spectrometry, LC–MS, or LC–MS/MS. Hence, label-free quantitation experiments need to be more carefully controlled than stable isotope methods to account for any experimental variations. Protein quantitation is performed using either ion peak intensity or spectral counting.

Relative quantitation by ion peak intensity relies on LC–MS only. The direct MS m/z values for all ions are detected and their signal intensities at a particular time recorded. The signal intensity from electrospray ionization has been reported to highly correlate with ion concentration, therefore the relative peptide levels between samples can be determined directly from these peak intensities. Because of the large amount of data collected from these experiments, sensitive computer algorithms are required for automated ion peak alignment and comparison. Label-free relative quantitation by spectral counts entails comparing the sum of the MS/MS spectra from a given peptide across multiple samples, which has been shown to directly correlate with protein abundance. Besides relative quantitation, label-free methods can be used to determine the absolute concentration of proteins in a sample. One method entails determining the exponentially modified protein abundance index (emPAI), which estimates protein abundance based on the number of peptides detected and the number of theoretically observed tryptic peptides for each protein, is used to determine the approximate absolute protein abundance in large-scale proteomic analyses. Another method, absolute protein expression (APEX) is based on spectral counts and uses correction factors to make protein abundance proportional to the number of peptides observed.

8.4.4.4 Flourescence Resonance Energy Transfer

Flourescence resonance energy transfer (FRET) is an important tool to study protein–protein interactions, protein-DNA interactions, protein conformational changes and other molecular dynamics quantification. It is used to detect protein binding, providing spatial and temporal information about the interaction. It is a type of Fluorescence Spectroscopy using two fluorescent dyes with overlapping emission and absorption spectra, which is used to indicate proximity of labeled molecules. This technique is useful for studying interactions of molecules and protein folding. It uses fluorescent energy transfer to visualize protein interactions. Fluorophores are fused to the proteins of interest and then bombarded with light at the excitation wavelength. The first fluorophore transfers some of the energy it absorbs from the light source to the second fluorophore, which in turn emits some of its energy into the environment where it is visible with the use of a fluorescent microscope.

Since the discovery of FRET by Theodor Forster in 1948, it has become useful tool in biology for three reasons: (1) FRET is really sensitive in the range of 100 Å and below, the scale at which transactions between biological macromolecules and complexes occur; (2) the instrumentation is extremely sensitive and is readily amenable to miniaturization, high throughput, and automation; and (3) cell permeable and genetically encodable FRET probes enable the real-time quantitation of dynamic cellular processes in live cells.

8.4.4.5 Mass Spectrometry

Mass spectrometry (MS) is an important emerging method for the characterization of proteins. MS is a technique in which gas phase molecules are ionized and their mass-to-charge ratio is measured by observing acceleration

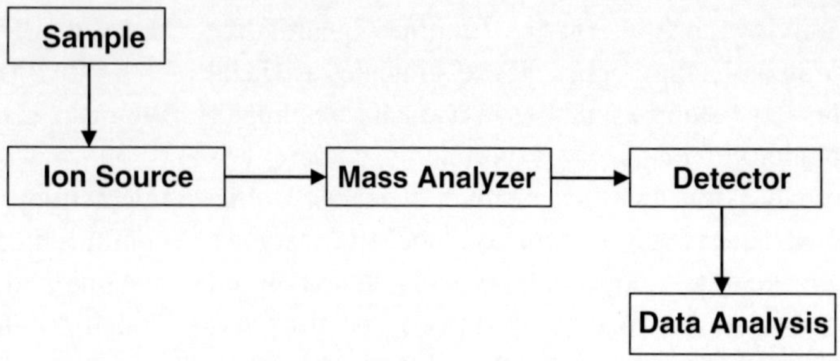

Fig. 8.6 Schematic diagram of a mass spectrometer

differences of ions when an electric field is applied (Fig. 8.6). Lighter ions will accelerate faster and be detected first. If the mass is measured with precision then the composition of the molecule can be identified. In the case of proteins, the sequence can be identified. Most samples submitted to MS are a mixture of compounds. A spectrum is acquired to give the mass-to-charge ratio of all compounds in the sample. MS throws light on molecular mechanisms within cellular systems. It is used for identifying proteins, functional interactions, and it further allows for determination of subunits. Several configurations of mass spectrometers that combine electrospray ionization (ESI) and matrix-assisted laser desorption ionization (MALDI) with a variety of mass analyzers (linear quadrupole mass filter [Q], time-of-flight [ToF], quadrupole ion trap, and Fourier transform ion cyclotron resonance [FTICR] instrument) are routinely used (Yates et al. 2009).

Whole protein mass analysis is primarily conducted using either TOF, MS, or FTICR. These two instruments are preferable because of their wide mass range and in the case of FTICR, its high-mass accuracy. Mass analysis of proteolytic peptides is a much more popular method of protein characterization, as cheaper instrument designs can be used for characterization. Additionally, sample preparation is easier once whole proteins have been digested into smaller peptide fragments. The most widely used instrument for peptide mass analysis are the MALDI time-of-flight instruments as they permit the acquisition of peptide mass fingerprints (PMFs) at high pace. Multiple stage quadrupole-time-of-flight and the quadrupole ion trap also find use in this application.

The relative abundance of an ion can also be measured using MS. Different compounds have differential ionization capabilities, therefore intensity of an ion is not in a direct correlation to concentration. It is an analytical method which has a variety of uses outside of proteomics, such as isotope and dating, trace gas analysis, atomic location mapping, pollutant detection, and space exploration. This technique was discovered during studies of gas excitation in a charged environment, more than 100 years ago by J. J. Thomson in 1913.

The ionization methods used for the majority of biochemical analyses are:

1. **ESI**

ESI is one of the atmospheric pressure ionization (API) techniques and is well suited to the analysis of polar molecules ranging from less than 100 Da to more than 1,000,000 Da in molecular mass. In ESI the ions of interest are formed from solution by applying a high electric field to the tip of a capillary, from which the solution will pass through. The sample will be sprayed into the electric field along with a flow of nitrogen to promote desolvation. Droplets will form and will evaporate in a vacuumed area. This causes an increase in charge on the droplets and the ions are now said to be multiply charged. These multiply charged ions can then enter the analyzer (Andersen et al. 1996). ESI is a method of choice because of the following properties: (1) The "softness" of the phase conversion process allows very fragile molecules to be ionized intact and even in some non-covalent interactions to be preserved for MS analysis. (2) The eluting fractions through liquid chromatography can then be sprayed into the mass spectrometer, allowing for the further analysis of

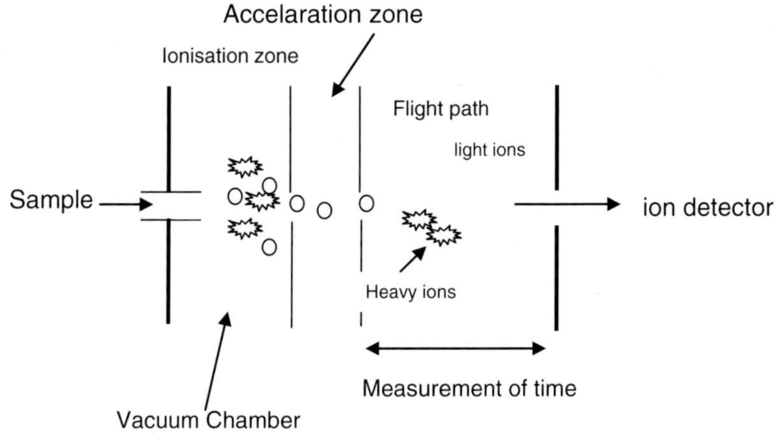

Fig. 8.7 MALDI-TOF mass spectrometry

mixtures. (3) The production of multiply charged ions allow for the measurement of high-mass biopolymers. Multiple charges on the molecule will reduce its mass to charge ratio when compared to a single charged molecule. Multiple charges on a molecule also allow for improved fragmentation which in turn allows for a better determination of structure.

2. **MALDI**

MALDI (Hillenkamp et al. 1991) deals with thermolabile, non-volatile organic compounds especially those of high molecular mass and is used successfully in biochemical areas for the analysis of proteins, peptides, glycoproteins, oligosaccharides, and oligonucleotides. It is relatively straightforward to use and reasonably tolerant to buffers and other additives. The mass accuracy depends on the type and performance of the analyzer of the mass spectrometer, but most modern instruments are capable of measuring masses to within 0.01 % of the molecular mass of the sample, at least up to ca. 40,000 Da. In MALDI, the molecular ions of interest are formed by pulses of laser light impacting on the sample isolated within an excess of matrix molecules. This enables the determination of masses of large biomolecules and synthetic polymers greater than 200,000 Daltons without degradation of the molecule of interest. The advantages of MALDI are its robustness, high speed, and relative immunity to contaminants and biochemical buffers. A type of mass spectrometer often used with MALDI is TOF or Time-of-Flight mass spectrometry. This enables fast and accurate molar mass determination along with sequencing repeated units and recognizing polymer additives and impurities. This technique is based on an ultraviolet absorbing matrix where the matrix and polymer are mixed together along with excess matrix and a solvent to prevent aggregation of the polymer. This mixture is then placed on the tip of a probe; then the solvent is removed while under vacuum conditions. This creates co-crystallized polymer molecules that are dispersed homogeneously within the matrix. A pulsing laser beam is set to an appropriate frequency and energy is shot to the matrix, which becomes partially vaporized. As a result the homogeneously dispersed polymer within the matrix is carried into the vapor phase and becomes charged.

Finally, in the TOF analyzer the molecules from a sample are imparted identical translational kinetic energies because of the electrical potential energy difference. These ionic molecules travel down an evacuated tube with no electrical field and of the same distance. The smallest ions arrive first at the detector, which produces a signal for each ion (Fig. 8.7). The cumulative data from multiple laser shots yield a TOF mass spectrum, which translates the detector signal into a function of time, which can be used to calculate the mass of the ion. These analyzers measure the mass/charge ratio of

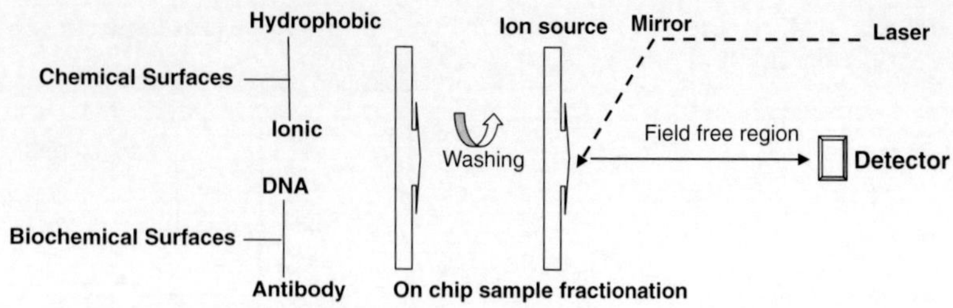

Fig. 8.8 The technique of SELDI-ToF MS

intact ionized biomolecules, as well as their fragmentation spectra, with high accuracy and high speed.

3. **Surface-enhanced laser desorption ionization (SELDI)–ToF MS**

This is a technique for the enrichment of proteins with specific chemical characteristics and combines chromatography with MS. Protein chip arrays have been designed that contain chemically or biochemically treated surfaces. The sample, consisting of crude extracts or mixtures of whole proteins, is applied to the surface. After a series of washes, the targeted proteins are selectively retained. An energy-absorbing solution is added to the surface and the sample is subjected to laser desorption ionization (Fig. 8.8). The formed ions are measured using a ToF mass analyzer as previously described. Characteristic features of spectra can be used for prognostic purposes.

The technique has at its heart a means of enriching for proteins with certain chemical characteristics that determine their interaction with a set of specific surfaces used as a laser desorption ionization target. One advantage of this chip-based technique is that crude samples can readily and rapidly be analyzed with high throughput. The key disadvantage is that the mass spectrum obtained does not enable identification of the proteins analyzed and further work is therefore required. The data obtained are analyzed to recognize critical features that may be used as markers for diagnosis or prognosis. For example, analysis of serum of ovarian cancer by an iterative algorithm that identified a proteomic pattern to discriminate healthy patients from those with malignant disease, using the minimum amount of information. A study on prostate cancer and lung cancer reveals the general applicability of this technique. Furthermore, there are indications that determination of a serum protein level (e.g., high-density lipoproteins) can be monitored effectively with SELDI-ToF and therefore, indicators of hematologic disease could theoretically be measured. Recently, study of non-small-cell lung tumors showed the potential of this technique for the detection of disease markers.

4. **MS/MS**

MS/MS is used to produce structural information about a compound by fragmenting specific sample ions inside the mass spectrometer and identifying the resulting fragment ions to generate structural information regarding the intact molecule. It also enables specific compounds to be detected in complex mixtures on account of their specific and characteristic fragmentation patterns.

MS/MS are instruments that have more than one analyzer and so can be used for structural and sequencing studies. The two analyzers are separated by a collision cell into which an inert gas (e.g., argon, xenon) is admitted to collide with the selected sample ions and bring about their fragmentation. The analyzers can be of the same or of different types, the most common combinations being e.g., quadrupole–quadrupole, magnetic sector-quadrupole, magnetic sector-magnetic sector, quadrupole-time-of-flight. Fragmentation experiments can also be performed on certain single analyser mass spectrometers such as ion trap and time-of-flight instruments.

The most common usage of MS/MS in biochemical areas is the product or daughter ion scanning experiment which is particularly successful for peptide and nucleotide sequencing. It involves more than one step of mass selection or analysis, and fragmentation is usually induced between the steps. One example of an application of tandem mass spectrometry is protein identification. The first mass analyzer isolates ions of a particular m/z value that represent a single species of peptide among many introduced into and then emerges from the ion source. Those ions are then accelerated into a collision cell containing an inert gas such as argon to induce fragmentation. This process is designated collisionally induced dissociation (CID) or collisionally activated dissociation (CAD). The m/z values of the fragment ions are then measured in a second mass analyzer to produce amino acid sequence information.

Uses of MS

Mass spectrometry can be used to identify proteins through various mass spectrometry techniques and also in protein quantification.

Top-down proteomics is a method of protein identification that uses an ion trapping mass spectrometer to store an isolated protein ion for mass measurement and tandem mass spectrometry analysis. Recently, a top-down approach has been developed for mapping the connectivity of the disulfide-rich peptide *hedyotide B2*. This method allows rapid characterization of the disulfide pattern of cystine-knot miniproteins such as cyclotides, conotoxin, knottin, and plant defensins. In this approach the complete proteins are directly analyzed by using mass spectrometer without solution digestion. The advantages of the top-down approach are that it can sometime provide the complete coverage of the protein. But since whole proteins are hard to handle biochemically compared to small peptide pieces, it makes top-down approach difficult to analyze.

Bottom-up proteomics is a common method to identify proteins and characterize their amino acid sequences and post-translational modifications by proteolytic digestion of proteins prior to analysis by mass spectrometry. The proteins may first be purified by a method such as gel electrophoresis resulting in one or a few proteins in each proteolytic digest. Alternatively, the crude protein extract is digested directly, followed by one or more dimensions of separation of the peptides by liquid chromatography coupled to mass spectrometry, a technique known as shotgun proteomics. By comparing the masses of the proteolytic peptides or their tandem mass spectra with those predicted from a sequence database or annotated peptide spectral in a peptide spectral library, peptides can be identified and multiple peptide identifications assembled into protein identification.

Another use of mass spectrometry in proteomics is protein quantification. By labeling proteins with stable heavier isotopes the relative abundance of proteins can be determined. Commercial kits such as iTRAQ (Applied Biosystems) are available for this to obtain a high-throughput level. One of the most powerful ways to identify a biological molecule is to determine its molecular mass together with the masses of its component building blocks after fragmentation. *In silico* proteome analysis can also be done to facilitate proteomics experiments using MS (Cagney et al. 2003).

8.4.4.6 Peptide Mass Fingerprinting

This technique was developed in 1993 independently by many groups. Protein identification by MS can be performed using sequence-specific peptide fragmentation or PMF, also known as peptide mass mapping. The standard approach to identify proteins includes separation of proteins by gel electrophoresis or liquid chromatography. Subsequently, the proteins are cleaved with sequence specific endoproteases, generally trypsin. Following digestion, the generated peptides are investigated by determination of molecular masses or generation of peptide fragments. For protein identification, the experimentally obtained masses are compared with the theoretical peptide masses of proteins stored in databases. MALDI-MS is the most commonly used technique to perform PMF. The predominant detection of singly charged peptide molecules by MALDI-MS facilitates the evaluation of PMFs significantly. Although PMF is an effective tool for the identification of relatively pure proteins, it often fails to identify protein mixtures. Separation of complex protein samples by high-resolution 2D gel electrophoresis (2DE) is well adapted to protein identification with PMF. On the other hand, the application of PMF in combination with one-dimensional gel electrophoresis or liquid chromatography must be adjusted to the separation capacity. A frequently used strategy to identify proteins by MS is to first generate a PMF because of the simplicity of the method. Peptide fragments can be generated directly by MALDI-MS using post-source decay (PSD)-MALDI-MS or collision-induced dissociation (CID)-MALDI-MS/MS. The relative new MALDI-ion traps, MALDI-Q-TOF, and MALDI-TOF/TOF instruments are particularly suitable for this purpose. Alternatively, nano-electrospray ionization tandem mass spectrometry (nano-ESI-MS/MS) is used frequently to achieve sequence information. Furthermore, the power and limitations of PMF are illustrated by the identification of phosphorylation sites, splicing variants, proteins with disulfide bridges, large proteins, and protein fragments. Another focus is the MS-Screener software which can be used for the advanced evaluation of PMFs.

The advantage of this method is that only the masses of the peptides have to be known. Time-consuming *de novo* peptide sequencing is then unnecessary. A disadvantage is that the protein sequence has to be present in the database of interest. Additionally most PMF algorithms assume that the peptides come from a single protein.

8.4.4.7 Shot Gun Proteomics

Shotgun proteomics refers to the use of bottom-up proteomics techniques in identifying proteins in complex mixtures using a combination of high-performance liquid chromatography combined with mass spectrometry. In this method, the proteins in the mixture are digested and the resulting peptides are separated by liquid chromatography. Tandem mass spectrometry is then used to identify the peptides. The advantage of Shotgun proteomics allows global protein identification as well as the ability to systematically profile dynamic proteomes. It also avoids the modest separation efficiency and poor mass spectral sensitivity associated with intact protein analysis. The disadvantage is dynamic exclusion filtering that is often used in shotgun proteomics maximizes the number of identified proteins at the expense of random sampling. This problem may be exacerbated by the undersampling inherent in shotgun proteomics.

8.4.4.8 Nuclear Magnetic Resonance

Nuclear magnetic resonance (NMR) is a physical phenomenon in which nuclei in a magnetic field absorb and re-emit electromagnetic radiation. This energy is at a specific resonance frequency which depends on the strength of the magnetic field and the magnetic properties of the isotope of the atoms.

The principle of NMR usually involves two sequential steps:

- The alignment (polarization) of the magnetic nuclear spins in an applied, constant magnetic field.
- The perturbation of this alignment of the nuclear spins by employing an electro-magnetic, usually radio frequency pulse. The required perturbing frequency is dependent upon the static magnetic field and the nuclei of observation.

The two fields are usually chosen to be perpendicular to each other as this maximizes the NMR signal strength. The resulting response by the total magnetization of the nuclear spins is the phenomenon that is exploited in NMR spectroscopy and magnetic resonance imaging. Both use intense applied magnetic fields in order to achieve dispersion and very high stability to deliver spectral resolution. NMR structure determination is limited currently by size constraints and lengthy data collection and analysis times (often months), and the method is best applied to proteins smaller than 250 amino acids. On the other hand, NMR experiments do not require crystals and samples appropriate for structure determination can be identified within minutes of protein purification.

Structural proteomics elucidates structure–function relationships of uncharacterized gene products based on the 3D protein structure. It proposes biochemical and cellular functions of unannotated proteins and thereby identifies potential drug design and protein engineering targets. NMR-based structural proteomics together with X-ray crystallography provides a comprehensive structural database to predict the basic biological functions of hypothetical proteins identified by the genome projects.

8.4.4.9 X-ray Crystallography

In order to determine the function of proteins the determination of the 3D structure of proteins is required. The X-ray crystallography is a major tool to explore the structure–function of macromolecule. X-ray crystallography currently is perceived as the potential workhorse for structural proteomics, because if provided with a well-diffracting crystal it is possible to determine a 3D structure in hours. The three components in an X-ray crystallographic analysis are a protein crystal, a source of X-rays, and a detector. X-ray crystallography is used to investigate molecular structures through the growth of solid crystals of the molecules they study. Crystallographers aim high-powered X-rays at a tiny crystal containing trillions of identical molecules. The crystal scatters the X-rays onto an electronic detector. After each blast of X-rays, lasting from a few seconds to several hours, the crystal is rotated by entering its desired orientation into the computer that controls the X-ray apparatus. This enables the scientists to capture in three dimensions how the crystal scatters, or diffracts, X-rays. The intensity of each diffracted ray is fed into a computer, which uses a mathematical equation to calculate the position of every atom in the crystallized molecule. The result is a 3D digital image of the molecule.

Crystallographers measure the distances between atoms in angstroms. The X-rays used by crystallographers are approximately 0.5–1.5 Å long. The scattering of X-rays is also known as "X-ray diffraction". Such scattering is the result of the interaction of electric and magnetic fields of the radiation with the electrons in the atoms of the crystal.

The patterns are a result of interference between the defracted X-rays governed by Bragg's Law: $2d \sin \theta = n * \lambda$, where d is the distance between two regions of electron density, θ is the angle of defraction, λ is the wavelength of the defracted X-ray, and n is an integer. If the angle of reflection satisfies the following condition:

$$\sin \theta = \frac{(n * \lambda)}{2d}$$

the defracted X-rays will interfere constructively. Otherwise, destructive interference occurs.

Drawbacks of X-ray crystallography are that the sample needs to be in a solid form, the sample must be present in a large enough quantity to be studied and the sample is often destroyed by the X-ray radiation used to study it. No sample in the gas or liquid state can be analyzed *via* X-ray crystallography. Also, rare or hard to synthesize samples may be difficult to study, because there may not be enough of the sample for the radiation to provide a clear image. Further, studying biological samples can be problematic because the radiation used to study the samples will harm or destroy the living tissues. X-ray crystallography also takes a lot of time to complete one protein structure.

8.4.4.10 X-ray Tomography

A new branch of X-ray microscopy is being used in proteomics analysis, called X-ray tomography. This method uses projected images to calculate and reconstruct a 3D object. This technology is being used in proteomics to determine the location of labeled proteins or large complexes within a cell. This technique can also be used in conjunction with images of cells from light-based microscopes to help identify where a protein is located and how this location factors into its function and identification. X-ray tomography is used to determine the location of labeled proteins or protein complexes in an intact cell.

8.4.4.11 Laser Capture Microdissection

Laser capture microdissection or LCM is a process that isolates and removes distinct populations from a tissue. This facilitates the comparison of diseased tissue with normal tissue from an organism. In LCM, an infrared laser beam melts a thermosensitive polymer film that traps a specific group of cells. This polymer film is then extracted and moved to a test tube where an extraction buffer is used to remove the groups of cells for more advanced proteomics analysis such as 2D-PAGE, Ion Chromatography, etc. This technology will become more useful as systems with higher sensitivity for analysis of smaller amounts of tissue are developed and realized.

8.4.4.12 Proteomic Complex Detection Using Sedimentation

Proteomic complex detection using sedimentation (ProCoDeS) is a technique for the high-throughput identification of both soluble and membrane proteins that are found in stable complexes. Relative sizes of protein complexes are estimated *via* their sedimentation in a gradient. In this case, a rate zonal gradient (RZG) is used to estimate the relative size of protein complexes. The distribution of a protein of interest in this sedimentation can be detected using classic techniques such as Western Blotting or ICAT. This can be done for a large number of proteins. Thus, ProCoDeS can be used to identify stable protein complexes and is especially well suited for the screening of unrefined cellular material to help find new proteins that cannot be discovered because they exist in protein complexes such as proteins found in protein membranes.

8.4.4.13 Protein Interactions

Most proteins function in collaboration with other proteins and one goal of proteomics is to identify which proteins interact. This is especially useful in determining potential partners in cell signaling cascades. Several methods are available to probe protein–protein interactions. The traditional method is yeast two-hybrid analysis. New methods include protein microarrays, immunoaffinity chromatography followed by mass spectrometry, dual polarization interferometry, and experimental methods such as phage display and computational methods. Protein interaction is crucial for every organism. Some proteins are composed of more than one polypeptide chain, and the interactions between the different peptides are necessary for the whole protein to function.

In many experiments and computational studies, the focus is on interactions between two different proteins. However, one protein can interact with other copies of itself (oligomerization), or three or more different proteins interacting. The stoichiometry of the interaction is also important—that is, how many of each protein involved are present in a given reaction. Some protein interactions are stronger than others, because they bind together more tightly. The strength of binding is known as affinity. Proteins will only bind each other spontaneously if it is energetically favorable. Energy changes during binding are another important aspect of protein interactions. Many of the computational tools that predict interactions are based on the energy of interactions. Predicting the interactions can help scientists predict pathways in the cell, potential drugs, and antibiotics and protein functions.

8.4.4.14 Protein chips

Protein chips, also referred to as protein arrays or protein microarrays, are similar to DNA microarrays. Protein microarrays are another approach for analyzing large sets of proteins. Protein chips enable researchers to quickly and easily survey the entire proteome of a cell within an organism and to automate protein experiments. Protein chips look identical, except each spot corresponds to one of the organism's thousands of proteins, instead of one of its genes. The intensity of the dot indicates the amount of protein present.

Protein chips were first developed in 2000 by researchers at Harvard University. Today there are many companies manufacturing protein chips using many types of techniques including spotting and gel methods. The types of protein chips available include "lab on a chip", antibody arrays and antigen arrays, as well as a wide range of chips containing "alternative capture agents" such as proteins, substrates, and nucleic acids. Analysis of protein chips comes with many challenges including dynamic protein concentrations, the sheer number of proteins in a cell's proteome, and the understanding of the probes for each protein. Steps include the reading of the protein levels off the chip, and then the use of computer software to analyze the massive amounts of data collected (Schena et al. 1995; MacBeath and Schreiber 2000). To examine expression at protein level and determine quantitative and qualitative information on proteins protein microarray (chips) two components are required: spatially addresses molecules that recognize individual protein–protein moieties and second a method to detect the interaction of the individual proteins in the mixture with their corresponding recognition molecules. In this technique, purified ligands or recognition molecules such as proteins, peptides, antibodies, antigens, aptamers, carbohydrates, or small molecules are spotted on a derivatized surface and generally used for examining the protein expression level for protein profiling and clinical diagnostics. Synthetic phage-displayed antibodies are used for this purpose but apart from antibodies any molecule that can specifically recognize individual protein with appropriate affinity and avidity can do so in immobilized form.

Applications of protein chip experiments include identifying biomarkers for diseases, investigating protein–protein interactions, and testing for the presence of antibodies in a sample. Protein chips also have applications in cancer research, medical diagnostics, homeland security, and proteomics. Many companies, including Biacore, Invitrogen, and Sigma-Aldrich, produce industrial level protein array systems

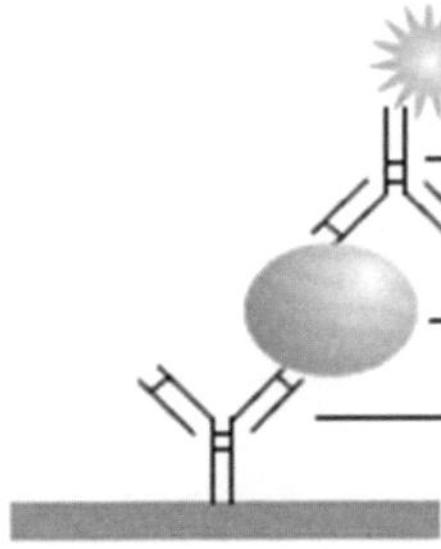

Fig. 8.9 Sandwich-style antibody-pair microarray

that can be used for drug discovery and basic biological research.

Antibody-Pair Protein Arrays

This microarray consists of arrayed capture antibodies and appropriate control and orientation elements. The assay is performed by adding an antigen standard or test sample followed by a detector antibody which is modified with a directly detectable label-enzyme, fluorescent molecule, isotope, etc., or it is biotinylated for detection after subsequent probing with labeled streptavidin. Antibody-pair microarrays are a type of multiplexed ELISAs. Sandwich-style antibody pair microarrays (Fig. 8.9) can be used for qualitative or comparative detection of protein analytes or for protein quantification when appropriate standards are available. Limitation of this technique is that it is difficult to produce specific antibodies for large number of human proteins.

Single Antibody/Labeled Microarray Sample Microarray

This technique is useful for protein targets, such as poorly characterized cell signaling proteins, for which paired antibodies are not available. Drawback of this technique is its lack of antibody redundancy, which helps to ensure specific antigen recognition (Fig. 8.10).

Cellular Lysate Protein Arrays

Microaarays of cellular proteins are either arrayed as complex protein mixtures or arrayed as purified or over-expressed proteins. Complex protein mixture arrays are dot blots of cellular lysates. These microarrays consisting of libraries

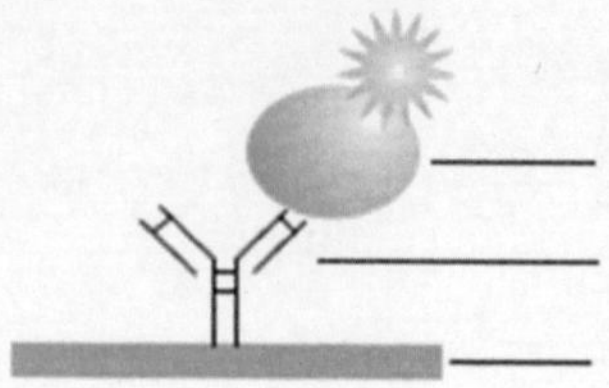

Fig. 8.10 Single antibody/labeled microarray sample microarray

of purified or over-expressed proteins allow screening for protein–protein interactions and kinase activities.

8.5 Post-translational Modifications

The human proteome is more complex than the human genome and single genes encode multiple proteins. Genomic recombination, transcription initiation at alternative promoters, differential transcription termination, and alternative splicing of the transcript are mechanisms that generate different mRNA transcripts from a single gene. The increase in complexity from the genome to the proteome is further facilitated by protein post-translational modifications (PTMs). PTMs are chemical modifications that play a key role in functional proteomics, because they regulate activity, localization and interaction with other cellular molecules such as proteins, nucleic acids, lipids, and cofactors.

PTM increases the functional diversity of the proteome by the covalent addition of functional groups or proteins, proteolytic cleavage of regulatory subunits, or degradation of entire proteins. These modifications include phosphorylation, glycosylation, ubiquitination, nitrosylation, methylation, acetylation, lipidation, and proteolysis and influence almost all aspects of normal cell biology and pathogenesis. PTMs occur at distinct amino acid side chains or peptide linkages and are most often mediated by enzymatic activity. It is estimated that 5 % of the proteome comprises enzymes that perform more than 200 types of post-translational modifications. These enzymes include kinases, phosphatases, transferases, and ligases, which add or remove functional groups, proteins, lipids, or sugars to or from amino acid side chains, and proteases, which cleave peptide bonds to remove specific sequences or regulatory subunits. Many proteins can also modify themselves using autocatalytic domains, such as autokinase and autoprotolytic domains. Post-translational modification can occur at any step in the "life cycle" of a protein. Other modifications occur after folding and localization are completed to activate or inactivate catalytic activity or to otherwise influence the biological activity of the protein. Proteins are also covalently linked to tags that target a protein for degradation. Besides single modifications, proteins are often modified through a combination of post-translational cleavage and the addition of functional groups through a step-wise mechanism of protein maturation or activation.

Protein PTMs can also be reversible depending on the nature of the modification. For example, kinases phosphorylate proteins at specific amino acid side chains, which is a common method of catalytic activation or inactivation. Conversely, phosphatases hydrolyze the phosphate group to remove it from the protein and reverse the biological activity. Proteolytic cleavage of peptide bonds is a thermodynamically favorable reaction and therefore permanently removes peptide sequences or regulatory domains. Consequently, the analysis of proteins and their post-translational modifications is particularly important for the study of heart disease, cancer, neurodegenerative diseases, and diabetes. The characterization of PTMs provides insight into the cellular functions underlying etiological processes.

8.5.1 Types of PTMs

Some of the post-translational modifications have been described in detail.

8.5.1.1 Phosphorylation

Reversible protein phosphorylation, principally on serine, threonine, or tyrosine residues, is one of the most important and well-studied post-translational modifications. Phosphorylation plays critical roles in the regulation of many cellular processes including cell cycle, growth, apoptosis, and signal transduction pathways.

8.5.1.2 Glycosylation

Protein glycosylation has significant effects on protein folding, conformation, distribution, stability, and activity. Glycosylation encompasses a diverse selection of sugar-moiety additions to proteins that ranges from simple monosaccharide modifications of nuclear transcription factors to highly complex branched polysaccharide changes of cell surface receptors. Carbohydrates in the form of aspargine-linked (N-linked) or serine/threonine-linked (O-linked) oligosaccharides are major structural components of many cell surface and secreted proteins.

8.5.1.3 Ubiquitination

Ubiquitin is an 8-kDa polypeptide consisting of 76 amino acids that is appended to the ε-amino group of a lysine residue on a target protein *via* the C-terminal glycine of ubiquitin. Following an initial monoubiquitination event, the formation of a ubiquitin polymer may occur and polyubiquitinated proteins are then recognized by the 26S proteasome that catalyzes the degradation of the ubiquitinated protein and the recycling of ubiquitin.

8.5.1.4 S-Nitrosylation

Nitric oxide (NO) is produced by three isoforms of nitric oxide synthase (NOS) and is a chemical messenger that reacts with free cysteine residues to form S-nitrothiols (SNOs). S-nitrosylation is a critical PTM used by cells to stabilize proteins, regulate gene expression, and provide NO donors, and the generation, localization, activation, and catabolism of SNOs are tightly regulated. S-nitrosylation is a reversible reaction, and SNOs have a short half life in the cytoplasm because of the host of reducing enzymes, including glutathione (GSH) and thioredoxin, that denitrosylate protein. Therefore, SNOs are often stored in membranes, vesicles, the interstitial space, and lipophilic protein folds to protect them from denitrosylation. Only specific cysteine residues are S-nitrosylated.

8.5.1.5 Methylation

The transfer of one-carbon methyl groups to nitrogen or oxygen (N- and O-methylation, respectively) to amino acid side chains increases the hydrophobicity of the protein and can neutralize a negative amino acid charge when bound to carboxylic acids. Methylation is mediated by methyltransferases, and S-adenosyl methionine (SAM) is the primary methyl group donor. Methylation occurs so often that SAM has been suggested to be the most-used substrate in enzymatic reactions after ATP. Additionally, while N-methylation is irreversible, O-methylation is potentially reversible. Methylation is a well-known mechanism of epigenetic regulation, as histone methylation and demethylation is influencing the availability of DNA for transcription. Amino acid residues can be conjugated to a single methyl group or multiple methyl groups to increase the effects of modification.

8.5.1.6 N-Acetylation

N-acetylation, or the transfer of an acetyl group to nitrogen, occurs in almost all eukaryotic proteins through both irreversible and reversible mechanisms. N-terminal acetylation requires the cleavage of the N-terminal methionine by methionine aminopeptidase (MAP) before replacing the amino acid with an acetyl group from acetyl-CoA by N-acetyltransferase (NAT) enzymes. This type of acetylation is co-translational, in that N-terminus is acetylated on growing polypeptide chains that are still attached to the ribosome. While 80–90 % of eukaryotic proteins are acetylated in this manner, the exact biological significance is still unclear. Lysine acetylation on histone N-termini is a common method of regulating gene transcription. It is reversible and reduces chromosomal condensation to promote transcription, and the acetylation of these lysine residues is regulated by transcription factors that contain histone acetyletransferase (HAT) activity. While transcription factors with HAT activity act as transcription co-activators, histone deacetylase (HDAC) enzymes are co-repressors that reverse the effects of acetylation by reducing the level of lysine acetylation and increasing chromosomal condensation. Protein acetylation can be detected by chromosome immunoprecipitation (ChIP) using acetyllysine-specific antibodies or by mass spectrometry, where an increase in histone by 42 mass units represents a single acetylation.

8.5.1.7 Lipidation

Lipidation is a method to target proteins to membranes in organelles (endoplasmic reticulum [ER], Golgi apparatus, mitochondria), vesicles (endosomes, lysosomes), and the plasma membrane. There are four types of lipidation: C-terminal glycosyl phosphatidylinositol (GPI) anchor, N-terminal myristoylation, S-myristoylation, and S-prenylation. Each type of modification gives proteins distinct membrane affinities, although all types of lipidation increase the hydrophobicity of a protein and thus its affinity for membranes. The different types of lipidation are also not mutually exclusive, in that two or more lipids can be attached to a given protein. GPI anchors tether cell surface proteins to the plasma membrane. This type of modification is reversible, as the GPI anchor can be released from the protein by phosphoinositol-specific phospholipase C. This lipase is used in the detection of GPI-anchored proteins to release GPI-anchored proteins from membranes for gel separation and analysis by mass spectrometry. N-myristoylation is a method to give proteins a hydrophobic handle for membrane localization. S-palmitoylation adds a C16 palmitoyl group from palmitoyl-CoA to the thiolate side chain of cysteine residues *via* palmitoyl acyl transferases (PATs). S-prenylation covalently adds a farnesyl (C15) or geranylgeranyl (C20) group to specific cysteine residues within 5 amino acids from the C-terminus *via* farnesyl transferase (FT) or geranylgeranyl transferases.

8.5.1.8 Proteolysis

Peptide bonds are indefinitely stable under physiological conditions, and therefore cells require some mechanism to break these bonds. Proteases comprise a family of enzymes that cleave the peptide bonds of proteins and are critical in antigen processing, apoptosis, surface protein shedding, and cell signaling. The family of over 11,000 proteases varies in substrate specificity, mechanism of peptide cleavage, location in the cell, and the length of activity. Proteases can generally be separated into groups based on the type of proteolysis. Degradative proteolysis is critical to remove unassembled protein subunits and misfolded proteins and to maintain protein concentrations at homeostatic concentrations by reducing a given protein to the level of small peptides and single amino acids. Proteases also play a biosynthetic role in cell biology that includes cleaving signal peptides from nascent proteins and activating zymogens, which are inactive enzyme precursors that require cleavage at specific sites for enzyme function.

Proteolysis is a thermodynamically favorable and irreversible reaction. Therefore, protease activity is tightly regulated to avoid uncontrolled proteolysis through temporal and/or spatial control mechanisms including regulation by cleavage in *cis* or *trans* and compartmentalization (e.g., proteasomes, lysosomes). The diverse family of proteases can be classified by the site of action, such as aminopeptidases and carboxypeptidase, which cleave at the amino or carboxy terminus of a protein, respectively. Another type of classification is based on the active site groups of a given protease that are involved in proteolysis, e.g., Serine proteases, Cysteine proteases, Aspartic acid proteases, and Zinc metalloproteases.

8.5.2 Methods for Study of PTMs

Methods such as phosphoproteomics and glycoproteomics are used to study post-translational modifications.

8.5.2.1 Glycoproteomics

It is now widely accepted that correct protein glycosylation is important for the normal function of many cellular and secreted proteins. Glycosylation has been shown to be necessary for many diverse aspects of protein function, for example, to ensure correct protein folding, to protect proteins from proteolytic attack, as recognition molecules for protein clearance from the serum, and as adhesion molecules. In parallel to the role of glycosylation in normal protein function, the number of examples in which changes occur to the glycosylation of proteins in disease expands.

The standard approach for the identification of (glyco) proteins after separation of the proteome by 1D- or 2D-PAGE is generally by MALDI-MS fingerprint analysis. The separated proteins are digested, in-gel, with a protease such as trypsin, analyzed on a mass spectrometer and the resulting masses are compared with those generated by the theoretical digestion of protein sequences in databases (SWISSPROT and NCBI). Using this approach, a proportion of peptide masses may not be identified as originating from the proteolytic digestion of the protein. These difficulties in protein identification are often explained by the occurrence of post-translational modifications such as glycosylation and phosphorylation, which are not predicted by the majority of databases whose sequences, in the main, have been determined from the characterization of cDNAs. Computational tools, such as FindMod, Peptldent, and Peptide-Mass (http://www.expasy.ch/tools/) can calculate the difference between the observed and predicted masses giving a useful insight into the types of possible modifications that may have occurred to the protein. Another approach to identify likely O-link glycosylation sites utilizes a neural network algorithm and incorporates into an O-GlycBase database (http://www.cbs.dtu.dk/databases/OGLYCBASE/Oglyc.base.html), which provides free access and information regarding the type of glycan, the glycosylation site, species from which the glycoprotein was derived, etc. To determine whether or not a protein is glycosylated is to use the periodic acid-Schiff (PAS) stain. Combining the periodic acid reaction with fluorescent dye technology enables the detection of glycoproteins at nanogram sensitivity. Proteins separated by 2D-PAGE can be transferred to inert membranes, usually made of either nitrocellulose or polyvinylidene difluoride, by a process known as Western blotting. The glycosylation of the proteins can then be examined by detecting the monosaccharides that are present by using carbohydrate-binding proteins, for example, lectins, or antibodies. The sensitivity of detection of glycan residues may be enhanced to nanomol concentrations by the use of biotin conjugated lectins and/or antibodies using streptavidin-horse radish peroxidase, to produce either a colored substrate or a light reaction. Most of the methods for detailed analysis of N-linked glycans involve the removal of the glycan moiety from the protein backbone first, although there have been recent reports of analysis of intact glycoproteins after separation by 2D-PAGE.

The most frequently used approach for glycan release is endoglycosidase digestion and for this PNGase-F (peptide N-glycosidase F) is used. Enzymatic treatment is favored over chemical methods, such as hydrazinolysis or j-elimination, for the release of N-linked sugars because the glycans are less extensively modified and the protein can also be recovered intact. Chemical methods such as hydrazinolysis and, more usually, non-reductive j-elimination or trifluoromethanesulfonic acid are employed. Glycan analysis is usually attempted for glycoproteins that are available in microgram quantities. Once released, various analytical techniques are available for the characterization of carbohydrate structures, e.g., chromatography, electrophoresis, and mass spectrometry. Web-based Bioinformatic analysis tools are available that enable the interpretation of sugar mass spectrometric data, e.g., GlycoMod and GlycoSuiteDB (www.expasy.ch/tools/glycomod and http://tmat.proteomesystems.com/glycosuite/).

8.5.2.2 Phosphoproteomics

Phosphoproteomics deals with the identification and quantification of phosphorylated proteins and identification of phosphorylation sites. Protein kinases control every basic cellular process including metabolism, growth, division, differentiation, motility, immunity learning, and memory *via* regulated protein phosphorylation. Mass spectrometry is the best method used for the identification of phosphorylation sites. One most common strategy for enrichment of phosphoproteomes is a combination of chromatography with affinity-based enrichment. These chromatographic techniques include immobilized metal affinity chromatography (IMAC), immunoprecipitation, metal oxide affinity chromatography (MOAC). In addition antibodies against phosphoaminoacid epitopes, magnetic materials, and nanoparticles as well as metal ion phosphopeptide precipitations are used for affinity-based enrichment of phosphopeptides. IMAC uses metal chelators to immobilize metal ions.

Identification of phosphopeptides and phosphorylation sites is based on database searches. Quantification of site-specific phosphorylation is measured in an MS assay. Occupancy of phosphorylation sites describes the level of phosphorylation compared with the unmodified sites.

8.6 Methods to Find Proteins that Function Together

Some proteins act alone, and the function of many of these has been known for years. But probably the majority of the proteins in a cell act in concert with others, e.g., the 79 proteins in the eukaryotic ribosome, the ~145 proteins in the spliceosomes, and the myriad transcription factors in the cell.

8.6.1 Affinity Chromatography

For affinity chromatography, the proteins whose partners are to be found are attached to a solid matrix in a glass column. A solution containing a mix of possible partners is run through the column. Those proteins that can bind to the target will stick; the others will flow through. A buffer is passed through the column which weakens the binding interactions. The partners are washed out and can be identified.

8.6.2 The Yeast Two-Hybrid System

The yeast two-hybrid system is an *in vivo* system that utilizes the binding domain (BD) and activation domain (AD) of genes that are used in transcription, to determine protein interactions. Proteins are fused to both the BD and the AD of a reporter gene, one to each. When the protein attached to BD interacts with the protein fused to the AD, transcription is initiated, and the reporter gene is transcribed. The effect is quantifiable by measuring the levels of reporter gene expressed. The budding yeast, *Saccharomyces cerevisiae* provides an excellent tool for discovering protein partners as it can easily be transformed with plasmids containing foreign DNA sequences, can live in either the haploid or diploid condition and haploid cells can fuse to form diploid cells if they are of opposite mating types. Using the two-hybrid method, it has been possible to identify many sets of interacting proteins in yeast and other organisms.

For this, using recombinant DNA methods, a plasmid is created containing (1) the DNA encoding the DNA-binding domain of a transcription factor needs to turn on expression of a reporter gene such as the *lacZ* gene that encodes the enzyme β-galactosidase coupled to (2) the DNA encoding the "target" protein, i.e., the protein whose possible partners are to be identified.

The plasmid is inserted into living haploid yeast of one mating type.

1. Using the same methods, many different plasmids are created each containing (a) the DNA encoding the activation domain of the transcription factor, (b) the DNA encoding a possible partner ("bait") protein, and (c) each of these plasmids is inserted into α yeast cells and grown as separate clones.
2. Each α clone is mated with the target clone.
3. If the fusion protein produced by the transcription and translation of a "bait"-containing plasmid can bind to the fusion protein containing the target, the two domains of the transcription factor can interact to turn on expression of the reporter gene (*lacZ*).
4. They are then grown on an indicator substrate as a result the colonies turn blue.
5. The DNA in these colonies can then be isolated and sequenced.
6. Finally the identification of the proteins that can associate with the target protein can be done.

8.6.3 Phage Display

This method involves a DNA bacteriophage that infects *E. coli* and its ability to remain infectious even if one of its coat proteins contains segments of a foreign protein. The method involves transformation of bacteriophages with a random mix of DNA from the organism of interest coupled to the DNA encoding one of the viral coat proteins. Then the *E. coli* is infected with these phages. As the viruses replicate, they not only propagate the recombinant gene but also express it as a coat protein. Both will be incorporated into new virions. The mix of viruses is infected. The mixture is passed through an affinity chromatography column to which the "target" protein has been fixed. Those viruses that display a piece of foreign protein (peptide) that can bind to the target will stick to it. The bound phage is eluted with a buffer. *E. coli* is infected and separate colonies (clones) are grown. The coat protein gene is sequenced to find the sequence of the foreign DNA inserted in it. Using the genetic code dictionary, the amino acid sequence of the peptide is determined. The databases are searched for a protein containing this sequence. As a result a protein that associates with the target protein is obtained.

8.6.4 Three-Dimensional Structure

The clearest picture of how different proteins interact with one another to form functional complexes will come from determining the 3D structure of the complex. There are two methods used for this, X-ray crystallography and NMR spectroscopy. X-ray crystallography requires crystallization of the protein. This is often difficult especially for complexes of two or more proteins. NMR spectroscopy has been especially useful in producing 3D images of proteins that cannot be crystallized.

8.7 Proteoinformatics/Bioinformatics in Proteomics

Proteoinformatics is the use of bioinformatics and computational biology techniques solely within the realm of protein identification and proteomics. Proteoinformatics is currently in its infancy and the largest work being done is on standardizing databases and data submission. Other proteoinformatic work is being done on the image analysis of 2D gels and other images in proteomics used to help identify and annotate proteins in the proteome. An increased ability to generate large amounts of data from the use of high-throughput methods has led to an increased reliance on computers for data acquisition, storage, and analysis. The internet has also enabled collaboration and sharing of data that would not have been possible, leading to the development of large public databases with contributors all over the world. Many databases exist for protein-related information, e.g., protein data bank (PDB) which handles structure and sequence information for proteins with a determined crystal structure. Expasy is a popular and well-curated resource for proteomics databases and tools, including resources such as the Prosite protein feature and domain database, protein BLAST (Basic Local Alignment and Search Tool, for similarity searching) and structure prediction. NCBI also provides many resources for proteins (Binz et al. 2004). As with other bioinformatics resources, "in silico" discovery is not meant as a replacement for lab techniques, but a supplement to work done in a wet lab.

8.7.1 Protein Identification Database

Protein Identifications Databases such as ProFound at Rockefeller University and Protein Prospector at the UCSF Mass Spectrometry Facility are used to identify proteins found with proteomics techniques such as mass spectrometry. Digestion of proteins into peptide fragments allows each protein to break apart in a different way, resulting in a unique peptide fingerprint that can be used to identify the protein. The masses of these fragments as well as the molecular weights and isoelectric points are stored in many of these databases. This data can be used to perform high-throughput protein identification. Protein databases are more specialized than primary sequence databases. They contain information derived from the primary sequence databases. Some contain protein translations of the nucleic acid sequences. Some contain sets of patterns and motifs derived from sequence homologs (Tables 8.2 and 8.3).

8.8 Applications of Proteomics

The goal of proteomics is to analyze the varying proteomes of an organism at different times, in order to highlight differences between them. Proteomics analyzes the structure and function of biological systems. For example, the protein content of a cancerous cell is often different from that of a healthy cell. Certain proteins in the cancerous cell may not be present in the healthy cell, making these unique proteins good targets for anti-cancer drugs. Both purification and identification of proteins in any organism is affected by a multitude of biological and environmental factors.

8.8.1 Drug Discovery

The process of drug discovery is quite complex, integrating many disciplines, including structural biology, metabolomics, proteomics, and computer science. The process includes target selection, lead identification, and preclinical and clinical candidate selection (Okerberg et al. 2005). Proteomics can be used for identifying proteins involved in disease pathogenesis and physiological pathway reconstruction facilitates the ever increasing discovery of new, novel drug targets, their respective modes of action mechanistically, and their biological toxicology. While the causes of many clinical problems vary greatly in their nature and origin, in some cases, the cause is found at the protein level, involving protein function, protein regulation, or protein–protein interactions. For example alkaptonuria is characterized by a defect in the gene coding for the enzyme homogentisic acid oxidase, inhibiting the metabolism of homogentisic acid to maleylacetoacetic acid, within the phenylalanine degradation pathway. The cause of this inborn disease is due to a single gene genetic defect, the clinical manifestations, which include excretion of black urine, are a function of the built up of homogentisic acid resulting from a defective enzyme (protein).

Virtual drug libraries are being developed, both in the public and private sectors. These databases contain potential drug compounds; these compounds may or may not exist outside of a computer database and new compounds developed through various methods of synthesis are continually added. Methods of modifying existing database entries to create new isomers and derivatives are also used, to more

Table 8.2 Some protein databases available on the Internet

S.N.	Database	Description
1.	PIR	Protein information resource—a comprehensive, non-redundant, expertly annotated, fully classified, and extensively cross-referenced protein sequence database
2.	SWISS-PROT and TrEMBL	SWISS-PROT is a curated protein sequence database. TrEMBL is a computer-annotated supplement of SWISS-PROT that contains all the translations of EMBL nucleotide sequence entries not yet integrated in SWISS-PROT
3.	TIGR	A collection of curated databases containing DNA and protein sequence, gene expression, cellular role, protein family, and taxonomic data for microbes, plants and humans
4.	ALIGN	A compendium of sequence alignments: it is a companion resource to PRINTS
5.	BLOCKS	Multiply aligned ungapped segments corresponding to the most highly conserved regions of proteins
6.	DOMO	A database of homologous protein domain families
7.	HOMSTRAD	A curated database of structure-based alignments for homologous protein families
8.	Pfam	A database of multiple alignments of protein domains or conserved protein regions. The alignments represent some evolutionary conserved structure which has implications for the protein's function. Profile hidden Markov models (profile HMMs) built from the Pfam alignments can be very useful for automatically recognizing that a new protein belongs to an existing protein family, even if the homology is weak
9.	ProSite	A database of protein families and domains consisting of biologically significant sites, patterns, and profiles
10.	Library of Protein Family Cores	Structural alignments of protein families and computed average core structures for each family. Useful for building models, threading, and exploratory analysis
11.	ModBase	A database of 3D protein models calculated by comparative modeling
12.	RCSB protein databank	Single international repository for the processing and distribution of 3D macromolecular structure data primarily determined experimentally
13.	Protein loop classification	Conformational clusters and consensus sequences for protein loops derived by computational analysis of their structures
14.	CluSTr	Offers an automatic classification of UniProtKB/Swiss-Prot + UniProtKB/TrEMBL
15.	CSA	The Catalytic Site Atlas (CSA) is a resource of catalytic sites and residues identified in enzymes using structural data
16.	InterPro	The InterPro database is an integrated documentation resource for protein families, domains, and functional sites
17.	IPI	International Protein Index contains a number of non-redundant proteome sets of higher eukaryotic organisms constructed from UniProtKB/Swiss-Prot, UniProtKB/TrEMBL, Ensemb,l and RefSeq
18.	PANDIT	PANDIT—protein and associated nucleotide domains with Inferred Trees. PANDIT is a collection of multiple sequence alignments and phylogenetic trees covering many common protein domains.
19.	Patent Data Resources	Patent data resources at the EBI contain patent abstracts, patent chemical compounds, patent sequences, and patent equivalents. There are various ways of accessing and searching the patent data
20.	UniProt	The Universal Protein Resource for protein sequences and is the central hub for the collection of functional information on proteins, with accurate, consistent, and rich annotation, the amino acid sequence, protein name or description, taxonomic data, and citation information
21.	UniProt Archive	A non-redundant archive of protein sequences extracted from public databases and contains only protein sequences

(continued)

Table 8.2 (continued)

S.N.	Database	Description
22.	UniProt/UniRef	Features clustering of similar sequences to yield a representative subset of sequences. This produces very fast search times
23.	UniProtKB-GOA	Provides assignments of proteins in UniProtKB/Swiss-Prot, UniProtKB/TrEMBL, and IPI to the Gene Ontology resource
24.	UniProtKB/Swiss-Prot	An annotated protein sequence database. Part of the UniProtKB
25.	UniProtKB/TrEMBL	A computer generated protein database enriched with automated classification and annotation. Part of the UniProtKB
26.	UniProt/UniMES	A repository specifically developed for metagenomic and environmental data
27.	UniSave	The UniProtKB sequence/annotation version archive (UniSave) is a repository of UniProtKB/Swiss-Prot and UniProtKB/TrEMBL entry versions
28.	Phospho.ELM	Database of experimentally verified phosphorylation sites in eukaryotic proteins
29.	DMPC	Database of membrane protein Contacts
30.	BioGRID	Database of protein and genetic interactions
31.	PROSITE	Database of protein families and domains
32.	IPI	International protein index
33.	HPRD	Human protein reference database
34.	CDD	NCBI conserved domain database
35.	ConSurf-DB	Evolutionary conservation profiles of protein structures

adequately cover a range of potential drug compounds. Docking and scoring are implemented using known and hypothetical drug targets on a protein, coupled with the databases of virtual chemical compounds. After docking, scoring is carried out using mathematical models. These models determine the chemical binding strength and energy state of the drug-protein complex. Those hits with high ranking scores are subsequently subjected to *in vivo* tests; hits with positive scores in both areas are then known to be leads. Evaluation of docked and scored complexes are then made, selecting an arbitrary number of top hits to be further screened manually. The first two steps are done entirely *in silico*; however, the best complexes now need to be examined using software visualization, often in 3D setups. The compounds that make it through docking, scoring, and evaluation become drug leads, and are then passed on to undergo drug testing techniques in a wet lab, to ensure that only compounds with effects relatively unique to the target system and safe to the rest of organism are considered.

8.8.2 Protein Biomarkers

Biomarkers of drug efficacy and toxicity are becoming a key need in the drug development process. Mass spectral-based proteomic technologies are ideally suited for the discovery of protein biomarkers in the absence of any prior knowledge of quantitative changes in protein levels. The success of any biomarker discovery effort will depend upon the quality of samples analyzed, the ability to generate quantitative information on relative protein levels, and the ability to readily interpret the data generated. Some diseases which have protein biomarkers that show promise as a screening tool are breast cancer, Alzheimer's, leukemia, Amyotrophic lateral sclerosis, and Parkinson's. A series of six steps must be accomplished in order to successfully validate a biomarker or set of biomakers: discovery, qualification, verification, assay optimization, validation, and commercialization. Once a biomarker is found and accepted, it can be used to possibly predict and prevent the disease it is related to.

Table 8.3 Some protein databases available on the Internet used for performing specific functions

S.N.	Database	Description
Identification and characterization with peptide mass fingerprinting data		
1.	FindMod	Predict potential protein post-translational modifications and potential single amino acid substitutions in peptides. Experimentally measured peptide masses are compared with the theoretical peptides calculated from a specified Swiss-Prot entry or from a user-entered sequence, and mass differences are used to better characterize the protein of interest
2.	FindPept	Identify peptides that result from unspecific cleavage of proteins from their experimental masses, taking into account artifactual chemical modifications, post-translational modifications (PTM), and protease autolytic cleavage
3.	Mascot	Peptide mass fingerprint from Matrix Science Ltd., London
4.	PepMAPPER	Peptide mass fingerprinting tool from UMIST, UK
5.	ProFound	Search known protein sequences with peptide mass information from Rockefeller and NY Universities [or from Genomic Solutions]
Identification and characterization with MS/MS data		
6	QuickMod	Open modification spectral library search tool for identification of MS/MS data
7.	Phenyx	Protein and peptide identification/characterization from MS/MS data from GeneBio, Switzerland
8.	Mascot	Sequence query and MS/MS ion search from Matrix Science Ltd., London
9.	OMSSA	MS/MS peptide spectra identification by searching libraries of known protein sequences
10.	PepFrag	Search known protein sequences with peptide fragment mass information from Rockefeller and NY Universities
11.	ProteinProspector	UCSF tools for fragment-ion masses data (MS-Tag, MS-Seq, MS-Product, etc.)
Identification with isoelectric point, molecular weight, and/or amino acid composition		
12.	AACompIdent	Identify a protein by its amino acid composition
13.	AACompSim	Compare the amino acid composition of a UniProtKB/Swiss-Prot entry with all other entries
14.	TagIdent	Identify proteins with isoelectric point (pI), molecular weight (Mw) and sequence tag, or generate a list of proteins close to a given pI and Mw
15.	MultiIdent	Identify proteins with isoelectric point (pI), molecular weight (Mw), amino acid composition, sequence tag, and peptide mass fingerprinting data
Other prediction or characterization tools		
16.	ProtParam	Physico-chemical parameters of a protein sequence (amino acid and atomic compositions, isoelectric point, extinction coefficient, etc.)
17.	Compute pI/Mw	Compute the theoretical isoelectric point (pI) and molecular weight (Mw) from a UniProt Knowledgebase entry or for a user sequence
18.	PeptideCutter	Predicts potential protease and cleavage sites and sites cleaved by chemicals in a given protein sequence

(continued)

Table 8.3 (continued)

S.N.	Database	Description
19.	PeptideMass	Calculate masses of peptides and their post-translational modifications for a UniProtKB/Swiss-Prot or UniProtKB/TrEMBL entry or for a user sequence
20.	xComb	Computes all possible crosslinks based on proteins of interest for further interrogation using standard search engine
21.	xQuest	Search machine to identify cross-linked peptides from complex samples and large protein sequence databases
22.	SmileMS	Small molecule identification
Other proteomics tools-glycotools		
23.	GlycanMass	Calculate the mass of an oligosaccharide structure
24.	GlycoMod	Predict possible oligosaccharide structures that occur on proteins from their experimentally determined masses (can be used for free or derivatized oligosaccharides and for glycopeptides)
25.	GlycospectrumScan	an analytical tool independent of MS-platform that accurately identifies and assigns the oligosaccharide heterogeneity on glycopeptides from MS data of a mixture of peptides and glycopeptides
26.	Glycoviewer	a visualization tool for representing a set of glycan structures as a summary figure of all structural features using icons and colors recommended by the Consortium for Functional Glycomics (CFG)
Other tools for MS data (vizualization, quantitation, analysis, etc.)		
27.	HCD/CID spectra merger	a tool to merge the peptide sequence-ion m/z range from CID spectra and the reporter-ion m/z range from HCD spectra into the appropriate single file, to be further used in identification and quantification search engines
28.	MALDIPepQuant	Quantify MALDI peptides (SILAC) from Phenyx output
29.	MSight	Mass spectrometry imager
Other tools for 2-DE data (image analysis, data publishing, etc.)		
30.	ImageMaster/Melanie	Software for 2-D PAGE analysis
31.	Make2D-DB II	A package to build a web-based proteomics database
DNA → Protein		
32.	Translate	Translates a nucleotide sequence to a protein sequence
33.	Transeq	Nucleotide to protein translation from the EMBOSS package
34.	(Reverse)-Transcription and translation tool genewise	Compares a protein sequence to a genomic DNA sequence, allowing for introns and frameshifting errors
35.	Reverse translate	Translates a protein sequence back to a nucleotide sequence
36.	BCM search launcher	Six frame translation of nucleotide sequence(s)

8.8.3 Study of Tumor Metastasis

Tumor metastasis is the dominant cause of death in cancer patients. However, the molecular and cellular mechanisms underlying tumor metastasis are still elusive. The identification of protein molecules with their expressions correlated to the metastatic process help to understand the metastatic mechanisms and thus facilitate the development of strategies for the therapeutic interventions and clinical management of cancer. Proteomic technology has been widely used in biomarker discovery and pathogenetic studies including tumor metastasis.

8.8.4 Neurotrauma

Proteomics is a burgeoning field that may provide a valuable approach to evaluate the post-traumatic central nervous system (CNS) and the newer proteomic technologies that are available to most researchers in neurotrauma. Proteomics will likely be very useful for developing diagnostic predictors after CNS injury and for mapping changes in proteins after injury in order to identify new therapeutic targets. Neurotrauma results in complex alterations to the biological systems within the nervous system, and these changes evolve over time. Exploration of the "new nervous system" that follows injury will require methods that can both fully assess and simplify this complexity.

8.8.5 Renal Disease Diagnosis

In the diagnosis and treatment of kidney disease, a major priority is the identification of disease-associated biomarkers. Proteomics, with its high-throughput and unbiased approach to the analysis of variations in protein expression patterns (actual phenotypic expression of genetic variation), is most suitable for biomarker discovery. Combining such analytical techniques as 2D gel electrophoresis with MS, has enabled considerable progress to be made in cataloguing and quantifying proteins present in urine and various kidney tissue compartments in both normal and diseased physiological states.

8.8.6 Neurology

In neurology and neuroscience, many applications of proteomics have involved neurotoxicology and neurometabolism, as well as in the determination of specific proteomic aspects of individual brain areas and body fluids in neurodegeneration. Investigation of brain protein groups in neurodegeneration, such as enzymes, cytoskeleton proteins, chaperones, synaptosomal proteins, and antioxidant proteins, is in progress as phenotype-related proteomics. The concomitant detection of several hundred proteins on a gel provides sufficiently comprehensive data to determine a pathophysiological protein network and its peripheral representatives. An additional advantage of advanced proteomics technology is that unknown proteins have been identified as brain proteins.

8.8.7 Fetal and Maternal Medicine

The expression levels of many proteins strongly depend on complex, but well-balanced regulatory systems. The proteome, unlike the genome, is highly dynamic. This variation depends on the biological function of a cell, but also on signals from its environment. In (bio) medical research it has become increasingly apparent that cellular processes, in particular in disease, are determined by multiple proteins. Hence, it is important not to focus on one single gene product (one protein), but to study the complete set of gene products (the proteome). In this way the multifactorial relations underlying certain diseases may be unraveled potentially identifying therapeutical targets. For many diseases, characterization of the functional proteome is crucial for elucidating alterations in protein expression and modifications. Abnormalities in splicing or post-translational modifications can cause a disease process but they can also be a consequence. An example is that patients with diabetes have high blood glucose which glycosylates hundreds or even thousands of proteins, including HbA1c which is used to monitor diabetic control.

8.8.8 Urological Cancer Research

Proteomic analysis allows the comparison of the proteins present in a diseased tissue sample with the proteins present in a normal tissue sample. Any proteins, which have been altered either quantitatively or qualitatively between the normal and diseased sample are likely to be associated with the disease process. These proteins can be identified and may be useful as diagnostic markers for the early detection of the disease or prognostic markers to predict the outcome of the disease or they may be used as drug targets for the development of new therapeutic agents. There are potential future applications of proteomic analysis in urological cancer research.

8.8.9 Autoantibody Profiling for Study and Treatment of Autoimmune Disease

Proteomics technologies enable profiling of autoantibody responses using biological fluids derived from patients with autoimmune disease. They provide a powerful tool to characterize autoreactive B-cell responses in diseases including rheumatoid arthritis, multiple sclerosis, autoimmune diabetes, and systemic lupus erythematosus. Autoantibody profiling may serve purposes including classification of individual patients and subsets of patients based on their 'autoantibody fingerprint', examination of epitope spreading and antibody isotype usage, discovery and characterization of candidate autoantigens, and tailoring antigen-specific therapy. In future, proteomics technologies will broaden understanding of the underlying mechanisms of and will further our ability to diagnose, prognosticate, and treat autoimmune disease.

8.8.10 Cardiovascular Research

The development of proteomics is a timely one for cardiovascular research. Analyses at the organ, subcellular, and molecular levels have revealed dynamic, complex, and subtle intracellular processes associated with heart and vascular diseases. Establishment of species- and tissue-specific protein databases provides a foundation for subsequent proteomic studies. Evolution of proteomic techniques has permitted more thorough investigation into molecular mechanisms underlying cardiovascular disease, facilitating identification not only of modified proteins but also of the nature of their modification. Continued development should lead to functional proteomic studies, in which identification of protein modification, in conjunction with functional data from established biochemical and physiological methods, has the ability to further our understanding of the interplay between proteome change and cardiovascular disease.

8.8.11 Diabetes Research

Proteomics is the investigation of all the proteins and their various modifications making up a system, a cell, tissue, or organism. The techniques involved in proteomics allow the global screening of complex samples of proteins and provide qualitative and quantitative evidence of altered protein expression. This lends itself to the investigation of the molecular mechanisms underlying disease processes and the effects of treatment. This can thus be applied to diabetes research.

8.8.12 Nutrition Research

Proteomics holds great promise for discoveries in nutrition research, including profiles and characteristics of dietary and body proteins; digestion, absorption, and metabolism of nutrients; functions of nutrients and other dietary factors in growth, reproduction and health; biomarkers of the nutritional status and diseases; and individualized requirements of nutrients. The proteome analysis is expected to play an important role in solving major nutrition-associated problems in humans and animals, such as obesity, diabetes, cardiovascular disease, cancer, aging, and intrauterine fetal retardation.

8.8.13 Plant Physiology

Although significant advances in the comprehensive profiling, functional analysis, and regulation of proteins has occurred in model organisms such as yeast (*Saccharomyces cerevisiae*) and in humans, proteomics research in plants has not advanced at the same pace. The availability of the complete Arabidopsis (*Arabidopsis thaliana*) genome, which is small compared to that of other plants, along with an increasingly comprehensive catalog of protein-coding information from large-scale cDNA sequencing and transcript mapping experiments, set it as a complex model organism to study plant proteomics. The application of proteomic approaches to plants involves (1) comprehensive identification of proteins, their isoforms, and their prevalence in each tissue; (2) characterizing the biochemical and cellular functions of each protein; (3) the analysis of protein regulation and its relation to other regulatory networks.

8.9 Conclusion

The post-genomic era requires the complete understanding of the protein structure and function. Proteomics is the study of the set of proteins produced (expressed) by an organism, tissue, or cell and the changes in the protein expression patterns in different environments and conditions. It has been classified into structural, expression, and functional proteomics. A number of analytical techniques used in proteomics study are 2D gel electrophoresis, chromatography, MS, NMR, X-ray crystallography, ICAT, FRET, Yeast two-hybrid system, etc. The sequencing of the human genome has increased interest in proteomics because DNA sequence information provides only a view of the various ways in which the cell might use its proteins, whereas proteomics includes identification of proteins in biological tissues, characterization of their physicochemical properties, and description of their behavior. This new data set holds great promise for proteomic applications in science, medicine, and mainly pharmaceuticals.

References

Andersen JS, Svensson B, Roepstorff P (1996) Electrospray ionization and matrix assisted laser desorption ionization mass spectrometry: powerful analytical tools in recombinant protein chemistry. Nat Biotechnol 14:449–457

Binz PA, Gasteiger E, Sanchez J, Bairoch A, Hochstrasser DF, Appel, RD (2004) Proteome analysis. In: Lengauer Th (ed) Bioinformatics. from Genomes to Drugs. Wiley-VCH Verlag GmbH & Co. KGaA, Weinheim, FRG. ISBN:3-527-29988-2, pp 69–118

Cagney G, Amiri S, Premawaradena T, Lindo M, Emili A (2003). In silico proteome analysis to facilitate proteomics experiments using mass spectrometry. Proteome Science 1:5. http://www.proteomesci.com/content/1/1/5 (Open Access article)

Garret RH, Grisham CM (1995) Proteins: their biological functions and primary structure. In: Biochemistry Gareet RH, Grisham CM (eds) Saunders College Publishing, Harcourt Brace College Publishers, Philadelphia, pp 81–131

Graves PR, Haystead AJ (2002) Molecular biologists guide to proteomics. Microbiol Mol Biol Rev 66(1):39–63

Hillenkamp F, Karas M, Beavis RC, Chait BT (1991) Matrix-assisted laser desorption/ionization mass spectrometry of biopolymers. Anal Chem 63(24):1193A–1203A

MacBeath G, Schreiber S (2000) Printing proteins as microarrays for high-throughput function determination. Science 289(5485):1760–1764

Okerberg ES, Wu J, Zhang B, Sammi B, Blackford K, Winn DT, Shreder KR, Burbaum JJ, Palricelli MP (2005) High resolution functional proteomics by active site peptide profiling. PNAS 102(14):4996–5001

Schena M, Shalon D, Davis RW, Brown PO (1995) Quantitative monitoring of gene expression patterns with a complementary DNA microarray. Science 270(5235):467–470

Yates JR, Ruse CI, Nakorchevesky A (2009) Proteomics by mass spectrometry: approaches, advances and applications. Annu Rev Biomed Eng 11:49–79

9 Recent Advances in Stem Cell Research

Shweta Kulshreshtha and Pradeep Bhatnagar

Abstract

Stem cells are totipotent progenitor cells which are capable of self-renewal and differentiation into multilineage cell types. The stem cells divide unlimitedly; therefore these are ideal targets for *in vitro* manipulation. Researchers have been studying the biology of stem cells to find new ways of treating various diseases. They provide nearly limitless potential for medical applications due to having ability of producing multilineage cell types. Although early researches have focused on hematopoietic stem cells (HSCs), these have also been found to be present in various other tissues. This chapter focuses on the stem cell technology and the potential of stem cells in treating various diseases. Also, the current researches and the clinical status of treatments based on stem cells are discussed.

9.1 Introduction

In the new millennium, biological and biotechnological approaches have been extensively explored to find out "biological solutions to biological problems." With the extraordinary advances taking place in the field of cellular and molecular biology, we are on the verge of finding simple mechanical care to consider biological solutions to health promotion, risk assessment, diagnosis, treatment, and prognosis. It is only in the last decade that an information explosion occurred in the area of stem cell research. Stem cells are likely to revolutionize the entire health care delivery. This is a high time to familiarize ourselves with stem cells, their characteristics, potential applications, current research and therapy, and possible barriers of their applications.

Stem cells, found mostly in multicellular organisms, can be defined as *primitive, generic, and undifferentiated form of cells* which are located in fetus, embryo, and adult body and are characterized by their ability to renew themselves through mitotic cell division and differentiate into a diverse range of specialized cell types. They can also build special tissues in the

S. Kulshreshtha (✉)
Amity Institute of Biotechnology, Amity University Rajasthan, Jaipur, India
e-mail: shweta_kul17@rediffmail.com

P. Bhatnagar
Department of Life Sciences, The IIS University, Mansarovar, Jaipur, India
e-mail: pradeepbhatnagar1947@yahoo.com

I. Ravi et al. (eds.), *Advances in Biotechnology*,
DOI: 10.1007/978-81-322-1554-7_9, © Springer India 2014

body and can be grown *in vitro* for use in research or medicine.

Stem cells are important for living organisms for many reasons. Different types of stem cells such as embryonic, induced pluripotent, fetal, and adult are useful in understanding human development and diseases. Human development can be understood by determining how human embryonic stem (ES) cells give rise to 220 types of cells in the body. Before using human ES cells, mouse was used as research model for developmental studies. However, developmental studies conducted in mice cannot be useful in understanding human development due to the differences in events of early development. Using ES cells, scientists can witness the very first events in human development to understand how normal development unfolds and how the process may go awry and lead to pregnancy loss, birth defects, and other problems.

Stem cells are also helpful in developing novel therapies and screening new therapeutic compounds. They offer enormous promise to get rid of diseases that rob the joy of life and even life. Besides, these cells provide a powerful source of cells for basic experimental studies that can be done in a laboratory, e.g., studying and characterizing the particular genes and specialized cell types.

9.2 Historical Developments

In 1981, a British research team succeeded for the first time in isolating and culturing cells from early mouse embryos and also to keep them in undifferentiated form. Embryonic cells, isolated from a brown mouse, were introduced into 4 to 7-days-old embryo of a white mouse and it was found that offsprings were born with a white coat with brown tufts, which were called as *chimeras*. This proved that cells of all organs originated from the injected embryonic cells which have totipotency. These cells can also be induced to differentiate into many different cell types in *in vitro* tissue culture and provide potentially unlimited source of raw material for damaged tissue. They act as an ideal model system to investigate the factors influencing the differentiation of cells in embryo. In 1998, a method was discovered to derive stem cells from human embryos (human embryonic stem cells—hESCs) and grow the cells in the laboratory. These embryos were produced for reproductive purposes through *in vitro* fertilization (IVF) procedures. Thus, pluripotent stem cells were isolated from the murine embryo in 1981 and from human embryo in 1998. Now, these pluripotent embryonic stem cells can be cultured beyond the Hayflick limit without changing their phenotype and genotype and can be used as an infinite source for *in vitro* production of all cell types of our body or for the generation of chimeric animals.

Another breakthrough came in 2006 when conditions for genetically reprogramming of some specialized adult cells and converting them into a stem cell-like state were identified by researchers. This new type of stem cell was named as induced pluripotent stem cells (iPSCs). In 2007, Mario R. Capecchi, Martin J. Evans, and Oliver Smithies were awarded the *Noble Prize in Medicine/Physiology* for their discovery regarding the principles in introducing specific gene modifications in mice by the use of embryonic stem cells. The collaboration of these scientists initially aimed to repair defects in the 'hprt gene'. They identified and selected cells that have undergone homologous recombination, thereby eliminating the defective gene. These cells were then implanted into a surrogate mother who gave birth to a knockout mice strain that was homozygous to the inert gene. In this way, these scientists developed a method to replace defected gene by homologous recombination and ensure heritability of the genetic material. After this major discovery the stem cell research got the new direction.

9.3 Properties

All stem cells regardless of their source have some general properties that help in their identification among the population of normal cells. These major properties include their capability of dividing and renewing themselves for long periods, their unspecialized nature and their ability to give rise to specialized cell types.

9.3.1 Stem Cells are Unspecialized

The fundamental characteristic of a stem cell is its unspecialized structure without any tissue specificity. Therefore, stem cells cannot perform specialized functions. However, unspecialized stem cells can be converted into specialized cells, including heart muscle cells, blood cells, or nerve cells that can perform special functions. All stem cells are characteristically of the same family type (lineage).

A starting population of stem cells that proliferates for many months in the laboratory can yield millions of cells. If the resulting cells continue to be unspecialized, like the parent stem cells, the cells are said to be capable of long-term self-renewal. The exact controlling factors and molecular events that allow stem cells to be unspecialized are still mystery to researchers.

It has taken scientists many years of trial and error to learn to grow stem cells in the laboratory without allowing them to differentiate into specific cell types. Therefore, an important area of research is focused on understanding the signals in a mature organism that causes a stem cell population to proliferate and remain unspecialized until the cells are needed for repair of a specific tissue. The internal signals are controlled by genes of cells, which are interspersed across long strands of DNA, and carry coded instructions for all the structures and functions of a cell. The process of differentiation of stem cells is complex and involves precise signals and the regulation of gene expression.

9.3.2 Ability to Self-Renew

Stem cells have a special ability to *divide* and *renew* themselves for extended periods of time. Stem cells repeatedly proliferate and produce copies of them. Unlike mature cells, which are permanently committed to their fate, stem cells can both renew themselves as well as create new cells of whatever tissue they belong to (and other tissues). An initial population of stem cells can produce millions of cells within months in a laboratory setting.

Embryonic stem cells are able to proliferate for a year or even longer without ever developing into specialized cells in contrast to adult cells. Researchers are contemplating on the different aspects that regulate stem cell renewal and the cells' proliferation and differentiation. Bone marrow stem cells, for example, are the most primitive cells in the marrow and from them various types of blood cells are descended. Bone marrow stem cell transfusions (or transplants) were originally given to replace various types of blood cells.

To ensure self-renewal, stem cells undergo two types of cell division. Symmetric division gives rise to two identical daughter cells, both endowed with stem cell properties, whereas asymmetric division produces only one stem cell and a progenitor cell with limited self-renewal potential (Fig. 9.1). Progenitors can go through several rounds of cell division before finally differentiating into a mature cell. It is believed that the molecular distinction between symmetric and asymmetric divisions lies in differential segregation of cell membrane proteins (such as receptors) between the daughter cells.

9.3.3 Ability to Give Rise to Specialized Cell

The ability of stem cells to give rise to specialized cells is a crucial one. In this process of differentiation, unspecialized stem cells produce specialized cells. It is believed that genes regulate the internal signals that trigger this process. These genes carry the specific code or instructions for all the parts and functions of a cell. External signals are those outside of the cell, which include chemicals released from other cells, physical connections with nearby cells,

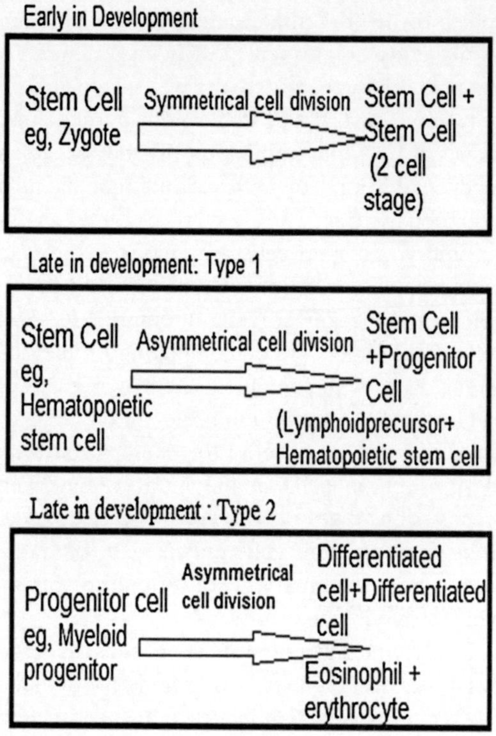

Fig. 9.1 Symmetric and asymmetric divisions (modified from http://stemcells.nih.gov/info/scireport/chapter1.asp)

and various other molecules in the surrounding area. Currently, scientists are searching for similarities and differences between the signals from one stem cell to another. Ultimately, answers will allow researchers to find ways to control stem cell differentiation. With control, scientists can more easily grow cells and tissues for specific functions.

9.3.4 Plasticity of Stem Cells

Plasticity is an important property of stem cells which allows these cells to be committed to generating a fixed range of progeny. When they have been relocated, to make other specialized sets of cells appropriate to their new niche, they can switch. The evidence for stem cell plasticity in rodents and man was explained by Poulsom et al. (2002).

Adult stem cells are surprisingly flexible in their differentiation repertoires due to the existence of different pathways. Adult stem cells typically generate the cell types of the tissue in which they reside. A blood-forming adult stem cell in the bone marrow, for example, normally gives rise to the many types of blood cells such as red blood cells, white blood cells, and platelets. Until recently, it had been thought that a blood-forming cell in the bone marrow (hematopoietic stem cell) could not give rise to the cells of a very different tissue, such as nerve cells in the brain. However, a number of experiments over the last several years have raised the possibility that stem cells from one tissue may give rise to cell types of a completely different tissue. This phenomenon is known as plasticity.

9.4 Classification

Stem cells can be classified according to their plasticity. Different types of stem cells vary in their degree of plasticity or developmental versatility. The categories into which they fall include: the totipotent stem cell, pluripotent stem cell, multipotent stem cell, and oligopotent and nullipotent stem cells (Fig. 9.2).

9.4.1 Totipotent Stem Cells

"Toti" comes from the Latin word meaning "whole" or "Total", so totipotent stem cells have total potential to develop into any cell type or organ/tissues including extra embryonic tissues. Totipotent stem cells develop during sexual reproduction when male and female gametes fuse during fertilization to form a zygote. The zygote is totipotent because its cells can become any type of cell and they have limitless replicative abilities. After fertilization, zygote divides into identical totipotent cells, which can later develop into any of the three germ layers of a human, i.e., endoderm, mesoderm, or ectoderm. This stage is

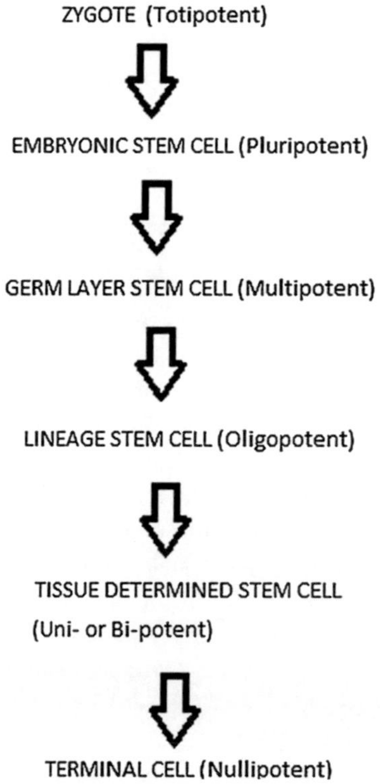

Fig. 9.2 Hierarchy of stem cells during differentiation showing that differential potential decreases and specialization increases at each stage of development

called morula stage (16-cell stage); the totipotent cells of morula start shifting toward pluripotent stem cells called blastocyst.

9.4.2 Pluripotent Stem Cells

Pluripotent stem cells are those cells which are able to form all the different types of cells in the body. Specialization in pluripotent stem cells is minimal, and therefore they can develop into almost any type of cell. Pluripotent cells show the formation of tumor-like mass called *teratoma*. Teratoma formation is a hallmark of true pluripotency. It results from the aggressive proliferation and differentiation of transplanted stem cells.

The best example of pluripotent stem cell is embryonic stem cell (ESC). ESCs seem to be more flexible compared to adult stem cells, because they have the potential to produce every cell type in the human body. ESCs are derived from inner cell mass, called blastocyst, of 5–6-days-old embryo. After the fertilization of an egg by a sperm, blastocyst which is a tiny and hollow sphere of 30–34 cells develops generally in 5 days that has potential to develop into a complete human being. In normal development, the blastocyst implants in the wall of the uterus to become embryo and then develops into a mature organism. Outer layer of cells initiates the formation of the placenta and the inner cell mass starts differentiating into the progressively more specialized cells of the body. During the early stages of development, ESCs present in specific part of embryo may develop into eye, blood, muscle, nerve, liver cells, etc. When ESCs which may become liver, are isolated and transferred to another section of the embryo, can be developed into eye or blood. This shows that differentiation of stem cell depends on their position in embryo.

ESCs can be induced to replicate in an undifferentiated state for very long periods of time and then produce specialized cells only after differentiation, which can be used for therapeutic and transplantation purposes. However, undifferentiated ESCs can form a teratoma and therefore, could not be used directly for tissue transplants. In an experiment, hESCs have been successfully differentiated into tissues derived from the three germ layers (Cai et al. 2007). This proves that hESCs can differentiate into a broad spectrum of cell types in culture.

The differentiation of hESCs is efficient and has advantages and disadvantages. The main advantages of hESCs are that these cells can develop into most of the cells/tissues of the body and are easy to cultivate in laboratory. These cells have great potential for developing future therapies to cure diseases. The major disadvantage of hESCs is that blastocyst must be destroyed when cells are removed. Moreover, egg donation is an important and serious issue which creates ethical problems.

Recently, iPSCs are developed which are genetically altered adult stem cells. These cells are induced or prompted in a laboratory to take the characteristics of ESCs. Although iPS cells behave like ESCs and express some of the same genes that are expressed normally in ESCs, they are not exact duplicates of ESCs.

9.4.3 Multipotent Stem Cells

Multipotent stem cells have the ability to differentiate into a limited number of specialized cell types. Multipotent stem cells typically develop into any cell of a particular group or type. For example, bone marrow stem cells can produce any type of blood cell. However, bone marrow cells do not produce heart cells. Mesenchymal stem cells (MSCs) are multipotent cells of bone marrow that have the ability to differentiate into several types of specialized cells related to them. These stem cells give rise to cells that form specialized connective tissues, as well as cells that support the formation of blood. Fetal stem cells, adult stem cells, and umbilical cord stem cells are examples of multipotent cells.

Fetal stem cell is the main site of hematopoiesis and is rich source of hematopoietic stem cells (HSCs) that leads to the generation of multiple cell types in blood. Tissues extracted from the fetal pancreas, when transplanted into diabetic mice, are reported to stimulate insulin production. The possible reason of this is due to a true stem cell, a more mature progenitor cell, or to the presence of fully mature insulin-producing pancreatic islet cells themselves.

Primordial germ cells are multipotent cells that have been isolated from the gonadal ridge, a structure that arises at an early stage of the fetus that will eventually develop into eggs or sperms in the adult. Germ cells can be cultured *in vivo* and have been shown to form multiple cell types of the three embryonic tissue layers.

Adult stem cells or somatic stem cells are undifferentiated cells, found throughout the body after embryonic development. These cells multiply by cell division to replenish dying cells and regenerate damaged tissues. Adult stem cells are present in the depth of organs along with millions of ordinary cells and restock some of the body's cells. These stem cells are tissue specific in their location and are derived from adult tissues such as bone marrow, blood, eye, brain, skeletal muscle, dental pulp, liver, skin, and the lining of the gastrointestinal tract and pancreas. Usually, they have very limited power of division due to which these can develop into the cells of only one particular tissue, therefore considered as multipotent stem cells. In some tissues, these cells sustain turnover and repair throughout life. For example, stem cells of skin will give rise to new skin cells and those obtained from the brain can differentiate into blood cells and muscle tissue.

Adult stem cells are not only difficult to identify and purify, but also difficult to maintain in the undifferentiated state during growth in culture medium. Culturing of adult stems cells in a definitive medium is a high priority of stem cell research. The one important feature of stem cell is that these stem cells can be manipulated easily and effectively.

An adult stem cell is a completely committed stem cell. This concept is one of the exciting discoveries of last millennium. There is now evidence that some of blood stem cells apparently are committed adult stem cells that are able to become a stem cell in a different organ. Committed blood forming stem cell can produce different cells performing different functions. There are experiments performed for bone marrow transplantation in rats with damaged livers which help the liver to re-grow partially with bone marrow-derived stem cells. Bone marrow-derived stem cells in muscle connective tissue, satellite cells in adults, and stem cell marker expressing cells located in muscle connective tissue can be derived from bone marrow in adulthood (Dreyfus et al. 2004).

9.4.4 Oligopotent Stem Cells

These stem cells have the ability to differentiate into just a few types of cells. A lymphoid stem cell is an example of an oligopotent stem cell. This type of stem cell cannot develop into any type of blood cell as bone marrow stem cells can do. They only give rise to blood cells of the lymphatic system such as *T* and *B* cells.

9.4.5 Unipotent Stem Cells

These stem cells have unlimited reproductive capabilities, but can only differentiate into a single type of cell or tissue. Unipotent stem cells are derived from multipotent stem cells and formed in adult tissue. Skin cells are one of the most prolific examples of unipotent stem cells. These cells must readily undergo cell division to replace damaged cells.

9.5 Stem Cell Technology

A stem cell has the potential to develop into all kinds of body cells. This versatility of stem cells can be used as an important tool to solve many kinds of problems in biology and medicine and treatment of numerous diseases to save lives. With further research, scientists will hopefully be able to understand more about the intriguing properties of stem cells and can then find ways to isolate, cultivate, and manipulate stem cells for therapeutic value. This process is called *stem cell technology* (Fig. 9.3). Researches on stem cells provide advance knowledge about the development of organism from a single cell and replacement of damaged cells with healthy cells in adult organisms.

9.5.1 Sources of Stem Cell

Stem cells can be isolated from different sources. These can be efficiently obtained from immature tissues possessing undifferentiated cells such as embryonic cells (Fig. 9.4). Different populations of stem cells exist in embryo and are arranged hierarchically. This arrangement makes the isolation of stem cells easy, different ESCs can thus be isolated from different stages of developing organism such as morula, blastocyst, placenta, amniotic fluid, organ primordia, and umbilical cord blood. The stem cells are also present in the adult organisms revealing the nondisappearing of adult cells after birth. However, these cells dilute among differentiated cells and also change their properties in response to the change of niche and new demands of the organism. Tissue-specific stem cells can be harvested from different tissues and at different time-points: from children tissues, adult tissues, tissues of elderly people, and even postmortem.

9.5.1.1 Embryonic Stem Cell Sources

(a) *In vitro* **fertilization**

The main potential source of blastocysts for stem cell research is IVF clinics. On administrating fertility drugs, stimulated ovaries start producing multiple mature eggs. These eggs are retrieved by surgical procedure and all donated eggs are fertilized by sperms in order to increase their chance of producing a viable *blastocyst* that can be used for implantation in the womb. However, all blastocysts are not implanted; remaining can be stored in freezers which can be a major source of ESCs for use in medical research.

IVF facilitates the formation of blastocysts by fusion of sperm (specific genetic traits) with eggs of donor. This will help us in the study of particular diseases. However, the creation of stem cells/blastocysts specifically for research using IVF is ethically problematic for some people, because it involves intentionally creating a blastocyst that will never develop into a human being.

(b) **Nuclear transfer**

Another potential way to produce ESCs is nuclear transfer. Nuclear transfer can be done by inserting the nucleus of an already differentiated adult cell into a donated enucleated egg. Due to insertion of nucleus, enucleated egg becomes nucleated which is then stimulated to form a blastocyst. This blastocyst is a source of ESC which are copies or clones of adult cell due to the presence of nucleus of adult stem cell. The ESCs created by nuclear transfer are genetically matched to the person needing a transplant,

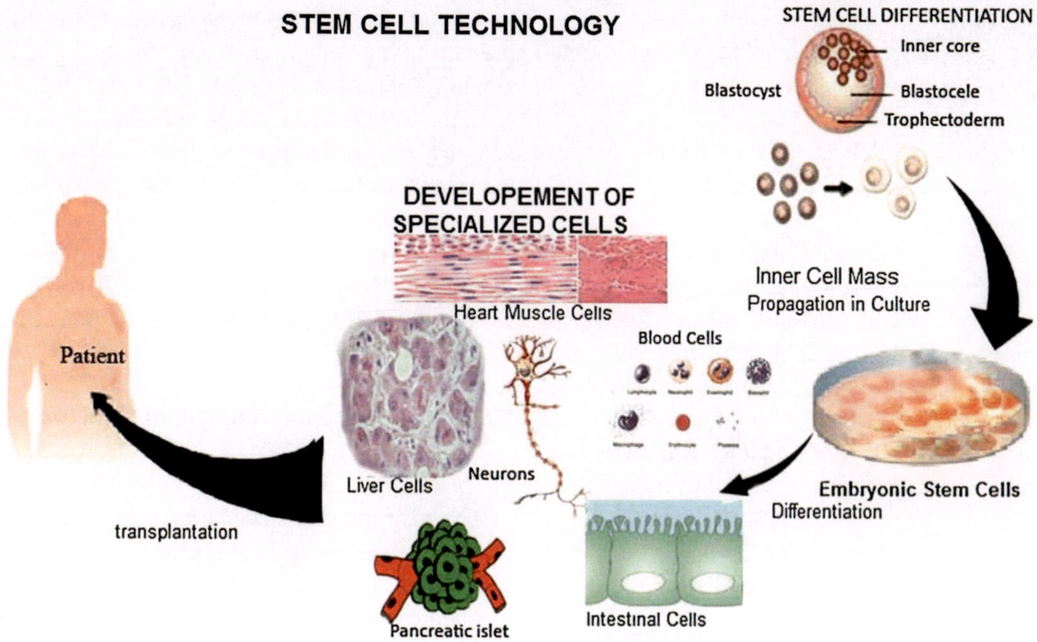

Fig. 9.3 Stem cell technology for replacing the damaged cells of human body (modified from http://www.stemcellsforhope.com)

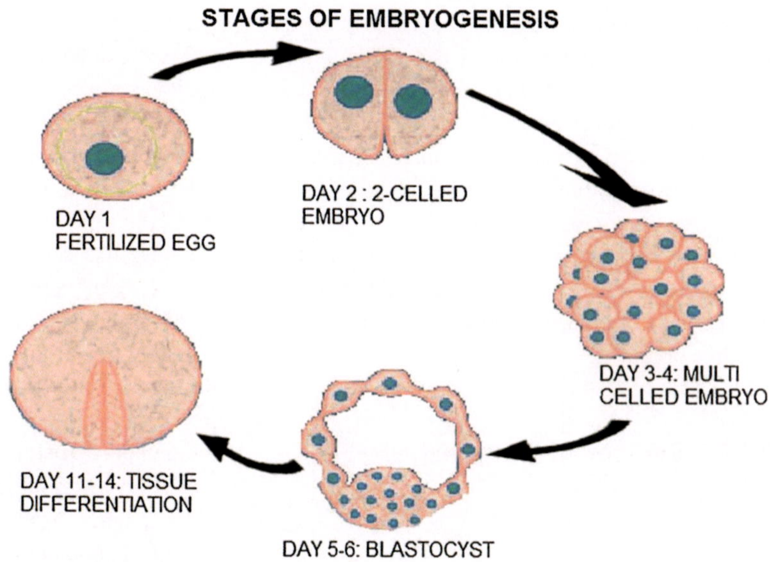

Fig. 9.4 Different stages of embryo development as a possible source of stem cells (modified from http://www.godandscience.org/slideshow/stem006.html)

therefore shows the less chance of rejection by immune system.

There is a misconception that nuclear transfer technique is similar to reproductive cloning. The use of nuclear transfer to develop disease-specific stem cells can be called research cloning, and the use of this technique for tissue transplants is called therapeutic cloning. These terms are distinguished from reproductive cloning in which the intent is to implant a cloned embryo in a female's womb and allow it to develop fully into an individual. This was the technique by which

"Dolly" the sheep was made and is now widely used for reproductive cloning in animals. In humans, however, reproductive cloning has been discouraged by most of researchers.

Unfortunately, nuclear transfer has not been successful in the production of hESCs. Scientists may use this technique to develop human stem cells in the future for studying the development and progression of specific diseases. This method may be useful in creating the stem cells that contain only DNA of a patient suffering from a specific disease.

(c) **Cord blood and aborted fetus**

Recent researches have shown that ESCs can be obtained from cord blood which may increase chances of the acceptance of stem cell research. Besides cord blood, stem cells can be harvested from aborted fetus.

(d) **Amniotic fluid**

In the womb, fetus is bathed with amniotic fluid which contains fetal cells including MSCs possessing ability to make a variety of tissues. Many pregnant women elect to have amniotic fluid drawn to test for chromosome defects, the procedure known as amniocentesis. This fluid is normally discarded after testing, but Dr Dario Fauza, a surgeon at Children's Hospital, Boston and a member of the stem cell program, have been investigating the idea of isolating MSCs and using them to grow new tissues for babies who have birth defects detected while they are still in the womb, such as congenital diaphragmatic hernia. These tissues would match the baby genetically, hence face less chances of rejection by the immune system, and can be easily implanted either in utero or after the baby is born.

9.5.1.2 Adult Stem Cell Sources

1. **Manipulating differentiated cells**

Specialized adult human cells can be reprogrammed genetically by introducing specific genes to create iPSCs (stem cell-like state). These iPSCs can be differentiated to any cell type.

2. **Somatic cells**

Stem cells are thought to reside as quiescent (nondividing) cells in a specific area of each tissue called a "stem cell niche". These cells are also called somatic stem cells and begin to divide and create new cells only when they are activated by tissue injury, disease, or anything else that makes the body need more cells. Many tissues have been reported to possess somatic stem cells such as bone marrow, blood vessels, and skeletal muscles.

3. **Other adult stem cell sources**

Many researchers have reported the presence of certain kind of adult stem cells which can transform, or differentiate, into apparently unrelated cell types. This phenomenon is called transdifferentiation and has been reported in some animals. Differentiation of brain stem cells into blood cells and blood forming cells into cardiac muscle cells is the best example of transdifferentiation process. Many questions are unanswered with this process such as how versatile they are, whether this is present in human stem cells, can these be synthesized reliably in laboratory. Stem cells are also present in tumors which are responsible for the growth of cancers. Hence, stem cells can be isolated from tumor which is removed from the body by surgical procedure.

Unlike ESCs, the use of adult stem cells in research and therapy is not controversial, because the production of adult stem cells does not require the creation or destruction of an embryo. However, the demand for autologous, patient-specific stem cells for regenerative therapies outstrip their supply. Several types of adult stem cells have been described which are given below:

(a) **Hematopoietic stem cells**

Blood stem cells have the ability to regenerate all types of blood cells; therefore can recreate an entire new blood system. Stem cells derived from bone marrow, when transplanted by intravenous injection to patients produce an entirely new blood system which helps the patients in recovering from cancer. There is a room for research that how and why stem cells find their way to the bone marrow site and regenerate a new blood system.

Recently, clinicians and scientists have focused on CXCR4 as the protein that provides the "navigation system" to allow human blood stem cells to find their way to the bone marrow and regenerate blood. The positive expression for CXCR4 has been widely hailed as a definitive marker for blood stem cells. Two distinct subgroups of blood stem cells called CXCR4-positive and CXCR4-negative have been identified. Their investigations showed that both subgroups possess the same blood cell regeneration capabilities but show other unique functional and physiological features including how they respond to growth factors. The two subgroups also preferentially differentiate into different types of blood cells: CXCR4-positive to disease fighting immune cells and CXCR4-negative to red blood cells (http://www.robarts.ca/discovery-new-stem-cell-properties-advance).

HSCs are very rare, hence, their isolation and tracking in patients is difficult. Blood stem cells are present in bone marrow stromal stem cells (BMSCs) that have become a standard in the field of adult stem cell biology and in regenerative medicine due to their high differentiation potentials and low morbidity during harvesting. Nevertheless, harvesting of BMSCs by bone marrow aspiration is a painful procedure and the number of cells acquired is usually low.

(b) **Mammary stem cells (MaSCs)**

MaSCs provide cells for growth of the mammary gland during puberty and gestation and play an important role in carcinogenesis of the breast. The mammary gland is a dynamic organ undergoing significant developmental changes during puberty, pregnancy, lactation, and involution. Numerous studies have provided strong evidence for the existence of MaSCs based on the fact that the mammary gland can be regenerated by transplantation of epithelial fragments in mice. Shackleton et al. (2006) isolated discrete populations of mouse mammary cells on the basis of cell-surface markers and identified a subpopulation (Lin-CD29hiCD24+) that is highly enriched for MaSCs. MaSCs have been isolated from human- and mouse- tissue as well as from cell lines derived from the mammary gland.

These stem cells capable of self-renewal and differentiation into the basal and luminal lineages comprise the functional mammary epithelium and have the ability to regenerate the entire organ in mice. The self-renewing capacity of these cells was demonstrated by serial transplantation of clonal outgrowths.

Recently in 2010, it was reported that MaSCs are highly responsive to steroid hormone signaling. However, they lack estrogen and progesterone receptors. Ovariectomy markedly diminished MaSC number and outgrowth potential *in vivo*, whereas MaSC activity increased in mice treated with estrogen plus progesterone (Labat et al. 2010).

(c) **Mesenchymal stem cell**

Mesenchymal stem cells (MSCs) are of stromal origin and may differentiate into a variety of tissues/cells including chondrocytes, which produce cartilage. MSCs have been isolated from placenta, adipose tissue, lung, bone marrow and blood, Wharton's jelly from the umbilical cord and teeth (perivascular niche of dental pulp and periodontal ligament).

MSCs cultured on regular flat surfaces such as petridishes, rapidly lose their multipotent properties which render them useless for stem cell experiments or clinical use. This problem is solved by using nanopatterned surfaces for the culturing of MSCs which helps in maintaining their phenotypic characteristics such as multipotency over long culture periods (McMurray et al. 2011). Generally, MSCs cultured with osteogenic media have very great efficiency of producing bone mineral *in vitro*. Similar efficiency can be achieved by the use of nanoscale disorder to stimulate human MSCs even in the absence of osteogenic supplements. However, it was reported that topographically treated MSCs have a distinct differentiation profile compared with those treated with osteogenic media, which has implications for cell therapies. This shows that stem cell differentiation to become osteoblasts occurs in the absence of chemical treatment without compromising material properties.

MSCs are involved in osteoblast differentiation. Metabolic acidosis can alter osteoblast differentiation from MSCs through its various effects on osteoblastic genes. Furthermore, osteoblastic genes produce different proteins which result in an impairment of bone formation.

Defect in MSCs can be corrected by genetic modification. After genetic modification, MSCs can be maintained in healthy and undifferentiated state. Furthermore, the genetically modified MSCs are able to engraft into many tissues of unconditioned transgenic mice making them an attractive therapeutic tool in a wide range of clinical applications. Chronic MSCs are also attractive for clinical therapy due to their ability to differentiate, provide trophic support, and modulate innate immune response. Researchers from Boston University's School of Dental Medicine proved that bone marrow-derived MSCs can relieve orofacial pain in rats within one day of treatment by either intravenous injection or direct injection of cells to the injured area.

(d) **Endothelial progenitor cells (EPCs)**

Endothelial progenitor cell is a type of *multipotent adult stem cell* found in the bone marrow. These cells are found to line blood vessels and originated from the hESCs. The differentiation of hESCs into endothelial cells avoids the formation of an embryoid-body intermediate.

Normal bone marrow sinusoidal endothelium, umbilical cord endothelium, and human telomerase reverse transcriptase (hTERT)-immortalized endothelial cell lines can support long-term proliferation and differentiation of human stem cells. On transducing these cells with adenovectors expressing various hematopoietic growth factors (thrombopoietin; erythropoietin; granulocyte–macrophage colony-stimulating factor, GM-CSF; and c-kit Ligand, flt3/flk-2 ligand), lineage appropriated expansion of mature hematopoietic cells (red cells, megakaryocytes, and neutrophils), as well as prolonged expansion of pluripotent stem cells is observed. These transduced endothelial cells are used by researchers for support of stem cell expansion on a clinical scale in a bioreactor system (http://www.mskcc.org/research/lab/malcolm-moore/lab/endothelial-cell-biology).

Bone marrow-derived endothelial stem cells (angioblasts) also exist in the circulatory system. In a study, at first dogs had been transplanted with allogeneic marrow and further grafted with artificial arteries which rapidly became lined with endothelium that was of the marrow donor genotype. This formation of new blood vessels is known as angiogenesis.

A rare population of cells comprising <1 % of the total CD34 positive population of human bone marrow, mobilized blood, or umbilical cord blood were found to express the vascular endothelial growth factor (VEGF) Receptor-2 (VEGFR2), CD34, and the HSCs antigen AC133 (http://www.mskcc.org/research/lab/malcolm-moore/lab/endothelial-cell-biology). Purified CD34 possessing hematopoietic progenitor cells, isolated from adults, can differentiate ex vivo to an endothelial phenotype. These cells were called as EPCs which showed expression of various endothelial markers, and incorporated into neovessels at sites of ischemia. Hence, EPCs are positive for both hematopoietic stem cell markers (CD34) and an endothelial marker protein (VEGFR2). However, CD34 is not an exclusive marker of HSCs and found to be present also on mature endothelial cells. In later studies, CD133, also known as prominin, was found to be suitable marker for immature hematopoietic stem cell which helps stem cells to differentiate in endothelial cells *in vitro* (Urbich and Dimmeler 2004). CD133 or AC133 is a highly conserved antigen with unknown biological activity. It is not expressed on mature endothelial cells and monocytic cells but present on HSCs.

(e) **Neural stem cells**

Neural stem cells exist not only in the developing mammalian nervous system but also in the adult nervous system of all mammalian organisms, including humans. The number of viable locations for neural stem cell is limited in the adult. Neural stem cells can also be derived from primitive ESCs. These are thought to be existed in the adult brain. The presence of stem cells in the mature primate brain was first reported in 1967. The clue for this has been found by the process of neurogenesis, the birth of new neurons that continues to adulthood in rats. Normally, adult neurogenesis is restricted to two areas of the brain—the subventricular zone, which lines the lateral ventricles, and the dentate gyrus of the hippocampus. However, the presence of true self-renewing stem cells has been debated. In ischemia, neurogenesis can be induced in other parts of brain such as neocortex following tissue damage. It has since been shown that new neurons are generated in adult mice, songbirds and primates, and including humans.

In vitro cultivation of neural stem cells leads to the formation of *neurospheres* which are floating heterogeneous aggregates of cells, containing a large proportion of stem cells. Neural cells can be propagated for extended periods of time and differentiated into both neuronal and glial cells and hence these are considered as stem cells. This behavior of neural cell is induced by the culture conditions in progenitor cells. The progeny of stem cell division normally undergoes a strictly limited number of replication cycles *in vivo*. Furthermore, neurosphere-derived cells do not behave as stem cells when transplanted back into the brain. Neurosphere-derived cells can differentiate into various cell types of the immune system if these are injected in blood. Potential uses of these stem cells in repair include transplantation to repair missing cells and the activation of endogenous cells to provide "self-repair."

(f) **Olfactory adult stem cells**

Olfactory stem cells have been successfully harvested from the human olfactory mucosa cells of the nose. These cells are also accessible in humans from nasal biopsies. If the appropriate chemical environment is provided to these cells, they are able to act as multipotent type of stem cells which can develop into many different cell types. Hence, they are easily grown in *in vitro* conditions and can be expanded to large numbers and stored frozen.

Olfactory stem cells hold the potential for therapeutic applications because these can be harvested with ease without harm to the patient. This means they can be easily obtained from all individuals, including older patients who might be most in need of stem cell therapies. The olfactory neuroepithelium undergoes continual neurogenesis and after extensive lesions or damage, fully regenerates to maintain sensory function. The olfactory epithelium is one of the few sites of adult neurogenesis, but the identity of its stem cell has been debated. Pathogens and other noxious substances constantly bombard the primary olfactory receptor neurons in the nose which kills them and results in the loss of the sense of smell. After extensive injuries resident neuronal precursors are depleted. This results in the loss of function of olfactory epithelium, if these neurons are not replaced. At this moment, horizontal basal cells (HBCs) transiently proliferate and their progeny fully reconstitutes the neuroepithelium. The pool of progenitor stem cells, that resides at the base of tissues, act as source of new neurons and can be used for the treatment of diseases.

(g) **Testicular cells**

Scientists at the School of Medicine, Stanford University and at University of California, San Francisco have succeeded in isolating stem cells from human testes. Previously, it was considered as these cells show resemblance to ESCs and they can differentiate into each of the three main types of tissues of the body (Science Daily 2009). However, it is proved that the testes stem cells are not pluripotent and have different patterns of gene expression and regulation. Firstly, due to the expression of genes differ from ESCs, testicular stem cells expressed many, but not all, genes associated with pluripotency. Detailed analysis of testicular stem cells reveals the different pattern of methylation. Testicular stem cells show methylation at different region compared to ESCs which leads to the modification to DNA that affects gene expression. Secondly, they do not proliferate and differentiate as aggressively as hESCs. Finally, testicular stem cells have limited ability to form *teratoma* when injecting into immune-compromised mouse. These cells differ in gene expression, methylation, and in their ability to form teratomas. Together, these results suggest that the stem cells isolated from male testes have some, but not all the characteristics of true pluripotent cells. Hence, these testicular stem cells are called as multipotent stem cells.

These multipotent testicular stem cells have been derived from spermatogonial progenitor cells found in the testicles of laboratory mice. Similar types of cells were isolated from the testicles of humans by the researchers of Germany and the United Kingdom. The extracted stem cells are known as human adult germline stem cells (Waters 2008).

(h) **Umbilical cord blood stem cells**

Recently, scientists have identified stem cells in umbilical cord blood and the placenta of newborn baby that are multipotent in nature. These cells can give rise to the various types of blood cells but not able to differentiate into all cell types of the body. These stem cells are found to be genetically similar to the cells of newborn. Researchers at the University of Minnesota recently found that they were able to reverse the effects of stroke in lab rats using stem cells found in human umbilical cord blood (http://news.nationalgeographic.com/news/2006/04/0406_060406_cord_blood.html). The transplanted stem cells took on properties of brain cells and seemed to be able to establish connection with brain of rat. These experiments show the possibility of using these cells in the future for therapeutic purpose. However, extensive research is required in this field. Therefore, umbilical cord blood is often banked, or stored for future use.

(i) **Adipose stem cells**

Adipose tissue is an attractive alternative source of stem cells. Stem cells can be collected in large quantities from adipose tissue fragments. Human adipose stem cells (ASCs) have therapeutic applicability in pre-clinical studies in diverse fields, due to their ability to readily expand and their capacity to undergo adipogenic, osteogenic, chondrogenic, neurogenic, and myogenic differentiation *in vitro*. Furthermore, ASCs have been shown to be immune privileged and appear to be more genetically stable in long-term culture compared to BMSCs. The safety and efficacy of ASCs for tissue regeneration or reconstruction is currently under assessment in clinical trials.

(j) **Lung stem cells**

Lung stem cells include basal cells and mucus-secreting cells in the trachea, Clara cells in the bronchioles, and type II pneumocytes in the alveoli. Mesenchymal cells predominate in fetal lung development. In postnatal life, the Clara cell is the most actively dividing cell in the tracheobronchial epithelium while type II pneumocytes are the progenitor cells for the alveolar epithelium. Under specific circumstances, basal cells and pulmonary neuroendocrine cells may also proliferate and act as stem cells.

(k) **Induced pluripotent stem cells**

These are adult cells that have been genetically reprogrammed and converted to an ESC-like state which can differentiate and form multiple fetal cell types due to force to express genes and factors important for maintaining the defining properties of ESCs. In animal studies, iPSCs have been shown to possess characteristics of pluripotent stem cells.

iPS and ESCs are very similar. Both can be used to grow nearly any cell type in the lab under precisely controlled conditions. They are also self-renewing—they can divide and produce their copies indefinitely. Both cell types can help to understand how specialized cell types develop from pluripotent cells. In the future, they might also provide an unlimited supply of replacement cells and tissues for patients with many different diseases.

In contrast to ESCs, making iPS cells does not depend on the destruction of an early embryo. Some genes in iPS cells may behave in a slightly different way to those in ESCs. So at the moment, ES cells cannot be replaced with iPS cells in basic research.

Reprogramming allows turning any cell of the body into a stem cell (Fig. 9.5). These were first developed for mouse in 2006 and for human beings in 2007. To develop these iPS cells, Yamanaka added four genes to skin cells isolated from a mouse. This leads to the initiation of reprogramming process which resulted in the conversion of these skin cells into iPSCs within 2–3 weeks. Finally, in May 2008, the first iPS cell line containing genetic disorder, a human

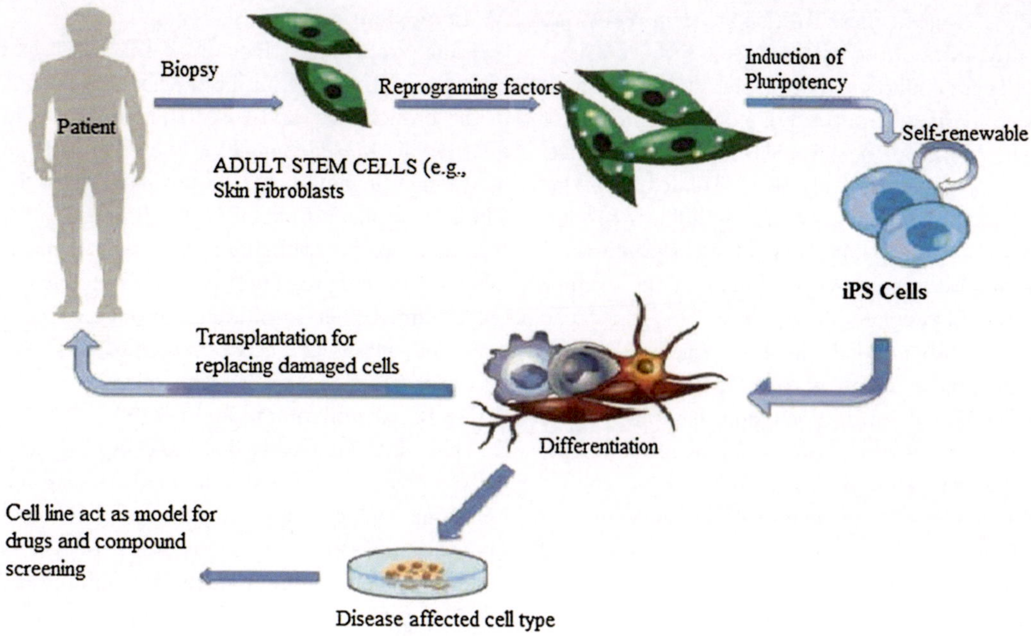

Fig. 9.5 Reprogramming of adult cells (modified from http://www.eurostemcell.org)

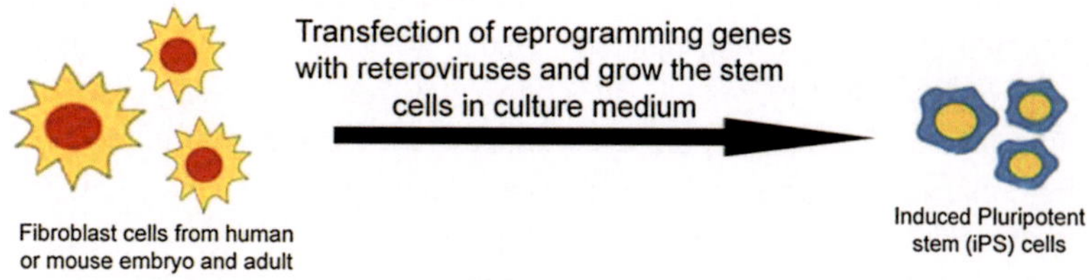

Fig. 9.6 Retroviral transfection of reprogramming genes

sickle cell anemia iPS cell line, was successfully established (Mali et al. 2008).

Moreover, this technique is used for *reprogramming of human cells* by adding only three genes. Mouse and human iPSCs possess important characteristics of pluripotent stem cells, including expressing stem cell markers, forming tumors containing cells from all three germ layers, and being able to contribute too many different tissues. Currently, viruses are used to introduce the reprogramming factors into adult cells (Fig. 9.6) which are found to cause cancers. Hence, this process must be carefully controlled and tested before the technique can lead to useful treatments for humans. Researchers are currently investigating nonviral delivery strategies. Tissues derived from iPSCs will be a nearly identical match to the cell donor, thus probably avoid the rejection by the immune system. The iPSC strategy creates pluripotent stem cells that together with studies of other types of pluripotent stem cells will help researchers to know how to reprogram cells to repair damaged tissues in the human body.

Recently, iPS cell lines have established the model for several diseases including adenosine deaminase deficiency-related severe combined immunodeficiency (ADA-SCID), Shwachman-Bodian-Diamond syndrome, Gaucher disease type III, Becker muscular dystrophy, Parkinson's disease (PD), Huntington's disease, juvenile-onset type 1 diabetes mellitus, Down syndrome/trisomy 21, and the carrier state of Lesch–Nyhan syndrome. The establishment of a diseased iPS cell bank would help to identify mechanisms for specific diseases and to discover effective treatments (Zaehres and Scholer 2007; Mali et al. 2008; Park et al. 2008).

9.5.2 Isolation of Stem Cells

There are different approaches for the isolation of stem cells largely depending on the tissue in which they are present. Generally, stem cells are present in the tissue which is made by these cells, e.g., bone stem cells can be taken from bone marrow. MSCs, which can make bone, cartilage, fat, fibrous connective tissue, and cells that support the formation of blood, can also be isolated from bone marrow. The stem cell isolation is greatly influenced by various factors depending on the source and type of cells and the donor age.

9.5.2.1 Use of Immunomagnetic Technique

The stem cells are diluted in the differentiated cells of adults. The enrichment of stem cells in the cell suspension by elimination of unwanted differentiated cells can be done rather than direct stem cell isolation. This may be achieved by the use of ferromagnetic nanoparticles labeled antibodies against the differentiated cells. When labeled antibodies bind to the surface markers of differentiated cells, the separation of these cells occur immunomagnetically. Sometimes, unlabeled antibodies are used against differentiated cells and cells-antibodies complex are incubated with complement proteins for their lysis.

9.5.2.2 Removal of Factors Supporting Growth of Differentiated Cells

Another method of isolation of stem cells is removal of factors that supports survival of differentiated cells. For example, removal of pro-differentiation factors in serum and use of plastic dishes that do not support cell adhesion (helps in the removal of differentiated cells). These methods are often combined with those that stimulate proliferation of stem cells while preserving their undifferentiated status, e.g., cell cultivation with the help of mitotic growth factors.

9.5.2.3 Amendment of Stem Cells Growth Supporting Factors

Since few stem cell-specific factors have also been identified, these may be recognized by antibodies using fluorescence-activated cell sorter. Stem cell-specific factors permit propagation of stem cells *in vitro* and their expansion. Under stimulation with growth factors, stem cells may undergo symmetric divisions and increase their numbers. As the growth of stem cells covers the dish in which they are cultured, replating them in a new culture dish is important after the cells reach a high density. The process of replating is called passaging.

9.5.2.4 Other Methods of Isolation

An interesting approach to enrich the stem cells is gradient centrifugation against different buoyant densities. Stem cells have different densities than that of adult stem cells due to having large nucleocytoplasmic ratio. Therefore, density gradient centrifugation enriches these cells in separators.

An interesting approach is to enrich stem cells as side population cells due to their capacity to efflux Hoechst dye. This method provides a novel method of selection of transfected target cells possessing stem cell-associated genes, which yields induced iPS cells.

9.5.3 Stem Cell Identification

9.5.3.1 Embryonic Stem Cells

During the process of generation, stem cells should be assessed for fundamental properties. This process is called characterization. This can be done by various methods at different stages of stem cell growth. One of the most important properties of stem cells is their self-renewable capability which can be assessed by growing and subculturing stem cells for many months. Subculturing of stem cells also helps in determining the ability of revival of these after freezing. Healthy and undifferentiated cells as well as the morphology of chromosomes can be observed under microscope. Presence of surface markers that are found only on undifferentiated cells (Oct-4) can also be used for the identification of stem cells. Marker Oct-4 is a transcription factor which turns genes on and off during the processes of cell differentiation and embryonic development.

The second important property of stem cells is their pluripotency which can be tested by allowing the cells to differentiate spontaneously in cell culture and by manipulating the cells, so that they will differentiate to form specific cell types. Stem cells are able to form teratoma on injecting the cells into an immunosuppressed mouse. This property of stem cells can be used for their identification.

9.5.3.2 Adult Stem Cells

In adults, stem cells are diluted amongst the mature and differentiated cells, therefore, are difficult to locate. In this case, the use of cell-specific molecular markers provides a good effective tool for locating the stem cells. Molecular markers are also applicable in labeling the cells in a living tissue and then used for determination of the specialized cell types generated by stem cells.

9.5.4 Culture of Stem Cells

The stem cells isolation is the first step in stem cell technology. Stem cells then need to be grown to large enough numbers to be useful for treatment purposes. When the blastocyst is used for stem cell research, cells from inner cell mass are isolated and placed in a culture dish with a nutrient-rich liquid. Over the course of several days, the cells of the inner cell mass proliferate and begin to crowd the culture dish. They are, therefore, removed gently and plated into several fresh culture dishes. This process is called subculturing and repeated many times for many months. After 6 months or more, the cells of the inner cell mass yield millions of ESCs. ESCs that have proliferated in cell culture for 6 or more months without differentiating are pluripotent and appear genetically normal, are referred as an ESC line. Once cell lines are established, or even before that stage, batches of that can be frozen and shipped to other laboratories for further culture and experimentation. These ESCs can be used for therapeutic purposes also. To be useful for producing medical therapies, cultured ESCs will need to be differentiated into appropriate tissues for transplantation into patients. Keeping this in view, researches are trying to achieve this type of differentiation. ES cell lines from patients with a variety of genetic diseases provide invaluable models to study those diseases.

Sometimes, stem cell cultivation requires the preparation of feeder layer in the culture dish for providing support to growing stem cells. To coat the culture dish with feeder layer, mouse skin cells can be used which is treated in this way that they will not divide after making a layer in the inner surface of culture dish. Now, human or mouse embryonic cells, isolated from inner cell mass, are placed in this culture dish and nutrient medium is poured over them. Feeder layer not only provides a sticky surface for the inner cell mass to which they can attach but also release nutrients into culture medium. However, culture media can be designed that does not require feeder layer.

9.5.5 Strategies for Differentiating Human ESC into Three Germ Layers

Human ESCs are an attractive cell source for transplantation therapy, regenerative medicine, and tissue engineering. To realize potential of stem cells, it is essential to control ESC

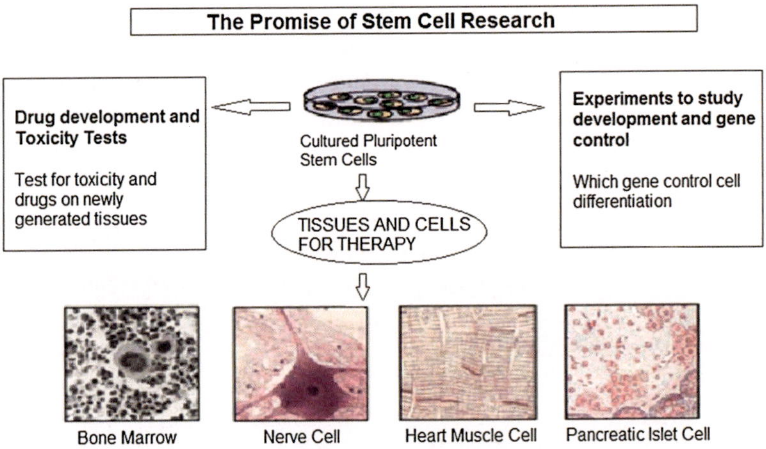

Fig. 9.7 Stem cell research in toxicology, drug design and genetic control (modified from http://science.jburroughs.org/mbahe/BioEthics/ppt_pdf/030%20StemCells.pdf)

differentiation and to direct the development of these cells along specific pathways. Basic science in the field of embryonic development, stem cell differentiation, and tissue engineering has offered important insights into key pathways and scaffolds that regulate hESCs differentiation, which have produced advances in modeling gastrulation in culture and in the efficient induction of endoderm, mesoderm, ectoderm, and many of their downstream derivatives.

The basic methods of hESC differentiation are divided into three categories:

1. Direct differentiation as a monolayer on extracellular matrix proteins.
2. Differentiation in co-culture with stromal cells.
3. The formation of 3-D spherical structures in suspension culture, termed embryoid bodies.

9.6 Applications

Stem cells have potential in different areas of health and medical research (Fig. 9.7). To start with, studying stem cells will help to understand how they transform into the dazzling array of specialized cells that make us what we are. Some of the most serious medical conditions such as, cancer and birth defects are due to problems that occur somewhere in this process. A better understanding of normal cell development will allow us to understand and perhaps correct the errors that cause these medical conditions.

Diseases treated by these nonembryonic stem cells include a number of blood and immune-system related genetic diseases, cancers, disorders, juvenile diabetes, Parkinson's, blindness, and spinal cord injuries. However, the use of this technology for treating various diseases is not very easy because of several problems associated with it. Besides the ethical problems of stem cell therapy, there is a technical problem of graft-versus-host disease associated with allogeneic stem cell transplantation. However, these problems associated with histocompatibility may be solved using autologous donor adult stem cells or *via* therapeutic cloning.

9.6.1 Therapy for Cancer

Cell therapy has the potential to offer novel treatment modalities for a number of diseases including cancer. MSCs play a particularly important role in the regulation of both solid tumor and hematological malignancies. MSCs have tumor-homing properties which may be important to use them as targeted delivery vehicles of gene products. However, conflicting reports have found that human MSCs can promote metastasis, while on the other hand other studies have shown that MSCs can stall the growth of metastatic lesions. About half of the tumor-burdened animals that were treated with murine and human MSCs, respectively harbored metastatic nodules significantly higher than controls (17 %) showing metastatic nodules. Hence, both human and mouse MSCs possess metastasis-promoting activity raising concerns about the safe use of MSCs. Although at the same time efforts are going on to make use of murine transgenic model systems feasible to study the role of MSCs in metastasis development and possibly finding ways of using them safely as cell therapeutic vehicles (Albarenque et al. 2011).

One can examine methods to take advantage of the unique immune properties of MSCs to improve therapy for cancer. Targeting MSCs might be more of a challenge due to the cells endogenous functions. MSCs have immune suppressive power which might partly explain reduced enthusiastic outcome of immune therapy for tumor.

9.6.2 Therapy for Cystic Fibrosis

The heterogeneous stem cell populations are present in bone marrow and lung tissue. Their presence in lung tissue may lead to stem cell manipulative and regenerative therapies for conditions such as idiopathic pulmonary fibrosis and cystic fibrosis transmembrane regulator gene therapy for cystic fibrosis.

9.6.3 Stem Cells in Tissue Repair

Organ-specific progenitors and HSCs are involved in repair of tissues. Epidermal stem cells and bone marrow progenitor cells both contribute for wound healing in the skin. Tissue regeneration is achieved by two mechanisms: (1) circulating stem cells divide and differentiate under appropriate signaling by cytokines and growth factors, e.g., blood cells; and (2) differentiated cells which are capable of division can also self-repair, e.g., hepatocytes, endothelial cells, smooth muscle cells, keratinocytes, and fibroblasts. These fully differentiated cells are limited to local repair. For more extensive repair, stem cells are maintained in the quiescent state, and can then be activated and mobilized to the required site.

Adult stem cells have shown promise as a cell source for a variety of tissue engineering and cell therapy applications. MSCs secrete a large spectrum of bioactive molecules. These molecules are immunosuppressive, especially for T-cells, thus allogeneic MSCs can be considered for therapeutic use. In this context, the secreted bioactive molecules provide a regenerative microenvironment for a variety of injured adult tissues to limit the area of damage and to mount a self-regulated regenerative response. This regenerative microenvironment is referred to as trophic activity; therefore, MSCs appear to be valuable mediators for tissue repair and regeneration. The natural titers of MSCs that are drawn to sites of tissue injury can be augmented by allogeneic MSCs delivered *via* bloodstream. Indeed, human clinical trials are now under way to use allogeneic MSCs for treatment of myocardial infarcts, graft-versus-host disease, Crohn's disease, cartilage and meniscus repair, stroke and spinal cord injury.

Recent reports indicate that umbilical cord blood stem cells may be beneficial for spinal cord injury and that lithium may promote regeneration and recovery of function after spinal cord injury. Both lithium and umbilical cord blood are widely available therapies that have long been used to treat diseases in humans.

9.6.4 Intervertebral DisK Regeneration

In humans, lower back pain is one of the most common causes of morbidity. In the normal intervertebral disk, the nucleus pulposus exerts a hydrostatic pressure against the constraining annulus fibrosus, which allows the disk to maintain flexibility between adjacent vertebrae, while absorbing necessary compressive forces. The nucleus pulposus performs this role because of its hydrophilic gel-like structure. The extracellular matrix of the nucleus pulposus is up to 80 % hydrated, as a result of large amounts of the aggregating proteoglycan, chondroitin sulfate proteoglycan. Intervertebral disk degeneration (IDD) is characterized by a progressive decrease in proteoglycan content and cell density in the nucleus pulposus.

Current therapies for IDD are aimed at treating the clinical symptoms arising from IDD rather than directly treating the underlying problem. Different strategies using stem cells to supplement/replenish the disk cell population have been proposed to slow, stop, or reverse the progressive loss of proteoglycan. Paracrine effects, cell–cell interactions are important characteristics that make stem cells as a source of cells for intervertebral disk (IVD) tissue engineering and cell therapy. Stem cell helps in the supplementation/replenishment of intervertebral disk cells which further leads to synthesis/maintenance of a more functional ECM (extra cellular matrix) in a degenerated disk. Transplanted stem cells survive and successfully engraft into the IVD tissue and are effective vehicles for exogenous gene delivery to the IVD. Thus, stem cells might be able to confer therapeutic effects in stem cell therapy of IVD degeneration.

Adult stem cells can be derived from muscle tissue. Muscle tissues serve as a good source of stem cells because of its vast abundance throughout the human body. Vadalà et al. (2008) reported a synergistic effect between muscles derived stem cells and NP cells resulting in an upregulated proteoglycan synthesis and NPCs proliferation *in vitro*.

Murrell et al. (2009) reported that olfactory stem cells could also differentiate into a nucleus pulposus chondrocyte phenotype *in vitro* and *in vivo* after transplantation into the injured intervertebral disk. This success provided a rationale for moving toward more extensive larger animal studies for assessment of regeneration and relief of symptoms. Chen et al. (2009) focused on the need to evaluate the *in vivo* effect on synovial-derived stem cells on disk regeneration.

9.6.5 Functional Genomic Studies

The ability to manipulate a specific gene of ES cells of mouse has been demonstrated since long. Mouse ES cells are able to produce transgenic animals. The study of the function of mammalian genes and proteins in the mouse was reported by Floss and Wurst (2002).

Stem cells help in the study of functional genes and the replacement of defective genes with the functional ones. Pelizaeus-Merzbacher disease (PMD) patients and myelin-deficient (md) rats carry mutations in the X-linked gene encoding myelin proteolipid protein (PLP). ES cell-derived oligodendrocytes can form myelin *in vivo*. The cells were grown in the presence of fibroblast growth factor (FGF2) and platelet-derived growth factor (PDGF) and then transplanted into the spinal cord of 1-week-old md rats. Two weeks after transplantation of ES cell-derived precursors into the dorsal columns of the spinal cord, numerous myelin sheaths were found in six of nine affected md-male rats. This experiment proposed the replacement of the defective gene with functional gene. It will also help in studying role of functional gene in an organism.

9.6.6 Study of Biological Processes

Studies of biological processes namely development of the organism is facilitated by the ability to trace stem cell fate. The spleen colony assay can be used in the study of the development of blood cells. In this method, single cells are injected into heavily irradiated mice so that all the hematopoietic cells are destroyed and new blood cells in these mice originate from the original colony.

9.6.7 Drug Discovery and Development

The isolation and purification of mouse ES cells and genetic engineering techniques have led to the use of murine ES cells in drug discovery.

With the sequencing of the human genome many potential targets of new drugs have been identified.

Normal and disease-specific iPS cell lines will prove extremely useful in studying how a drug influences a disease. BMSCs have become a standard in the field of adult stem cell biology and in regenerative medicine due to their high differentiation potentials and low morbidity during harvesting.

9.6.8 Stem Cells in Liver Disease

A number of animal and human studies demonstrated that either HSCs or MSCs could be applied to therapeutic purposes in certain liver diseases. The diseased liver may recruit migratory stem cells, particularly from the bone marrow, to generate hepatocyte-like cells either by transdifferentiation or cell fusion. Transplantation of BMSCs has therapeutic effects of restoration of liver mass and function, alleviation of fibrosis, and correction of inherited liver diseases. Several potential approaches for BMSCs delivery in liver diseases have been proposed in animal studies and human trials. BMSCs can be delivered *via* intra-portal vein, systemic infusion, intraperitoneal, intrahepatic, and intrasplenic. The optimal stem cell delivery should be easy to perform, less invasive and traumatic, with minimum side effects, and high cells survival rate (Xu and Liu 2008).

BMSCs were transplanted into liver-injured rats to test the therapeutic effect. BMSCs expressed the albumin mRNA and production of protein after cultivation with hepatocyte growth factor (HGF) for 2 weeks. The BMSCs appeared to differentiate into hepatocyte-like cells which were transplanted into carbon tetrachloride (CCl_4)-injured rats by injection through the caudal vein. Transplantation of the BMSCs into liver-injured rats restored their serum albumin level and suppressed transaminase activity and liver fibrosis. Therefore, BMSCs were shown to have a therapeutic effect on liver injury (Yagi et al. 2008). MSCs were more resistant to reactive oxygen species *in vitro*, reduced oxidative stress in recipient mice, and accelerated repopulation of hepatocytes after liver damage.

Hepatic progenitor cells from MSC differentiation in the growth-factor-free co-culture system may also contribute to the therapeutic effect for liver diseases *in vivo*. MSCs from bone marrow of green fluorescent protein-transgenic Sprague–Dawley rats were co-cultured with hepatocytes from normal Sprague–Dawley rats, sharing growth-factor-free media. After 14 days, cells were implanted into the spleen of CCl_4-injured rats and remarkably decreased fibrosis was observed. CCl_4/co-cultured MSC had reduced alpha-fetoprotein expression, CK18, and CK19 exhibited stronger expression. Albumin in CCl_4/co-cultured MSC increased in protein level (Li et al. 2010). Kuo et al. (2008) reported that both MSC-derived hepatocytes and MSCs, transplanted by either intrasplenic or intravenous route, engrafted recipient liver, differentiated into functional hepatocytes, and rescued liver failure. Intravenous transplantation was more effective in rescuing liver failure than intrasplenic transplantation. In this type of transplantation, paracrine effect was found to be effective.

Hepatic cirrhosis is the end-stage of chronic liver diseases. The majority of patients with hepatic cirrhosis die from life-threatening complications occurring at their earlier ages. The transplantation of autologous bone marrow-derived mesenchymal stem cells holds great potential for treating hepatic cirrhosis. MSCs can differentiate to hepatocytes, stimulate the regeneration of endogenous parenchymal cells, and enhance fibrous matrix degradation. Experimental and clinical studies have shown promising beneficial effects.

9.6.9 Acute Myocardial Infarction

A number of stem cell types, including ES cells, cardiac stem cells that naturally reside within the heart, myoblasts (muscle stem cells), adult bone marrow-derived cells including mesenchymal cells (bone marrow-derived cells that give rise to tissues such as muscle, bone, tendon, ligament, and adipose tissue), EPCs (cells that give rise to the endothelium, the interior lining of blood vessels), and umbilical cord blood cells, have been investigated as possible sources for regenerating damaged heart tissue.

Stem cells that are injected into the circulation or directly into the injured heart tissue appear to improve cardiac function and/or induce the formation of new capillaries. The mechanism for this repair remains controversial and the stem cells likely regenerate heart tissue through several pathways. Whether these cells can generate heart muscle cells or stimulate the growth of new blood vessels that repopulate the heart tissue, or help *via* some other mechanism is actively under investigation. For example, injected cells may accomplish repair by secreting growth factors, rather than actually incorporating into the heart.

By flow cytometry, myogenic stem cell populations expressing the neural cell adhesion molecule can be selected in the presence and absence of $\alpha 6$ integrin. The expression of $\alpha 6$ integrin showed an advantage in the formation of myotubes, possibly by an improved cell fusion capacity. It was possible to generate healthy heart muscle cells in the laboratory and then transplant those cells into patients with chronic heart disease. Preliminary research in mice and other animals indicated that BMSCs, transplanted into a damaged heart, can have beneficial effects (Wilschut et al. 2011).

Recent randomized clinical trials indicate that intracoronary delivery of bone marrow (stem) cells leads to an improvement in systolic function after acute myocardial infarction, thereby providing the first evidence that cell therapy may work in patients (Wollert 2008).

Dual stem cell therapy comprises granulocyte colony stimulating factor (GCSF)-based stem cell mobilization and diprotin A mediated dipeptidyl peptidase-IV (DPP-IV) inhibition. Dual stem cell treatment effectively stimulated the pool of resident cardiac stem cells which was reversed by AMD3100 treatment. According to Theiss et al. (2011), homing through the SDF-1/CXCR-4 axis is essential for the success of dual stem cell therapy.

Although adult stem cells with the capacity to transform into various cardiac cell types and to secrete cardioprotective cytokines have been identified, endogenous repair mechanisms in the adult heart are not sufficient for meaningful tissue regeneration.

9.6.10 Stem Cell Therapy for Diabetes

In type 1 diabetes mellitus (T1DM), the cells of the pancreas that normally produce insulin are destroyed by the patient's own immune system. Any potential stem cell-based cure for T1DM should address the need for beta-cell replacement, as well as control of the autoimmune response to cells which express insulin.

It may be possible to direct the differentiation of hESCs in cell culture to form insulin-producing cells that eventually could be used in transplantation therapy for persons with diabetes. These stem cells reconstitute a functional beta-cell mass that have molecular characteristics closely resembling bonafides insulin-secreting cells. These cells express insulin; however, these cells are often unresponsive to glucose. The use of mesenchymal stromal cells or umbilical cord blood to modulate the immune response is already in clinical trials; however, definitive results are still pending.

Ductal structures of the adult pancreas contain stem cells that differentiate into islets of Langerhans. Here, pancreatic ductal epithelial cells isolated from prediabetic adult nonobese diabetic mice were grown in long-term cultures. They were induced to produce functioning islets containing cells. These *in vitro* generated islets showed temporal changes in mRNA transcripts for islet cell-associated differentiation markers, responded to *in vitro* glucose challenge and reversed insulin-dependent diabetes after being implanted into diabetic nonobese diabetic mice

(Ramiya et al. 2000). The ability to control growth and differentiation of islet stem cells provides an abundant islet source for cell reconstitution in type I diabetes. Recent success in islet transplantation-based therapies for diabetes mellitus and the extreme shortage of pancreatic islets have motivated recent efforts to develop renewable sources of stem cells for islet-replacement tissue.

Increase in endothelial cell sloughing and diminished function of endothelial stem cell progenitors in diabetic subjects are well-known phenomena. The transplantation of BMSCs including MSCs into type 2 diabetic mice would restore insulin sensitivity and glucose tolerance. This approach, when combined with induction of HO-1 (a cytoprotective antioxidant system) in the recipient, would further improve bone marrow function. BMSC administered by the intrabone marrow–bone marrow transplantation (IBM–BMT) significantly increased BMSC function, serum adiponectin, and glucose tolerance. It has been observed that induction of HO-1 in the recipients greatly enhanced the ability of BMSC to prevent diabetes.

ESC can be differentiated into insulin-producing cells by manipulating culture conditions. *In vitro* differentiation of mouse ESC can generate embryoid bodies, which, after selection for nestin expressing ESC, were stimulated to differentiate toward different cell-like phenotype. For example, the addition of phosphoinositide kinase inhibitors promoted differentiation of larger numbers of ESC toward functional cells (Hussain and Theise 2004).

Pancreatic endocrine precursor cells on basement membrane extract (BME) differentiate into insulin-producing cells. A thin BME coating is sufficient to maintain an undifferentiated phenotype during embryonic stem cell expansion, while a thick BME hydrogel may be employed to induce stem cell differentiation. Stem cell differentiation *in vitro* can be used to identify cell types based on morphology and gene expression, and also to determine potential functionality *in vivo* (Arnaoutova et al. 2012).

9.6.11 Therapy for Parkinson's Disease

Parkinson's disease (PD) is a neurodegenerative disorder, symptomatized by tremors, rigidity, bradykinesia etc. In this disease, the majority of the signs and symptoms appear because of the progressive loss of cells in a small area known as the *substantia nigra*. These cells make dopamine, which is delivered to a part of the basal ganglia known as the *striatum*; when nigral neurons die and striatal dopamine diminishes, the signs and symptoms of Parkinson's disease become manifested. Thus, replenishing missing neurons in a limited area of the brain should in theory reverse parkinsonism, making this an attractive approach. But the challenge of actually replacing injured and/or lost neurons in the adult human nervous system has proven to be a daunting task. Stem cells might be effective in case of different forms and at all stages of parkinsonism.

Adult stem cells harvested from the noses of Parkinson's patients gave rise to dopamine-producing brain cells (http://www3.griffith.edu.au/03/ertiki/tiki-read_article.php?articleId=16841). When these stem cells were cultured and injected into the damaged area the rats re-acquired the ability to run in a straight line. Stem cells from the olfactory nerve in the nose are 'naive', having not yet differentiated into which sort of cells they will give rise to. They can still be influenced by the environment they are put into.

Dopaminergic neurons can also be obtained from fetal brain tissue. At stages I–III, when uni- and bi-lateral manifestations of the disease are observed, such as balance disturbances, tremor, mild and moderate rigidity etc., administration of fetal stem cells is effective in 85 % of cases. After the treatment, patients report improvement of stiffness, decrease in shaking while walking and tremor. The dose of antiparkinson drugs can be lowered. At stage IV (severe stiffness, but the patient still can walk or stand unassisted), stem cell treatment is effective in 65–70 % of Parkinson's disease cases. At stage V (bed- or wheelchair-bound), this therapy helps to improve patient's quality of life and ease care for the patient due to decreased spasticity and tremor syndrome and improved sleep, normalization of organ, and system functioning (heart, lungs, and bowels). In 100 % of cases of PD, treatment with fetal stem cells results in psychoemotional improvements. After the stem cell treatment, patients demonstrate positive emotions, improved thinking, more expressive, intelligible and louder speech, and more persistent memory and intellect. These trials provide proof of principle for the approach, since in a few of these trials major and long lasting improvements were seen in some patients. The trials also emphasized several issues that need to be resolved, one of which is the need to produce large amounts of pure, uniform cells for transplantation into patients. Recent findings also highlight a further concern about cell transplantation therapies. The fetal transplants that some patients received began to show signs of being affected by Parkinson's disease. This showed that the disease from the patient was transmitted to the transplanted fetal cells (http://www.eurostemcell.org/).

Mouse iPS cell-derived neurons could functionally integrate into donor brains, and symptoms of Parkinson's disease in donor mice were relieved after transplantation of iPS cell-derived dopaminergic neurons (Wernig et al. 2008).

Human-induced pluripotent stem cells (hiPSCs) are a potentially unlimited source of patient-specific cells for transplantation. Dopamine (DA) neurons derived from protein-based hiPSCs exhibited gene expression, physiological, and electrophysiological properties similar to those of midbrain dopamine (mDA) neurons. Transplantation of these cells into rats with striatal lesions, a model of PD, significantly rescued motor deficits. These data support the clinical potential of protein-based hiPSCs for personalized cell therapy of PD. However, it is critical to evaluate the safety of hiPSCs generated by different reprogramming methods. On comparing multiple hiPSC lines derived by virus- and protein-based reprogramming to hESCs, it was found that neuronal precursor cells (NPCs) and DA neurons delivered from lentivirus-based hiPSCs exhibited residual expression of exogenous reprogramming genes, but those cells derived from retrovirus- and protein-based hiPSCs did not. Furthermore, NPCs derived from virus-based hiPSCs exhibited early senescence and apoptotic cell death during passaging, which was preceded by abrupt induction of p53. In contrast, NPCs derived from hESCs and protein-based hiPSCs were highly expandable without senescence (Rhee et al. 2011). The use of viruses encoding the reprogramming factors represents a major limitation of the current technology since even low vector expression may alter the differentiation potential of the hiPSCs or induce malignant transformation. Soldner et al. (2009) reported that residual transgene expression in virus-carrying hiPSCs can affect their molecular characteristics and that factor-free hiPSCs, therefore represent a more suitable source of cells for modeling of human disease.

9.6.12 Therapy for Bone and Cartilage Repair

Bone and cartilage defects are common features of bone fracture and joint diseases, such as rheumatoid arthritis or osteoarthritis that have great social and economic impact on the aging occidental population. Injured cartilage tissue cannot spontaneously heal and, if not treated, can lead to osteoarthritis of the affected joints. Although a variety of procedures are being employed to repair cartilage damage, methods that result in consistent durable repair tissue are not yet available. Tissue engineering is a recently developed science that merges the fields of cell biology, engineering, material science, and surgery to regenerate new functional tissue. Three critical components in tissue engineering of cartilage are as follows: First, sufficient cell numbers within the defect, such as chondrocytes or multipotent stem cells capable of differentiating into chondrocytes; second, access to growth and differentiation factors that modulate these cells to differentiate through the chondrogenic lineage; third, a cell carrier or matrix that fills the defect, delivers the appropriate cells, and supports cell proliferation and differentiation (Gao et al. 2007). Stem cells that exist in the embryo or in adult somatic tissues are able to renew themselves through cell division without changing their phenotype and are able to differentiate into multiple lineages including the chondrogenic lineage under certain physiological or experimental conditions.

Recent investigations on the stromal MSC offer a new perspective for bone and cartilage tissue engineering. Xian and Foster (2006) proposed few points for exploring the potentials of MSCs in cartilage repair. These include (a) identifying readily available sources of and devising appropriate techniques for isolation and culture expansion of MSCs that have good chondrogenic differentiation capability; (b) discovering appropriate growth factors (such as TGF-beta, IGF-I, BMPs, and FGF-2) that promote MSC chondrogenic differentiation; (c) identifying or engineering biological or artificial matrix scaffolds as carriers for MSCs and growth factors for their transplantation and defect filling. Richardson et al. (2010) exposed implanted MSCs to traumatic physical loads and high levels of locally produced inflammatory mediators and catabolic cytokines. They also explored the potential of culture models of MSCs, fully differentiated cells and co-cultures as "proof of principle" ethically acceptable "3Rs" models for engineering articular cartilage.

Caplan (2005) used MSCs with site-specific delivery vehicles to repair cartilage, bone, tendon, marrow stroma, muscle, and other connective tissues. In the beginning of the twenty-first century, they have made substantial advances: the most important is the development of a *cell-coating technology*, called painting that allows us to introduce informational proteins to the outer surface of cells. These paints can serve as targeting addresses to specifically dock MSCs or other reparative cells to unique tissue addresses. The scientific and clinical challenge include to perfect cell-based tissue engineering protocols to utilize the body's own rejuvenation capabilities by managing surgical implantations of scaffolds, bioactive factors, and reparative cells to regenerate damaged or diseased skeletal tissues. Arinzeh (2005) evaluated the efficacy of MSC-loaded scaffolds in large bone defects as a potential substitute for autologous and allogeneic bone grafts.

Current strategies to engineer cartilage tissues from adult stem cells and human ESC-derived cells are also in developing state (Hwang and Elisseeff 2009). Earlier, Hwang et al. (2007) demonstrated that ESCs encapsulated within polyethylene glycol-based photopolymerizing hydrogels and cultured in an appropriate growth factor and medium conditions undergo chondrogenic differentiation with extracellular matrix deposition characteristic of neocartilage. Vinatier et al. (2009) evaluated the optimal combinations that will answer to the functional demand placed upon cartilage tissue replacement in animal models and in clinics.

Similar to those of bone marrow-derived mesenchymal stem cells (BMSCs), adipose-derived adult stem (ADAS) cells show characteristics of multipotent adult stem cells and under appropriate culture conditions, synthesize cartilage-specific matrix proteins that are assembled in a cartilaginous extracellular matrix. The growth and chondrogenic differentiation of ADAS cells is strongly influenced by factors in the biochemical as well as biophysical environment of the cells.

Autologous chondrocyte transplantation (ACT) is technically feasible and biologically relevant to repairing disk damage and retarding disk degeneration. In ACT, a cartilage biopsy is taken from the patient and articular chondrocytes are isolated. The cells are then expanded after several passages *in vitro* and used to fill the cartilage defect. A dog model was used to investigate the hypothesis that autologous disk chondrocytes can be used to repair damaged intervertebral disk. Disk chondrocytes were harvested and expanded in culture under controlled and defined conditions, returned to the same animals from which they had been sampled (autologous transplantation) *via* percutaneous delivery. The animals were analyzed at specific times after transplantation by several methods to examine whether disk chondrocytes integrated with the surrounding tissue, produced the appropriate intervertebral disk extracellular matrix, and might provide a formative solution to disk repair. Meisel et al. (2007) designed EuroDISC according to the German Drug Law (Arzneimittelgesetz) and good manufacturing practices. This has been certified according to internationally approved DIN EN ISO 9001 standards. To broaden the scope, better understanding and long-term efficacy, clinical trial was initiated to assess the potential autologous disk chondrocyte transplantation in a broader population. The clinical goals of the EuroDISC, provided satisfactory results such as long-term pain relief, maintain disk height, and prevent adjacent segment disease.

Despite promising experimental approaches, however, none of the current cartilage repair strategies have generated long-lasting hyaline cartilage replacement tissue that meets the functional demands placed upon this tissue *in vivo*. The reasons for this are diverse and can ultimately result in matrix degradation, differentiation or integration insufficiencies, or loss of the transplanted cells and tissues.

9.6.13 Stem Cells for Use in Regenerative Medicine

Stem cells are ideal candidates for use in regenerative medicine because of their ability to self-renew and to commit to multiple cell lineages. Stem cells for regenerative medicine applications should meet the following criteria:

1. These should be found in abundant numbers (millions to billions of cells).
2. These should be harvested by a minimally invasive procedure with minimal morbidity.
3. These should be differentiated along multiple cell lineage pathways in a controllable and reproducible manner.
4. These should be safely and effectively transplanted to either an autologous or allogeneic host.
5. These should be produced in accordance with current good manufacturing practice guidelines.

Recent advancements in tissue engineering and regenerative medicine have highlighted MSCs as a potential source of cells which would differentiate to a variety of tissue tailored to individual needs. Transplanted hMSCs have the potential to migrate into normal and injured liver parenchyma, particularly under conditions of chronic injury, but differentiation into hepatocyte-like cells is a rare event (di Bonzo et al. 2008).

9.6.14 Treatment of Immunological Disorders and Other Diseases

MSC can migrate to injured tissues and some of their reparative properties are mediated by paracrine mechanisms including their immune-modulatory actions. ASCs are good candidate for cell-based therapies, because these can survive in a low oxygen environment and secrete angiogenic cytokines. These cells also possess immunomodulatory properties and protective effects in immunological disorders. Reports have shown that the immunosuppressive capacity of the ASCs may in some cases favor the growth of tumor cells but contradictory results exist and the question is a matter of intense debate.

According to Lindroos et al. (2010), ASCs exert profound immune-modulatory properties and protective effects on acute graft-versus-host disease (GvHD) and experimental arthritis. By functional characterization, ASCs have been shown to be immune-privileged due to lack of HLA-DR expression (major histocompatibility complex class II) and the suppression of proliferation of activated allogeneic lymphocytes. Additionally, ASCs have been shown to inhibit the production of inflammatory cytokines by both CD4+ T helper cells and CD8+ T cytotoxic cells, stimulate the production of the anti-inflammatory/suppressive cytokine IL-10 by monocytes and T cells, and induce the generation of antigen-specific regulatory T cells. The use of ASCs for immunologic disorders, for example, in autoimmune-induced rheumatoid arthritis or inflammatory bowel disease, such as Crohn's disease and ulcerous colitis, is currently under investigation.

ASCs also protect from severe sepsis by reducing the infiltration of inflammatory cells in various target organs and by downregulating the production of various inflammatory mediators. Colitic and septic mice were treated intraperitoneally with human ASCs or murine ASCs, and diverse clinical signs of disease were investigated as well as the levels of various inflammatory cytokines and chemokines, T helper 1-type response and generation of regulatory T cells in affected organs. Systemic infusion of ASCs significantly reduced the severity of colitis by eliminating weight loss, diarrhea, and inflammation, thereby increasing survival (Gonzalez-Rey et al. 2009).

9.6.15 Stem Cells in Graft-Versus-Host Disease

Twenty clinical trials of BMSCs for treatment of GvHD have been reported (http://clinicaltrials.gov), of which 10 have already been completed. Yet, no clinical trials using ASCs for treating severe and acute GvHD are currently under way. Furthermore, the efficacy of ASCs is currently being studied in GvHD; however, their effect on alloreactivity in solid organ transplant patients is unknown.

9.6.16 Stem Cells in Veterinary Medicine

Stem cells play an important role in veterinary medicine in different ways. Currently, several stem cell therapies for animal patients are being developed and some, like the treatment of equine tendinopathies with MSCs, have already successfully entered the market. However, clinical applications of ES cells are not possible yet due to their *in vivo* teratogenic degeneration. MSCs from either extra embryonic or adult tissues are in the focus of attention in veterinary medicine and research.

9.7 Advantages of Stem Cell Therapy

Stem cell therapy is very promising and has several advantages: (1) it is apparently not toxic; (2) it may be administrated once without the need for repetition, (3) the progeny of transplanted cells assume phenotypes in the target tissue, (4) the therapy may be applied without the precise knowledge of mechanisms that control cell homing, their proliferation and differentiation, and (5) the incurable diseases can also be treated by this therapy.

9.8 Disadvantages

Although the BMSCs continue to be a viable option for a stem cell population for cell therapy applications, there are drawbacks to utilizing the source. A bone marrow harvest is a painful procedure with possible donor site morbidity as a result. Secondly, although MSCs grow well under standard tissue culture conditions, ex vivo expansion is necessary due to relatively low numbers of MSCs present in the harvested marrow.

There are several drawbacks which are associated with iPS. These are based on the use of retroviruses and lentiviruses that integrate into the DNA. In fact, there is correlation between the number of insertions of the transgenes and the quality of the reprogramming (Okita et al. 2007). Viral integration on its own either cause mutations or in the case of Klf4 and c-Myc, which are classified as oncogenes, the vectors become reactivated and cause tumors (Ton-That et al. 1997). Moreover, Oct-4 can cause dysplasia when overexpressed in adult tissues (Hochedlinger et al. 2005). Theoretically, use of nonintegrating viruses would be a potential solution; one clear example is adenoviruses, while for lentiviruses DNA integration can be drastically reduced or eliminated by simple mutation of the integrase viral gene. In both cases, the viruses will become progressively diluted with cell divisions and disappear.

9.9 Conclusion

Patients who are diagnosed with difficult to treat nervous system disorders, malfunctioning or damage of any organ suffer tremulously, and often desperately look for help and hope. In this new age of medical science, stem cell therapy can provide a ray of hope to these patients. Stem cells serve as sort of repair system. They can theoretically divide without limit to replenish other cells for as long as the person or animal is still alive. When a stem cell divides, each "daughter" cell has the potential to either remain a stem cell or become another type of cell with a more specialized function, such as a muscle cell, a red blood cell, or a brain cell. Some current therapies, such as bone marrow transplantation, already make use of stem cells and their potential for regeneration of damaged tissues. Other therapies are under investigation that involves transplanting stem cells into a damaged body part and directing them to grow and differentiate into healthy tissue. But, as it is with any new advancement, people, who are not too familiar with it, try to use it for their own gain. Some of the treatment strategies are under clinical trials and their successful completion will bring the technology to the field to treat various diseases and give new life to patients. Diseases and conditions, which were not possible to treat before are now treatable and to some extent even curable by stem cell therapy.

References

Albarenque SM, Zwacka RM, Mohr A (2011) Both human and mouse mesenchymal stem cells promote breast cancer metastasis. Stem Cell Res 7:163–171

Arinzeh TL (2005) Mesenchymal stem cells for bone repair: preclinical studies and potential orthopedic applications. Foot Ankle Clin 10:651–665

Arnaoutova I, George J, Kleinman HK, Benton G (2012) Basement membrane matrix (BME) has multiple uses with stem cells. Stem Cell Rev 8:163–169

Cai L, Ye Z, Zhou BY, Mali P, Zhou C, Cheng L (2007) Promoting human embryonic stem cell renewal or differentiation by modulating Wnt signal and culture conditions. Cell Res 17:62–72

Caplan AI (2005) Mesenchymal stem cells: cell-based reconstructive therapy in orthopedics. Tissue Eng 11:1198–1211

Chen S, Emery SE, Pei M (2009) Coculture of synovium-derived stem cells and nucleus pulposus cells in serum-free defined medium with supplementation of transforming growth factor-beta1: a potential application of tissue-specific stem cells in disc regeneration. Spine (Phila Pa 1976) 34:1272–1280

di Bonzo LV, Ferrero I, Cravanzola C, Mareschi K, Rustichell D, Novo E, Sanavio F, Cannito S, Zamara E, Bertero M, Davit A, Francica S, Novelli F, Colombatto S, Fagioli F, Parola M (2008) Human mesenchymal stem cells as a two-edged sword in hepatic regenerative medicine: engraftment and hepatocyte differentiation versus profibrogenic potential. Gut 57:223–231

Dreyfus PA, Chretien F, Chazaud B, Kirova Y, Caramelle P, Garcia L, Butler-Browne G, Gherardi RK (2004) Adult bone marrow-derived stem cells in muscle connective tissue and satellite cell niches. Am J Pathol 164:773–779

Floss T, Wurst W (2002) Functional genomics by gene-trapping in ES cells. In: Turksen K (ed) Embryonic stem cells methods and protocols. pp 347–379

Gao J, Yao JQ, Caplan AI (2007) Stem cells for tissue engineering of articular cartilage. Proc Inst Mech Eng H 221:441–450

Gonzalez-Rey E, Anderson P, González MA, Rico L, Büscher D, Delgado M (2009) Human adult stem cells derived from adipose tissue protect against experimental colitis and sepsis. Gut 58:929–939

Hochedlinger K, Yamada Y, Beard C et al (2005) Ectopic expression of Oct-4 blocks progenitor-cell differentiation and causes dysplasia in epithelial tissues. Cell 121:465–477

Hussain MA, Theise ND (2004) Stem-cell therapy for diabetes mellitus. www.thelancet.com. Accessed on July 2010

Hwang NS, Elisseeff J (2009) Application of stem cells for articular cartilage regeneration. J Knee Surg 22:60–71

Hwang NS, Varghese S, Elisseeff J (2007) Cartilage tissue engineering: directed differentiation of embryonic stem cells in three-dimensional hydrogel culture. Methods Mol Biol 407:351–373

Science Daily. Jan 8, 2009

Kuo TK, Hung SP, Chuang CH, Chen CT, Shih YR, Fang SC, Yang VW, Lee OK (2008) Stem cell therapy for liver disease: parameters governing the success of using bone marrow mesenchymal stem cells. Gastroenterology 134:2111–2121

Labat MLA, Vaillant F, Sheridan JM, Pal B, Wu D, Simpson ER, Yasuda H, Smyth GK, Martin TJ, Lindeman GJ, Visvade JE (2010) Control of mammary stem cell function by steroid hormone signaling. Nature 465:768–802

Li TZ, Kim JH, Cho HH, Lee HS, Kim KS, Lee SW, Suh H (2010) Therapeutic potential of bone-marrow-derived mesenchymal stem cells differentiated with growth-factor-free coculture method in liver-injured rats. Tissue Eng Part A 16:2649–2659

Lindroos B, Suuronen R, Miettinen S (2010) The potential of adipose stem cells in regenerative medicine. Stem Cell Rev Rep. doi 10.1007/s12015-010-9193-7

Mali P, Ye Z, Hommond H et al (2008) Improved efficiency and pace of generating induced pluripotent stem cells from human adult and fetal fibroblasts. Stem Cell 26:1998–2005

McMurray RJ, Gadegaard N, Tsimbouri PM, Burgess KV, McNamara LE, Tare R, Murawski K, Kingham E, Oreffo ROC, Dalby MD (2011) Nanoscale surfaces for the long-term maintenance of mesenchymal stem cell phenotype and multipotency. Nat Mater 10:637–644

Meisel HJ, Siodla V, Ganey T, Minkus Y, Hutton WC, Alasevic OJ (2007) Clinical experience in cell-based therapeutics: disc chondrocyte transplantation—a treatment for degenerated or damaged intervertebral disc. Biomol Eng 24:5–21

Murrell W, Sanford E, Anderberg L, Cavanagh B, Mackay-Sim A (2009) Olfactory stem cells can be induced to express chondrogenic phenotype in a rat intervertebral disc injury model. Spine J 9:585–594

Okita K, Ichisaka T, Yamanaka S (2007) Generation of germline-competent induced pluripotent stem cells. Nature 448:313–317

Park IH, Arora N, Huo H et al (2008) Disease-specific induced pluripotent stem cells. Cell 134:877–886

Poulsom R, Alison MR, Forbes SJ, Wright NA (2002) Adult stem cell plasticity. J Pathol 197:441–456 Stem cells

Ramiya VK, Maraist M, Arfors KE, Schatz DA, Peck AB, Cornelius JG (2000) Reversal of insulin-dependent diabetes using islets generated *in vitro* from pancreatic stem cells. Nat Med 6:278–282

Rhee YH, Ko JY, Chang MY, Yi SH, Kim D, Kim CH, Shim JW, Jo AY, Kim BW, Lee H, Lee SH, Suh W, Park CH, Koh HC, Lee YS, Lanza R, Kim KS, Lee SH (2011) Protein-based human iPS cells efficiently generate functional dopamine neurons and can treat a rat model of Parkinson disease. J Clin Invest 121(6):2326–2335

Richardson SM, Hoyland JA, Mobasheri R, Csaki C, Shakibaei M, Mobasheri A (2010) Mesenchymal stem cells in regenerative medicine: opportunities and challenges for articular cartilage and intervertebral disc tissue engineering. J Cell Physiol 222:23–32

Rob Waters (2008) Testicle stem cells become bone, muscle in German experiment. http://www.bloomberg.com/apps/news?pid=newsarchive&sid=aNmiXs8SPp4w&refer=home

Shackleton M, Vaillant FO, Simpson KJ et al (2006) Generation of a functional mammary gland from a single stem cell. Nature 439:84–88

Soldner F, Hockemeyer D, Beard C, Gao Q, Bell GW, Cook EG, Hargus G, Blak A, Cooper O, Mitalipova M, Isacson O, Jaenisch R (2009) Parkinson's disease patient-derived induced pluripotent stem cells free of viral reprogramming factors. Cell 136:964–977

Theiss HD, Vallastera M, Rischplera C, Kriega L, Zarubaa MM, Brunnera S, Vancheva Y, Fischera R, Gröbnera M, Hubera B, Wollenweber T, Assmann G, Mueller-Hoecker J, Hacker M, Franza WM (2011) Dual stem cell therapy after myocardial infarction acts specifically by enhanced homing *via* the SDF-1/CXCR4 axis. Stem Cell Res 7:244–255

Ton-That H, Kaestner KH, Shields JM et al (1997) Expression of the gut-enriched Kruppel-like factor gene during development and intestinal tumorigenesis. FEBS Lett 419:239–243

Urbich C, Dimmeler S (2004) Endothelial progenitor cells: characterization and role in vascular biology. Circ Res 95:343–353

Vadalà G, Sobajima S, Lee JY, Huard J, Denaro V, Kang JD, Gilbertson LG (2008) *In vitro* interaction between muscle-derived stem cells and nucleus pulposus cells. Spine J 8:804–809

Vinatier C, Bouffi C, Merceron C, Gordeladze J, Brondello JM, Jorgensen C, Weiss P, Guicheux J, Noël D (2009) Cartilage tissue engineering: towards a biomaterial-assisted mesenchymal stem cell therapy. Curr Stem Cell Res Ther 4:318–329

Wernig M, Zhao JP, Pruszak J et al (2008) Neurons derived from reprogrammed fibroblasts functionally integrate into the fetal brain and improve symptoms of rats with Parkinson's disease. Proc Natl Acad Sci USA 105:5856–5861

Wilschut KJ, Tola HTAV, Arkesteijn GJA, Haagsman HP, Roelen BAJ (2011) Alpha 6 integrin is important for myogenic stem cell differentiation. Stem Cell Res 7:112–123

Wollert KC (2008) Cardiovascular and renal-cell therapy for acute myocardial infarction. Curr Opin Pharmacol 8:202–210

Xian CJ, Foster BK (2006) Repair of injured articular and growth plate cartilage using mesenchymal stem cells and chondrogenic gene therapy. Curr Stem Cell Res Ther 1:213–229

Xu YQ, Liu ZC (2008) Therapeutic potential of adult bone marrow stem cells in liver disease and delivery approaches. Stem Cell Rev 4:101–112

Yagi K, Kojima M, Oyagi S, Ikeda E, Hirose M, Isoda K, Kawase M, Kondoh M, Ohgushi H (2008) Application of mesenchymal stem cells to liver regenerative medicine. Yakugaku Zasshi 128:3–9

Zaehres H, Scholer HR (2007) Induction of pluripotency: from mouse to human. Cell 131:834–835

10 Biosensors

Mayank and Rachana Arya

Abstract

A biosensor is an analytical tool used for qualitative and quantitative detection of analyte. It incorporates a biological recognition element such as enzyme, microbial cell, receptor, DNA, antibody, antigen, etc., which interacts with the analyte to produce primary response. This initial response is identified and converted into a suitable electronic form with the help of a suitable transducer device. Selection of transducer is based on the type of signal obtained from the interaction of biological recognition element and analyte. The commonly used devices are electronic, optical, mechanical, and thermal transducer. To insure accurate response, suitable immobilization techniques are required for biological recognition elements, which provide proper interaction of transducer with the signal obtained from primary sensing. Adsorption, cross-linking, entrapment, and covalent binding are some commonly applied techniques for immobilization. The electronic signal obtained from transducer may easily be modified, amplified, displayed, and recorded by suitable electronic devices. The use of biological component for sensing makes the biosensor very specific, fast, and reliable, thus biosensors are nowadays utilized in various industries for monitoring such as food quality control, medical research, clinical diagnosis, environmental monitoring, agriculture, bioprocess monitoring and control, and pharmaceutical industry, etc.

10.1 Introduction

In 1956, Leland C. Clark Jr., who is known as the father of biosensors, published his definitive paper on the oxygen electrode (Clark 1956). Based on this experience and addressing his desire to expand the range of analytes that could be measured in the body, he made a land mark address in 1962 at New York Academy of

Mayank (✉)
Biochemical Engineering Department, BT Kumaon Institute of Technology, Dwarahat, Uttarakhand 263653, India
e-mail: findmayank12@yahoo.co.in

R. Arya
Electronics and Communication Engineering Department, BT Kumaon Institute of Technology, Dwarahat, Uttarakhand 263653, India
e-mail: rachna009@gmail.com

I. Ravi et al. (eds.), *Advances in Biotechnology*,
DOI: 10.1007/978-81-322-1554-7_10, © Springer India 2014

Science Symposium in which he described "how to make electrochemical sensors (pH, polarographic, potentiometric, or conductimetric) more intelligent" by adding "enzyme transducers as membrane enclosed sandwiches". The concept was illustrated by an experiment in which glucose oxidase was entrapped at a Clark oxygen electrode using dialysis membrane. The decrease in measured oxygen was proportional to glucose concentration (Clark and Lyons 1962). Clark's ideas became commercial reality in 1975, with the successful launch of the Yellow Springs Instrument Company (Ohio) glucose analyzer, based on the amperometric detection of hydrogen peroxide. This was the first of many biosensor-based laboratory analyzers to be built by companies around the world.

The term biosensor has been applied to devices either used to monitor living systems, or incorporating biotic elements. The consensus, however, is that the term should be reserved for use in the context of a sensor incorporating a biological element such as an enzyme, antibody, nucleic acid, microorganism, or cell.

A typical biosensor consists of a biological component (primary sensing element) and an electronic device (transducer) that converts the biological signal into a measurable output. The biological part of the sensor reacts with a particular substance of interests (i.e., the analyte) to produce a physical or biochemical change that is detected and converted to an electrical signal by the transducer. The amplifier increases the intensity of the signal so that it can be readily measured. These components are usually housed within a single portable unit that can be placed at fixed strategic locations. The biosensors display can be tailored to meet the needs of the application and can range from a simple output such as switching on (or off), a light-emitting diode to a quantitative result displayed in graphical format. The general schematic diagram of a biosensor is given in Fig. 10.1.

Biosensor consists of a biological entity such as an enzyme, antibody or nucleic acid that interacts with an analyte and produces a signal that is measured electronically. Each biosensor, therefore, has a biological component that acts as the sensor and an electronic component to transduce and detect the signal. A variety of substances including nucleic acids, proteins (particularly antibodies and enzymes), lectins (plant proteins that bind sugar moieties) and complex materials (organelles, tissue slices, microorganisms) can be used as the biological components. In each case it is the specificity of the biological components for an analyte (or group of related analytes) that makes the biomolecules attractive as sensing component. For example, a single strand of DNA can be used as a biomolecular sensor that will hybridize only to its complementary strands under appropriate conditions. The signal, which can be electrical, optical, or thermal, is converted by means of a suitable transducer into a measurable electrical parameter such as current or voltage. Biosensor probes are attaining increasing sophistication because of the fusion of two technologies: microelectronics and biotechnology. Biosensors provide a useful means for measuring a wide spectrum of analytes (e.g., gases, ions, and organic compounds, or even bacteria) and are suitable for studies of complex microbial environments.

10.2 Overview of Biosensor Technology

The important components of biosensor technology and their interaction in the form of flow chart are represented in Fig. 10.2.

10.3 Biological Recognition Elements

The bioactive component could be an enzyme, an antibody, a cell (bacterial and fungal), a tissue slice, a receptor, nucleic acid, or an organelle. It is used to recognize and interact specifically with the analyte. Generally, there are two categories of biosensors, catalytic and affinity:

1. *Catalytic biosensors*: In this type of biosensors, biocatalysts such as enzyme, microbiological cells, animal and plant cells,

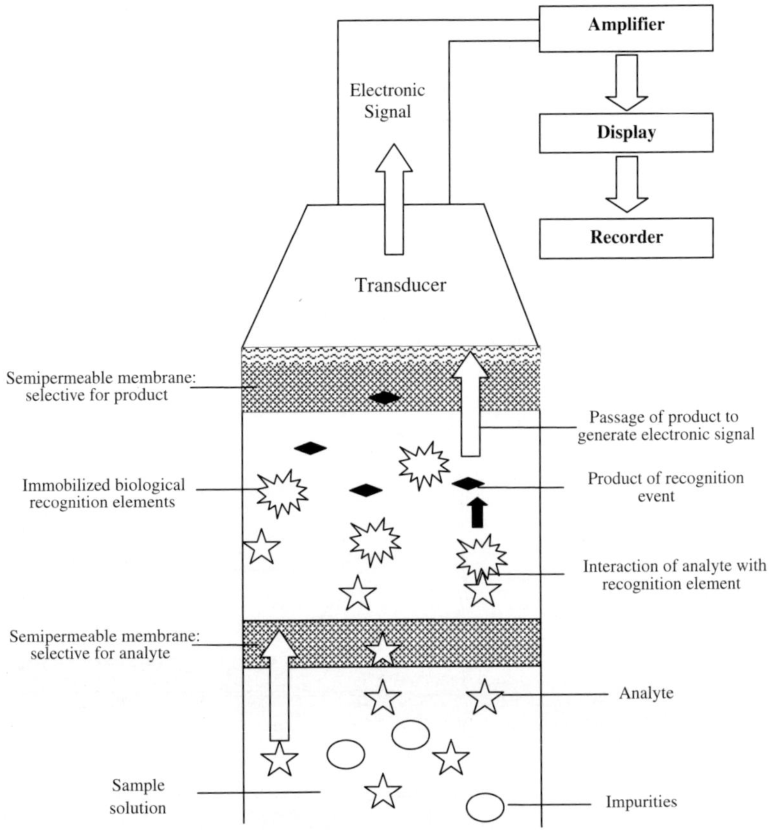

Fig. 10.1 Schematic diagram of biosensor

organelles and tissues are used to recognize, bind, and chemically convert a molecule identified by the transducer. These biorecognition elements can be used to analyze substrate, inhibitor, activator, co-factor, and enzyme activity.

2. *Affinity biosensors*: In this category the biological element (receptor) binds specifically to the analyte leading to a complex. The transducer reveals the complex and generates an electronic signal. Receptor molecule such as antibodies, nucleic acids, lectins, and hormone receptors are used to bind molecules irreversibly and non-catalytically (Table 10.1).

10.3.1 Enzymes

Enzymes were the first biological components used in biosensors (Clark 1956) and remain by far the most commonly employed elements. Enzymes are the large protein molecules that act as biological catalysts, i.e., they speed up chemical reactions at moderate temperature usually acting on a unique substrate or group of substrates. It is this remarkable specificity that makes enzymatic methods attractive in sensing devices. Enzymes are normally the preferred choice in catalytic-type biosensor. However, in many cases the use of enzymes is currently impractical or impossible. The enzyme may be unstable, may require soluble cofactors, may be difficult to purify, may not involve a readily detectable species, or the analyte may not be substrate for a specific enzyme.

10.3.2 Antibodies

In principle, an antibody-based assay can quantify analyte on the basis of binding reactions between antibody and antigen. It is done from a total of the

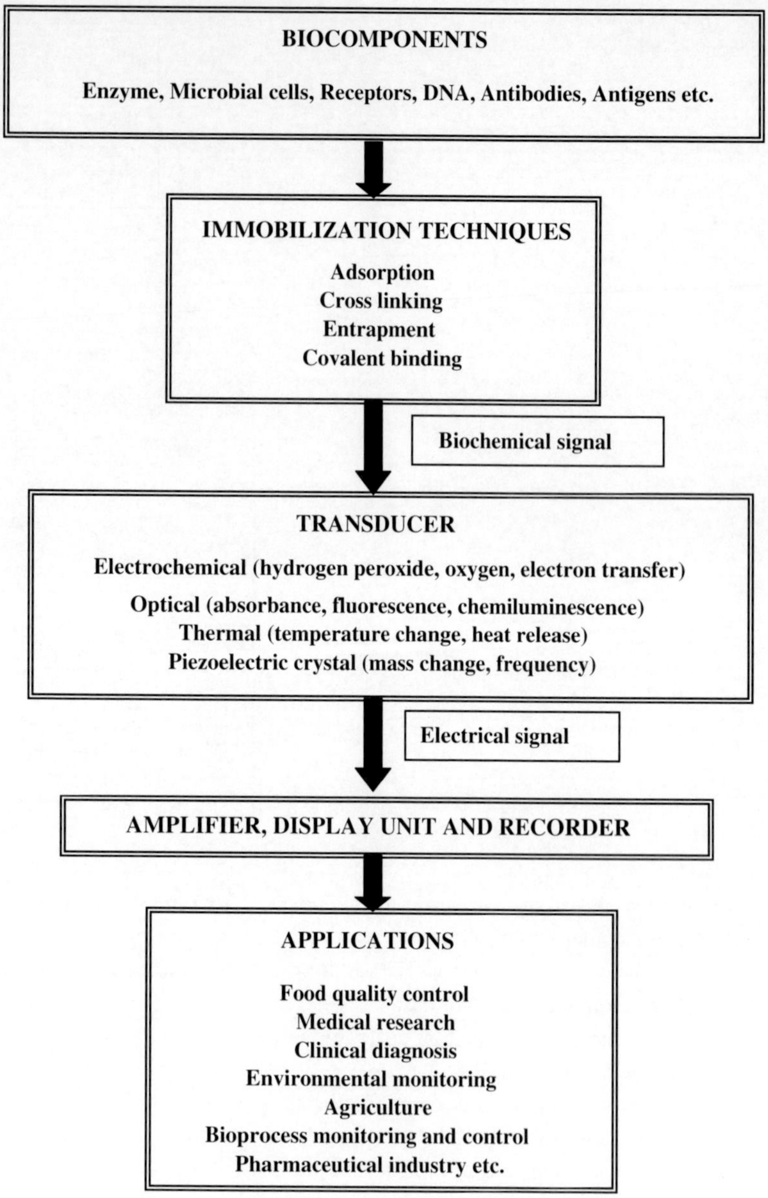

Fig. 10.2 An overview of biosensor technology

Table 10.1 Affinity biosensor

Device name	Bio-recognition element	Analyte
Immunoassay	Antibody	Antigen
–	Receptor	Hormone
–	Enzyme	Inhibitor
–	Membrane transport protein	Substrate analogue
Ion selective electrode	Ionophore	Ions
–	Lectin	Saccharide glycoprotein
–	Apoenzyme	Prosthetic group

antibody-antigen complex after 20 min assay operation. To shorten the time for the assay, a new format was developed. The binding reactions between antibody and antigen can be monitored as a time-dependent change of fluorescence signal. The signal change was proportional as a reaction ratio of antibody to the analyte. Thus, it was possible to monitor the analyte concentration in the sample from the signal. As a result of the monitoring, the analytical time was dramatically shortened to less than 2 min.

10.3.3 Receptors

An active biological receptor can be immobilized and stabilized in a polymeric film to construct receptor-based biosensor for determining an analyte of interest in a sample. The receptor-based biosensors include a polymeric film having a biological receptor capable of binding an analyte of interest immobilized therein according to the method of the present invention and an electrical means for determining the presence and quantity of the analyte. In particular, acetylcholine and opiate receptors have been immobilized in a polymeric film made by combining the receptor, a material capable of polymerization (e.g., bovine serum albumin, gelatin) and a polymerizing agent (e.g., glutaraldehyde).

10.3.4 Fragments of Nucleic Acids

An automated optical biosensor system was reported based on fluorescence excitation and detection in the evanescent field of a quartz fiber used to detect 16-mer oligonucleotides in DNA hybridization assays (Abel et al. 1996). For all hybridization experiments a biotinylated capture probe was immobilized using aminosilanized fibers.

10.3.5 Microorganisms

Microorganisms have a number of advantages as biological sensing materials in the fabrication of biosensors. They are present ubiquitously and are able to metabolize a wide range of chemical compounds. Microbes have a great capacity to adapt to adverse conditions and to develop the ability to degrade new molecules with time. They are also amenable for genetic modifications through mutation or recombinant DNA technology and serve as an economical source of intracellular enzymes. A microbial biosensor consists of a transducer in conjunction with immobilized viable or non-viable microbial cells (Riedel 1998). Non-viable cells obtained after permeabilization or whole cells containing periplasmic enzymes have mostly been used as an economical substitute for enzymes. Viable cells make use of the respiratory and metabolic functions of the cell, the analyte to be monitored being either a substrate or an inhibitor of these processes.

Microorganism-based biosensors tend to use one of three primary mechanisms. For the first mechanism, the pollutant is a respiratory substrate. Biosensors that detect biodegradable organic compounds measured as biological oxygen demand (BOD) are the most widely reported of the microorganism-based biosensors using this mechanism. Several of these devices are commercially available from vendors including: Nisshin Electric Co. Ltd., Tokyo; Autoteam, GmbH, Berlin; Prufgeratewrk, Medingen GmbH, Dresden; and Dr. Lange, GmbH, Berlin. The use of these devices has been incorporated into industrial standard methods in Japan.

Another mechanism used for microorganism-based biosensors involves the inhibition of respiration by the analyte of interest. Microbial respiration and its inhibition by various environmental pollutants have been measured both optically and electrochemically. Inherent advantages of these techniques apply primarily to the use of microorganisms as compared to isolated enzymes. These biosensors are relatively inexpensive to construct and can operate over a wide range of pH and temperature. General limitations involve the long assay times including the initial response and return to baseline. These characteristics are primarily determined by the cellular diffusion characteristics that can be modified by using genetically engineered microorganisms. The broad specificity of these biosensors to environmental toxins may be an advantage or disadvantage depending on the intended application. In this respect, these devices might be most applicable for general toxicity screening or in situations where the toxic compounds are well defined, or where there is a desire to measure total toxicity through a common mode of action.

Biosensors have also been developed using genetically engineered microorganisms (GEMs) that recognize and report the presence of specific environmental pollutants. The microorganisms used in these biosensors are typically produced with a constructed plasmid in which genes that code for luciferase or β-galactosidase are placed under the control of a promoter that recognizes the analyte of interest. Because the organism's biological recognition system is linked to the reporting system, the presence of the analyte results in the synthesis of inducible enzymes which then catalyze reactions resulting in the production of detectable products. With respect to environmental applications, the primary disadvantage for this type of biosensor is the limited number of GEMs which have been constructed to respond to specific environmental pollutants. Nevertheless, reported advances include the development of GEMs involving a variety of bioremediation pathways and mechanisms. GEMs that could report both the metabolic condition of the relevant microorganisms as well as the rates of pollutant breakdown could be very useful (Burlage and Kuo 1994).

10.3.6 Lectin

Lectins are proteins which are highly specific for their sugar moieties. They play a role in biological recognition phenomena involving cells and proteins. Specific mono- and oligosaccharides bind to lectins and provide their attachment to cell membranes. Microbial surface bear many of the sugar residues capable of interacting with lectins. The ability of lectins to react with microbial glycoconjugates renders them to be employed as probes and sorbents for whole cells, mutants, and numerous cellular constituents and metabolites. Lectins are attractive reagents for the clinical diagnostic laboratory because of their diverse specificity, commercial availability, a wide range of molecular weights, and their stability in standard buffers. A lectin-based biosensor is reported for electrochemical assay of glycan expression on living cancer cells which may facilitate the medical diagnosis and treatment in early process of cancer (Zhang et al. 2010).

10.3.7 Membrane Transport Protein

A membrane transport protein is a membrane protein involved in the movement of ions, small molecules, or macromolecules, such as another protein across a biological membrane. Transport proteins are integral membrane proteins; that is they exist within and span the membrane across which they transport substances. The proteins may assist in the movement of substances by facilitated diffusion or active transport. These mechanisms of action are known as carrier-mediated transport. A prototype lactose sensor is reported which consists of a quartz slide covered by a lipid membrane containing the protein lactose permease from *Escherichia coli*. This protein is a lactose/H^+ co-transporter, hence lactose in the external medium initiates lactose/H^+ co-transport across the lipid membrane. This leads to a rise in proton concentration in the small volume between the lipid membrane and the quartz surface which can be detected by a pH-sensitive fluorescence dye (Kiefer et al. 1991).

10.4 Transducers

The transducing element of a biosensor is used to convert the biological recognition step into an electrical signal that can be coupled to a

Table 10.2 Summary of transducer devices

Type	Device	Parameter measured
Electronic	Amperometric	Change in current
	Potentiometric	Change in potential
	Impedance	Change in impedance
Thermal	Thermistor	Change in temperature
Mechanical	Piezo-electric	Change in mass vibrations
Acoustical	Acoustical	Change in amplitude, phase, or frequency of acoustic wave
Optical (Colorimetric and Photometric)	Photo-conductive	Change in intensity, colour or emission
	Photo-emissive	Fluorescence or luminescence activation, quenching, and polarization
	Photo-diodes	
	Photo-transistors	
	Fluorescence or Chemiluminiscence	

microprocessor for control and display. Transducers are mainly divided into four major categories as summarized in Table 10.2.

10.4.1 Electrochemical Transducers

The most common transducer for the construction of a biosensor is electrochemical transducer. The electrochemical device uses a chemical change to measure the input parameter, whereas the output is a varying electrical signal proportional to the measured chemical change. On the basis of output electrical signal the electrochemical transducers may be classified as:

10.4.1.1 Amperometric Biosensors

These biosensors are based upon the resulting steady state limiting current between two electrodes under constant voltage. Most of the glucose biosensors are based upon this principle. It can be either directly or indirectly mediated using suitable redox mediator like ferrocene dye. The electrode potential E will control the concentration of the two redox forms according to Nernst equation:

$$\mathbf{E = E^0 + RT/nF\ ln\ \{[ox]/[red]\}}$$

E^0 is electrode potential at standard conditions

R is universal gas constant

T is absolute temperature

F is Faraday constant

n is electrons transferred in the reaction

[ox]/[red] is ratio of oxidized to reduced molecules

10.4.1.2 Potentiometric Biosensors

These biosensors are based upon non-faradic electrodes process with no net current flow. They are based on the rise in electrode potential due to the accumulation of the ionic species produced from the particular biochemical reactions and is measured with a reference electrode. This electrode potential is proportional to analyte concentration in log scale and is also governed by Nernst equation. These sensors have accordingly glass electrode, ion selective electrode, or field effect transistor (FET) as the transducer. Many biosensors for β-lactam antibiotic are based upon this principle.

10.4.1.3 Impedance Biosensors

Impedance biosensors may be used to measure the electrical impedance of an interface in AC steady state with constant DC bias conditions. Mostly this is accomplished by imposing a small sinusoidal voltage at a particular frequency and measuring the resulting current; the process can be repeated at different frequencies. The current–voltage ratio gives the impedance and this technique is known as electrochemical impedance spectroscopy (EIS). The impedance of the electrode-solution interface changes when the target analyte is captured by the probe. EIS can be utilized over a wide frequency range or at a single frequency to study a variety of electrochemical phenomena. Impedance measurement does not require special reagents or labels for operation.

10.4.2 Optical Transducers

Optical sensors, originally developed for oxygen, carbon dioxide, and pH, have been extended to the construction of fluorescent and luminescent opto-electrodes. The selective biocomponent is immobilized at one end of optical fiber, with both the excitation and detection components located at the other end. $NAD(P)^+/NAD(P)H$- dependent dehydrogenases are obvious candidates for these biosensors, because NAD(P)H is known to absorb light strongly at 340 nm and fluorescence at 460 nm. Several $NAD(P)^+/NAD(P)H$-dependent dehydrogenase enzymes exist for a number of food and beverage-related compounds, but the requirement for expensive and unstable cofactors is the major drawback of this technique. Hydrogen peroxide fiber-optic sensor based on absorbance or chemiluminescence has also been developed for various analytes using oxidases. Optical transducers can also take advantages of affinity interactions between antibodies and antigens to screen for microbial contaminants (Mehrvar et al. 2000).

10.4.3 Thermal Transducers

This device is regarded as a small microcalorimeter with immobilized biocomponents, packed in a small column in proximity to the heat sensing transducer, commonly a thermistor. Most enzyme or microorganism catalyzed reactions are accompanied by considerable heat evolution (20–100 kJ/mol), hence this device has a broader applicability than other transducers. The commercial enzyme thermistor claims to detect temperature change in the range 0.0001–0.05 °C. The sensitivity of this technique is equivalent to the substrate concentration of 10^{-5} M. Thermal transducers have also been applied for the detection of antibody-antigen interactions. This technique is known as the thermometric enzyme linked immunosorbent assay, however its use is sophisticated and expensive.

10.4.4 Piezoelectric Transducer (Quartz Microbalance)

A quartz crystal can oscillate when partially or completely immersed in a liquid, and if the physiochemical properties of the liquid (such as viscosity, density, and conductivity) are defined, a change in the weight of the quartz crystal can be detected by measuring its vibration frequency. Biologically active materials such as antibodies, enzymes, and antigens have been used as active coatings. For the past two decades, intensive research has been directed to developing piezoelectric sensors for a variety of applications. However, non-specific binding and poor sensitivity are the two major drawbacks of piezoelectric transduction (Sochaczewski et al. 1990).

10.5 Immobilization Techniques

An efficient and effective biosensor requires intimate contact between its component parts, normally achieved by restricting the biomolecule, cell, or organelles to a thin layer on or near the transducer surface. The processes employed in achieving this restriction are collectively known as immobilization techniques, defined according to the recommendation of standardization of nomenclature in enzyme technology as:

1. Entrapment with a membrane, gel, or microcapsule.
2. Physical or chemical adsorption.

3. Cross-linking between molecules.
4. Covalent bonding to soluble or insoluble supports.

The localization of biological components at the transducer confers several advantages:

1. Enables recovery of the biocomponent without dilution or contamination.
2. Permits continuous or repeated use of the sensor.
3. Maximizes response by concentrating the biocomponent.
4. Often results in improvement of enzyme stability due to increased rigidity (which maximizes unfolding), a favorable microenvironment, or diffusional effects (Klibnov 1983).

The original concept of biosensor used enzyme entrapment at an electrode with dialysis membrane. This simple method has generally been superseded by other techniques but is still employed with enzymes and microbial cells. Enzyme entrapment within a polyacrylamide gel matrix was successfully employed by Updike and Hicks (1967). This method has been widely adopted. Despite its popularity as an enzyme immobilization technique, gel entrapment is better suited for cells and organelles since enzyme molecules tend to leak from the matrix. Other polymers used for the occlusion of cells and/or enzymes include alginate, carrageenan, agar, cellulose acetate, polyvinyl alcohol (PVA), polyvinyl chloride, polycarbonate, and polypyrrole (Brooks et al. 1991). The procedure with their inherent advantages and disadvantages are presented in Table 10.3.

Adsorption is the oldest known form of immobilization, dating back to 1916 when Nelson and Griffin first immobilized invertase on alumina and charcoal. Adsorption in aqueous phase is a complex process involving hydrophobic interactions, physical adsorption, and chemisorption. Physical adsorption is a result of Van der Waals forces which are made up of London dispersion forces and electrostatic interaction. In contrast to physical adsorption where there is no net transfer of electrons from the interacting species, in chemisorption there is a sharing or transfer of electrons to form a chemical bond, for example *via* pi–pi bonding. Biosensor applications of adsorption techniques are limited by their reversibility, since the degree of attachment is often insufficient to maintain a constant coverage during repeated or continuous uses (Riedel 1998).

Copolymerization of enzymes can be accomplished by cross-linking with a low molecular weight bi-functional reagent, resulting in a soluble film which adheres to, or is retained at, a transducer. In biosensor applications, glutaraldehyde, which couples to lysine residues, is by far the most popular of the numerous bi-functional reagents that have been developed for cross-linking of proteins. A lysine rich protein, such as albumin or collagen, is often added to glutaraldehyde-enzyme preparations to increase the number of cross-linking sites and to provide physical strength. Bi-functional reagents are available for modification of other protein functional groups, including carboxyl, sulfhydryl, phenol, and imidazole groups (Deshpande et al. 1986; Svitel et al. 1998).

Some remarkably stable and durable immobilized enzyme preparations have been achieved by the covalent attachment of biological molecules to a solid surface, such as a membrane, bead or tube, which is positioned at or near the transducer surface. Support materials can be broadly classified into inorganics (glasses, titanium dioxide), polysaccharides (starch, agarose, cellulose), and other polymers (nylon, polyaminostyrene, and polymethacrylate). A bifunctional reagent is covalently bonded to the support and the enzyme subsequently reacted with the activated surface. The permanence of covalent attachment is a great advantage in the fabrication of enzyme sensors. However, such immobilization may be deleterious as the reagents used can cause an initial loss of activity due to inadvertent modification of amino acid residues at or near the site. Alternatively, glycoproteins such as glucose oxidase can be cross-linked or bound to a surface *via* activated carbohydrate residues with the advantage that the active site remains unchanged (Scouten 1987).

Table 10.3 Immobilization procedure for biosensor construction

Methods	Advantages	Disadvantages
Adsorption on insoluble materials (Van der waal's forces, hydrophobic forces, or ionic bonding)	Simple, mild conditions, and mild toward proteins	This technique is highly sensitive toward pH, solvent, and temperature
Cross-linking (using multifunctional groups)	Simple technique, strong chemical binding of the biomolecules, physically adsorbed proteins (covalently bound onto a support) are stabilized with this technique	Relatively low enzyme activity so requires more enzymes, reaction control is difficult, the protein layer is gelatinous, provides less rigidity
Entrapment (gel or semi permeable membrane)	Mild procedure and common technique for any enzyme	Mass transfer resistance is high, leakage of biocatalyst, biocatalyst may denature as a result of reaction
Covalent binding (membrane or insoluble supports)	Stable enzyme-support complex, leakage of the biomolecule is rare, ideal for mass production and commercialization	Intricate and time consuming procedure, activity may decrease due to reaction with groups essential for the biological activity (can be overcome by immobilization in the presence of the substrate or inhibitor of enzyme)

Table 10.4 Summary of immobilization techniques

	Adsorption	Entrapment	Covalent coupling	Cross-linking
Matrix material	Ion exchange resins, active charcoal, silica gel, clay, aluminum oxide, porous glass	Alginate, carageenan, collagen, polyacrylamide, gelatine, silicon rubber, polyurethanes	Agarose, cellulose, Polyvinyl Chloride, ion exchange resins, porous glass	Cross linking agents: glutaraldehyde, bis-isocyanate, bis-diazobenzidine
Nature of bonding	Reversible; changes in pH, ionic strength may detach the enzyme	Physical entrapment	Chemical bonding	Entrapment; functionally inert proteins are often used together (BSA, gelatin)
Cost	Inexpensive	Inexpensive	Expensive	Inexpensive

Matrix material, nature of bonding, the cost of immobilization is compared for various techniques in Table 10.4.

10.6 Characteristics of Biosensors

10.6.1 Response Time

When, there is a change in input parameter, biosensors do not change output state immediately. Rather, it will change to the new state over a period of time, called the response time. The response time can be defined as the time required for a sensor output to change from its previous state to a final-settled value within a tolerance band of the correct new value. This concept is somewhat different from the notion of the time constant of the system.

10.6.2 Precision and Accuracy

Accuracy is a measurement of how close the sensor comes indicating the actual value of the measured variable. Accuracy is always given in terms of inaccuracy such as ± 1 %. Precision is a measure of the consistency of a sensor in measuring the same value under the same operating conditions over a period of time.

10.6.3 Reproducibility

Reproducibility has the same definition for electrochemical biosensors, as any analytical device. It is the measure of the scatter or of the drift in a series of observation or result performed over a period of time. It is a generally determined for analyte concentrations with the usable range.

10.6.4 Lifetime

Although some biosensors have been reported usable under laboratory condition for periods of more than 1 year, their practical life time when incorporated into industrial processes or to biological tissue, such as glucose biosensors implanted *in vivo*, is either unknown or limited to days or weeks. While it is relatively easy to determine the laboratory bench stability of biosensors, both during storage and operation in the presence of analyte, procedure for assigning their behavior during several days of introduction into industrial reactors is much complex and difficult to handle. In both cases one should specify whether lifetime is a storage (shelf) or operational (use) one and what are the storage and operating conditions. Finally the mode of assessment of lifetime should be specified, i.e., by reference to initial sensitivity, within the linear range of the calibration curve, to decrease by 10 % (LT 10) or 50 % (LT 50). For the determination of the storage life time, it is suggested to compare sensitivities of different biosensors, issued from the same production batch of homogenous patterns, after different storage duration time under identical conditions. Term 'lifetime' is still the topic of ongoing discussion.

10.6.5 Stability

The operational stability of a biosensor response may vary considerably depending on the sensor geometry, method of preparation, as well as on the applied receptor and transducer. Furthermore, it is strongly dependent upon the response rate-limiting factor, i.e., substrate external/inner diffusion or biological recognition reaction. Finally, it may vary significantly depending upon the operational conditions such as continuous/discontinuous contact with substrate containing solution, substrate concentration, temperature, pH, presence of organic solvents, and the sample matrix.

Within the biosensor community the term "biosensor stability" is controversial. However, it has been agreed that a general definition would be useful and beneficial. Without a general agreement it would be almost impossible to directly compare 'stability data' obtained from different experimental setups at different research centers.

1. **Operational stability** depends upon following factors:
   (a) Response rate-limiting factor: mass transfer or catalytic limiting
   (b) Continuous versus discontinuous contact with substrate containing solution
   (c) Substrate concentration
   (d) Temperature
   (e) Buffer
   (f) Organic solvents
   (g) Sample matrix
2. **Storage stability** depends on following factors:
   (a) Dry or wet storage
   (b) Air or nitrogen atmosphere
   (c) Temperature
   (d) Buffer
   (e) Additives.

10.6.6 Drift Rate

Alternatively, biosensor stability measure may be quantified as a drift rate, when the sensitivity evolution is monitored during storage or operational conditions. Drift rate is defined as the slow change in output signal, independent of measured property. It is especially useful for biosensor where sensitivity evolution is either very slow or studied during rather short periods of time.

10.6.7 Specificity

Like other bioanalytical methods (such as immuno-assays and enzyme-assays), biosensors use a biologically derived compound as the sensing elements. The advantage of biological sensing elements is their remarkable ability to distinguish between the analyte of interest and similar substances. With biosensors, it is possible to measure specific analytes with great accuracy.

10.6.8 Speed

One characteristic of biosensors that distinguish them from other bioanalytical method is that the analyte tracers or catalytic products can be directly and instantaneously measured. There is no need to wait for results from lengthy procedures carried out in centralized laboratories.

10.6.9 Simplicity

The uniqueness of a biosensor is that the receptor and transducer are integrated into one single sensor. This combination enables the measurement of target analytes without using reagents. For example, glucose concentration in a blood sample can be measured directly by a biosensor (which is made specifically for glucose measurement) by simply dipping the sensor in the sample. This is in contrast to the conventional assay in which many steps are used and each step may require a reagent to treat the sample.

10.6.10 Online Monitoring

Another advantage that biosensors have over bioanalytical assay is that they can regenerate and reuse the immobilized biological recognition component. For enzyme-based biosensors, an immobilized enzyme can be used for repeated assays: this feature allows these devices to be used for continuous or multiple assays. By contrast, immunoassay including enzyme-linked immunosorbent assay (ELISA), are typically based on irreversible binding and are thus used only once and then discarded.

10.7 Applications of Biosensors

Biosensors are being developed for different applications including environmental and bioprocess monitoring and control, quality control in food industry, agriculture, military, and particularly, medical uses. In fact, most of the commercially available biosensor systems are applied in the clinical and pharmaceutical markets. A few applications of biosensors are summarized in Table 10.5.

10.7.1 Food and Pharmaceutical Quality Control

Food microbiologists are constantly seeking rapid and reliable automated systems for the detection of biological activity. Biosensors provide sensitive, miniaturized systems that can be used to detect unwanted microbial activity or the presence of a biologically active compound, such as glucose or a pesticide. The use of immunodiagnostics and enzyme biosensors has reduced the time for detection of pathogens such as *Salmonella* and has provided detection of biological compounds such as cholesterol or chymotrypsin. Improving the preservation or increasing the yield of agricultural products is a common aim for the producers, normally realized by using herbicides, insecticides, antibiotics, or hormones. The uses of these xenobiotics are a potential risk for consumers. The safety and quality of the food is a major concern these days. In novel foods such as functional foods, ingredients or constituents are present to tackle health and bad eating-habit-related issues. In all these cases there is the need of monitoring an increasing number of various analytes. The continued research in biosensor technology will

Table 10.5 Applications of microbial biosensors in food, fermentation, environmental, and allied fields

Analyte	Microorganism	Transducer/ immobilization	Reference
BOD	*Pseudomonas putida*	Oxygen electrode (adsorption on porous nitro cellulose membrane)	Chee et al. (1999)
BOD	*Trichosporum cutaneum*	Miniature oxygen electrode (UV cross-linking resin (ENT-3400)	Yang et al. (1996)
BOD	Activated sludge (mixed microbial consortium)	Oxygen electrode/flow injection system (entrapped in dialysis membrane)	Liu et al. (2000)
BOD	Salt tolerant mycelial yeast	Oxygen electrode (PVA)	Tag et al. (2000)
Anionic surfactants [linear alkyl benzene sulfonates (LAS)]	LAS degrading bacteria isolated from activated sludge	Oxygen electrode, (reactor type sensor, Ca-alginate)	Nomura et al. (1994)
Phenolic compounds	*Pseudomonas putida*	Oxygen electrode (reactor with cells adsorbed on PEI glass)	Nandakumar and Mattiasson (1999)
Nitrite	*Nitrobacter vulgaris*	Oxygen electrode (adsorption on Whatman paper)	Reshetilov et al. (2000)
Cyanide	*Saccharomyces cerevisiae*	Oxygen electrode (PVA)	Ikebukuro et al. (1996)
Chlorophenols	*Rhodococcus* sp.; *Trichosporon beigelii*	Oxygen electrode (PVA)	Riedel et al. (1993, 1995)
Organophosphate nerve agents (paraxon, methyl parathion, diazinon)	Genetically engineered *Escherichia coli* (organophosphorous hydrolase)	Potentiometric (adsorption on electrode surface)	Mulchandani et al. (1998)
Alcohol	*Candida vini*	Oxygen electrode (porous acetyl cellulose filter)	Mascini et al. (1989)
Glucose	*Aspergillus niger* (glucose oxidase)	Oxygen electrode (entrapment in dialysis membrane)	Katrlik et al. (1996)
Glucose, sucrose, lactose	*Gluconobacter oxydans* (D-glucose dehydrogenase), *Saccharomyces cerevisiae* (invertase), *Kluyveromyces marxianus* (β-galactosidase)	Oxygen electrode (gelatin)	Svitel et al. (1998)
Short chain fatty acids in milk (butyric acid)	*Arthrobacter nicotianae* (acyl-CoA oxidase)	Oxygen electrode (Polyvinyl alcohol)	Ukeda et al. (1992a, b)
Phosphate	*Chlorella vulgaris*	Oxygen electrode (polycarbonate membrane)	Matsunaga et al. (1984)
CO_2	CO_2 utilizing autotropic bacteria (*Pseudomonas*)	Oxygen electrode (bound on cellulose nitrate membrane)	Suzuki and Karube (1987)
Vitamin B-6	*Saccharomyces uvarum*	Oxygen electrode (adsorption on cellulose nitrate membrane)	Endo et al. (1995)

(continued)

Table 10.5 (continued)

Analyte	Microorganism	Transducer/ immobilization	Reference
Vitamin B-12	*Escherichia coli*	Oxygen electrode (trapped in porous acetyl cellulose membrane)	Karube et al. (1987)
Peptides (aspartame)	*Bacillus subtilis*	Oxygen electrode (filter paper strip and dialysis membrane)	Renneberg et al. (1985)
Phenylalanine	*Proteus vulgaris* (Phenylalanine deaminase)	Amperometric oxygen electrode (Ca-alginate)	Liu et al. (1996)
Pyruvate	*Streptococcus faecium* (Pyruvate dehydrogenase complex)	CO_2 gas sensing electrode (direct immobilization on sensor membrane)	Di Paolantonio and Rechnitz (1983)
Tyrosine	*Aeromonas phenologenes* (Tyrosine-phenol lyase)	NH_3 gas sensing electrode (direct immobilization on sensor membrane)	Di Paolantonio and Rechnitz (1982)
Enalapril maleate (angiotensin)	*Bacillus subtilis*	Oxygen electrode	Fleschin et al. (1998)

soon provide on-line quality control of food production, which will not only reduce the cost of food production but will also provide greater safety and increased food quality.

The pharmaceutical industry is constantly looking for new tools, techniques and practices to increase the productivity of research and development and to improve the quality of the innovative medicines they produce. Biosensors comprise a very specific biological component such as enzyme, antibody receptor etc., thus provide an advantage of quantitative study of the interaction between a drug compound and an immobilized biocomponent. Affinity biosensors are suitable for high-throughput screening of bioprocess produced antibodies and for candidate drug screening. Enzyme- based biosensors can be applied in the pharmaceutical industry for monitoring chemical parameters in the production processes.

10.7.2 Medical Research and Clinical Diagnosis

The initial thrust for advancing biosensor technology came from health care area, where diagnostics and monitoring have key roles to play in optimising health care. It is now generally recognized that measurements of blood gases, ions, and metabolites are often essential and allow a better estimation of the metabolic state of a patient. There is still a notable gap between laboratory biosensing and commercially viable medical or consumer diagnostic devices. The biosensor community needs to refine its work to the wider population for telemedicine or telehealthcare.

10.7.3 Environmental Monitoring

Another assay situation which may involve a considerable degree of the unknown is environmental monitoring. The primary measurement media here will be water or air, but the variety of target analytes is vast. At sites of potential pollution, such as factory effluent, it would be desirable to install on-line real-time monitoring and alarm, targeted at specific analytes, but in many cases random or discrete monitoring of both target species and general hazardous compounds would be sufficient. The possible analytes include biological oxygen demand (BOD) which provides a good indication of pollution, atmospheric acidity, and river water pH, detergent, herbicides, and fertilizers (organophosphates, nitrates, etc.). Biosensors can be excellent analytical tools for monitoring programs working to implement legislation.

10.7.4 Agriculture

Biosensor technology is a powerful technique, having the specificity and sensitivity of biological systems in small and low cost devices for measurement. Though various biosensors are developed in research laboratories, there are not many reports of applications in agricultural monitoring. The need for fast, on-line, and accurate sensing opens up opportunities for biosensors in many different agricultural areas such as *in situ* analysis of pollutants in crops and soils, detection and identification of diseases in crops and livestock, monitoring animal fertility and screening therapeutic drugs in veterinary testing and the detection of various compounds affecting the fertility of soil.

10.7.5 Bioprocess Monitoring and Control

While real-time monitoring with feedback control involving automated systems does exist, currently a few common variables are measured on-line (e.g., pH, temperature, CO_2, O_2) which are often indirectly related with the process under control. Real-time monitoring of carbon sources, trace metals, dissolved gases, important components of media, and products in fermentation processes could lead to optimization of the procedure giving increased yields at decreased materials cost.

10.8 Conclusion

In this chapter we have discussed various types of biosensing elements, transducers, and immobilization techniques utilized to develop biosensors. The diversity of these biosensor components provides opportunity to develop biosensors for a variety of analytes and thus can be used in diverse applications. The activity and specificity of biosensing elements make them suitable for analysis of complex solutions with short response time. Various biosensors are having the capability to be used for online analysis which enhances the use of biosensors in food and bioprocess industries to optimize production, in medical field for screening of the potential drug molecules and in environmental field to identify and quantify the potential pollutants. The efforts are continuously made to develop new biosensors and commercialize them but still few biosensors are commercially successful and there is a vast scope of research and development in this sector.

References

Abel AP, Weller MG, Duveneck GL, Ehrat M, Widmer HM (1996) Fiber-optic evanescent wave biosensor for the detection of oligonucleotides. Anal Chem 68:2905–2912

Brooks SL, Higgins IJ, Newman JD, Turnerp APF (1991) Biosensors for process control. Enzyme Microbial Technol 13:946–955

Burlage R, Kuo CT (1994) Living biosensors for the management and manipulation of microbial consortia. Annu Rev Microbiol 48:291–309

Chee GI, Nomura Y, Karube I (1999) Biosensor for the estimation of low biochemical oxygen demand. Anal Chim Acta 379:85–191

Clark LC Jr (1956) Monitor and control of blood and tissue oxygen tensions. Trans Am Soc Artif Intern Organs 2:41–48

Clark LC Jr, Lyons C (1962) Electrode systems for continuous monitoring cardiovascular surgery. Ann N Y Acad Sci 102:29–45

Deshpande A, D'Souza SF, Nadkarni GB (1986) Immobilization of microbial cells in gelatine using gamma-irradiation. Indian J Biochem Biophys 23:353–354

Di Paolantonio CL, Rechnitz GA (1982) Induced bacterial electrode for the potentiometric measurement of tyrosine. Anal Chim Acta 148:1–12

Di Paolantonio CL, Rechnitz GA (1983) Stabilized bacteria-based potentiometric electrode for pyruvate. Anal Chim Acta 148:1–12

Endo H, Kamata A, Hoshi M, Hayashi T, Watanabe E (1995) Microbial biosensor system for rapid determination of vitamin B-6. J Food Sci 60:554–557

Fleschin S, Bala C, Bunaciu AA, Panait A, Aboul-Enein HY (1998) Enalapril microbial biosensor. Prep Biochem Biotechnol 28:261–269

Ikebukuro K, Honda M, Nakanishi K, Nomura Y, Masuda Y, Yokoyama K, Yamauchi Y, Karube I

(1996) Flow-type cyanide sensor using an immobilized microorganism. Electroanalysis 8:876–879

Karube I, Wang Y, Tamiya E, Kawarai M (1987) Microbial electrode sensor for vitamin-B12. Anal Chim Acta 199:93–97

Katrlik J, Svorc J, Rosenberg M, Miertus S (1996) Whole cell amperometric biosensor based on *Aspergillus niger* for determination of glucose with enhanced upper linearity limit. Anal Chim Acta 331:225–232

Kiefer H, Klee B, John E, Stierhof YD, Jähnig F (1991) Biosensors based on membrane transport proteins. Biosens Bioelectron 6(3):233–237

Klibanov AM (1983) Stabilization of enzymes against thermal inactivation. Adv Appl Microbiol 29:1–28

Liu B, Cui Y, Deng J (1996) Studies on microbial biosensor for DL-phenylalanine and its dynamic response process. Anal Lett 29:1497–1515

Liu J, Bjornsson L, Mattiasson B (2000) Immobilized activated sludge based biosensor for biochemical oxygen demand measurement. Biosens Bioelectron 14:883–893

Mascini M, Memoli A, Olana F (1989) Microbial sensor for alcohol. Enzyme Microb Technol 11:297–301

Matsunaga T, Suzuki S, Tomoda R (1984) Photomicrobial sensor for selective determination of phosphate. Enzyme Microb Technol 6:355–357

Mehrvar M, Bis C, Scharer JM, Young MM, Luong JH (2000) Fiber-optic biosensors–trends and advances. Anal Sci 16:677–692

Mulchandani A, Mulchandani P, Kaneva I, Chen W (1998) Biosensor for direct determination of organophosphate nerve agents using recombinant *Escherichia coli* with surface-expressed organophophorous hydrolase. 1. Potentiometric microbial electrode. Anal Chem 70:4140–4145

Nandakumar R, Mattiasson B (1999) A microbial biosensor using *Psuedomonas putida* cells immobilized in an expanded bed reactors for the on-line monitoring of phenolic compounds. Anal Lett 32:2379–2393

Nomura Y, Ikebukuno K, Yokoyama K, Takeuchi T, Arikawa Y, Ohno S, Karube I (1994) A novel microbial sensor for anionic surfactant determination. Anal Lett 27:3095–3108

Renneberg R, Riedel K, Scheller F (1985) Microbial sensor for aspartame. Appl Microbiol Biotechnol 21:180–181

Reshetilov AN, Iliasov PV, Knackmuss HJ, Boronin AM (2000) The nitrite oxidising activity of *Nitrobacter* strains as a base of microbial biosensor for nitrite detection. Anal Lett 33:29–41

Riedel K (1998) Microbial biosensors based on oxygen electrodes. In: Mulchandani A, Rogers KR (eds) Enzyme and Microbial Biosensors: techniques and protocols. Humanae Press, Totowa, pp 199–223

Riedel K, Hensel J, Rothe S, Neumann B, Scheller F (1993) Microbial sensor for determination of aromatics and their chloroderivatives. Part-II: determination of chlorinated phenols using a *Rhodococcus* containing biosensor. Appl Microbiol Biotechnol 38:556–559

Riedel K, Beyersdorf R, Neumann B, Scheller F (1995) Microbial sensor for determination of aromatics and their chloroderivatives. Part-III: Determination of chlorinated phenols using a biosensor containing *Trichosporon beigelli* (cutaneum). Appl Microbiol Biotechnol 43:7–9

Scouten WH (1987) A survey of enzyme coupling techniques. Methods Enzymol 135:30–65

Sochaczewski EP, Luong JHT, Guilbault GG (1990) Development of a piezoelectric immunosensor for the detection of *Salmonella typhimurium*. Enzyme Microb Technol 12:173–183

Suzuki S, Karube I (1987) An amperometric sensor for carbon dioxide based on immobilized bacteria utilizing carbondioxide. Anal Chim Acta 199:85–91

Svitel J, Curilla O, Tkac J (1998) Microbial cell-based biosensorfor sensing glucose, sucrose or lactose. Biotechnol Appl Biochem 27:153–158

Tag K, Lehmann M, Chan C, Renneberg R, Riedel K, Kunze G (2000) Measurement of biodegradable substances with a mycelia-sensor based on the salt tolerant yeast *Arxula adeninivorans* LS3. Sens Actuators B 67:142–148

Ukeda H, Wagner G, Bilitewski U, Schmid RD (1992a) Flow injection analysis of short-chain fatty acids in milk based on a microbial electrode. J Agric Food Chem 40:2324–2327

Ukeda H, Wagner G, Weis G, Miller M, Klostermeyer H, Schmid RD (1992b) Application of a microbial sensor for determination of short-chain fatty acids in raw milk samples. Z. Lebensm Unters Forsch 195:1–2

Updike SJ, Hicks GP (1967) The enzyme electrode. Nature 214:986–988

Yang Z, Suzuki H, Sasaki S, Karube I (1996) Disposable sensor for biochemical oxygen demand. Appl Microbiol Biotechnol 46:10–14

Zhang X, Teng Y, Fu Y, Xu L, Zhang S, He B, Wang C, Zhang W (2010) Lectin-based biosensor strategy for electrochemical assay of glycan expression on living cancer cells. Anal Chem 82(22):9455–9460

11 Monoclonal Antibody Production and Applications

B. D. Lakhchaura

Abstract

Antiserum is a heterogeneous mixture of antibodies produced against an antigen. Antigens usually have multiple epitopes and separate antibodies are produced against each by an individual clone of B cell. Monoclonal antibody produced by hybridoma technique is a single homogeneous antibody population which is specific for one epitope and exhibits no cross-reaction. The technique involves fusion of B cell with the myeloma cell and cultivation of the fused clone for indefinite production of the desired antibody. Myeloma cells are the tumor cells and impart immortality to the clone while B cells contribute to the antibody production ability. Selection of the fused cell is carried out in hypoxanthine-aminopterin-thymidine (HAT) containing medium so that unfused myeloma cells are unable to grow by virtue of their HGPRT negative character. Hybridomas once produced can be cryopreserved in liquid nitrogen for indefinite storage. It is possible to antibodies in culture medium or as ascitic fluid in mice. Both procedures allow sizeable harvest of the antibody. Monoclonal antibodies have proven to be valuable tools in immunodiagnostic, immunotherapy, and in biological and biochemical research. Monoclonal antibody-based immunodiagnostic kits are available for detection of pregnancy, for diagnosing numerous pathogenic microorganisms, measuring level of drug in blood or urine, matching histocompatibility antigens, and detecting antigens shed by various tumors. A sizable number of monoclonal antibodies are available as therapeutic agents.

11.1 Introduction

Antibodies play an important role in body defence against infectious diseases. Antibodies are glycoprotein and are produced by B lymphocytes with the help of helper T lymphocytes and antigen presenting cells. An antigen, when

B. D. Lakhchaura (✉)
G. B. Pant University of Agriculture and Technology, Pantnagar, India
e-mail: lakhchaurabd@rediffmail.com

I. Ravi et al. (eds.), *Advances in Biotechnology*,
DOI: 10.1007/978-81-322-1554-7_11,

Table 11.1 Comparison between conventional serum and monoclonal antibodies

S. N.	Property	Conventional antiserum	Monoclonal antibody
1.	Determinant	Several	Single
2.	Specificity	Variable with animal and bleed	Standard
		Partial cross-reactions with common determinants	Unexpected cross-reactions may occur
		Seldom too specific	May be too specific for requirements
3.	Affinity	Variable with bleed	May be selected during cloning
4.	Yield of useful antibody	Up to 1 mg/ml	Up to 100 μg/ml in tissue culture, Up to 20 mg/ml in ascitic fluid
5.	Contaminating Immunoglobulin	Up to 100 %	None in culture, 10 % in ascetic fluid
6.	Purity of antigen	Either pure antigen or serum absorption	Some degree of antigen purification desirable but not essential.
7.	Approx. Minimum cost	Usually below £100	Capital cost £10,000, Running cost £10,000 p.a.

administered in an appropriate vertebrate host, results in eliciting an immune response. Basically there are two types of immune responses, humoral and cell mediated. In humoral response, B lymphocytes are activated by antigen presenting cells and helper T lymphocytes. On the other hand, cell-mediated immune response requires activation of cytotoxic T lymphocytes. An antigen molecule possesses a large number of antigenic determinants which are called epitopes. Each antibody is specific for a particular epitope. Thus the antisera raised against a given antigen are a mixture of antibodies, each specific for an epitope. Such antisrea are known as polyclonal antisera. As all antibody molecules are similar in their physical and chemical properties, it is impossible to separate epitope specific antibodies from each other. Antibodies produced from a single clone of B lymphocyte are called as monoclonal antibody (MAb).

The theory of monoclonal antibody production is based on the clonal selection hypothesis of F. Macfarlane Burnet (1959). Each mammalian B lymphocyte has the potential to make a monospecific antibody. The constant region of the antibody chain may alter during the differentiation of the lymphocyte clone but the variable region retains this singular specificity. The first report of Hybridoma production was in fact in 1970 (Sinkovics et al. 1970). Kohler and Milstein in 1975 developed a technique that made it possible to raise epitope specific antibodies; this technique was christened as Hybridoma Technology and the antibodies produced by this method are called Monoclonal antibodies. Table 11.1 gives a comparison between conventional serum and monoclonal antibodies.

11.2 Production of Monoclonal Antibodies

Antibodies are secreted by plasma cells which are programmed in lymph glands for producing antibodies against a particular epitope. The plasma cells can be cultured and made to secrete antibodies but they have a very short life span and die soon. The technique developed by Kohler and Milstein (1975) immortalizes the plasma cells which can thus produce monoclonal antibodies indefinitely. For this purpose they fused the plasma cells with cancerous cells called myeloma cells. Myeloma cells can be obtained from animals suffering from myeloma tumors or tumors can be induced in experimental animals by injecting mineral oil. They used an inbred strain of mice known as Balb/c mice for this purpose. A schematic illustration for monoclonal antibody production is given in Fig. 11.1. All the steps are also detailed subsequently.

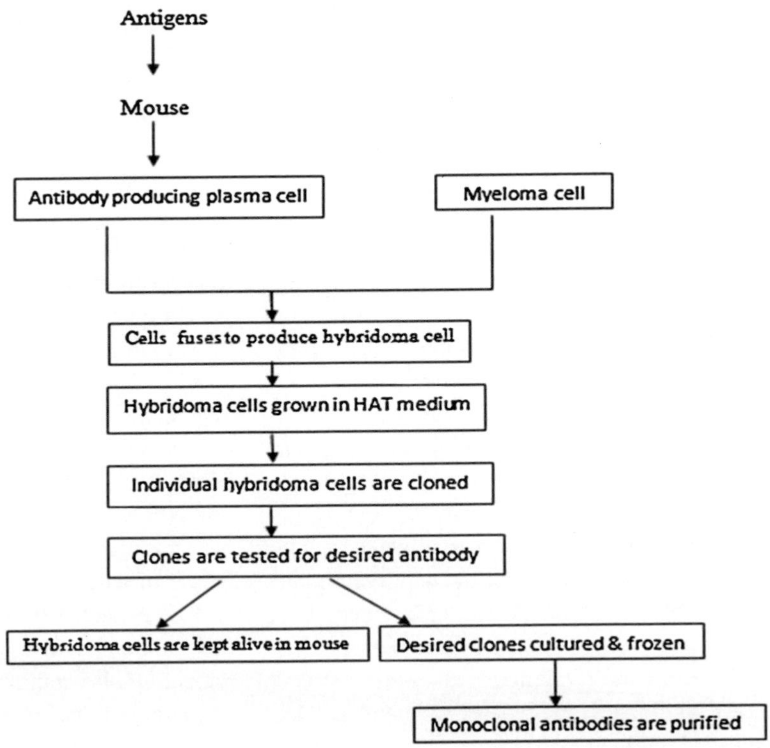

Fig. 11.1 Schematic representation of monoclonal antibody production

11.3 Monoclonal Antibody Assay Requirements

Assay is the most critical factor in production of good hybridomas and is very sensitive. Understanding of the theoretical background is essential for the production of a large number of hybridomas secreting antibody of the required characteristics.

a. **Rate of association**

The initial rate of association of an antibody with an antigen is described by the equation:

$$\text{Rate of formation of product} = K_1[\text{antibody}][\text{antigen}]$$

The number of epitopes on the antigen is much reduced and may be one in case of small protein. Thus the effective concentration of the antigen is quite low. The antibody will still be bivalent or decavalent for IgM but its concentration may be very low. The association rate is usually the parameter which determines the final equilibrium constant and the concentration of antigen in most assays remains constant.

b. **Rate of dissociation**

The rate of dissociation of an antibody-antigen complex is represented by the following equation:

$$\text{Initial rate of dissociation} = K_2([\text{antibody} - \text{antigen}])$$

The dissociation rate is nearly always very much lower than the association rate. With most antibody-antigen reactions, it is the dissociation rate rather than the association rate that determines the affinity as the dissociation rate can vary over 8–9 orders of magnitude. The dissociation rate is not only the most variable among different antibodies under a defined set of conditions but also the most variable in a single antibody with respect to environmental conditions such as pH, temperature, etc.

c. **Equilibrium concentration of reactants**

The equilibrium constant for an antibody-antigen reaction is the ratio of the forward to the backward rates i.e.,

$$\mathrm{Keq} = \mathrm{K_1/K_2} \text{ or}$$

$$\mathrm{Keq} = (\mathrm{Ab - Ag})/(\mathrm{Ab})(\mathrm{Ag})$$

The concentration of the reactants and the assay conditions may influence the possibility of detection of a suitable hybridoma. With polyclonal sera it can be assumed that an optimal equilibrium condition will be achieved by incubation for one or two hours at room temperature, biological pH, and ionic strength, etc. but for monoclonal antibodies several of the parameters will have to be varied and screening procedures will have to be extended if antibodies to more interesting or relevant epitopes are to be detected.

d. **Effect of multivalence**

All antibodies are at least divalent and IgA and IgM antibodies may have several idiotypes on the same molecule. Multivalent antibodies are helpful to screening procedures since the effective dissociation rate can be reduced if enough antigens are present. Antigens having multiple epitopes, mostly found in bacterial systems with symmetrical cell wall structures, can affect the reaction in many ways.

e. **Specificity and affinity**

Monoclonal antibodies against polymorphic antigens could exhibit extreme specificity in one assay and considerable epitope overlap in a second higher affinity assay. However, it is assumed that at least in the initial stages, an assay detecting the maximum numbers of positive clones is required. Suitable conditions for obtaining specific responses can then be achieved later.

f. **Number of assays**

In a typical fusion using 4 × 96 well plates some 400 samples must be assayed in a short time. Subcloning usually involves a similar number although a valuable clone may be subcloned with more. Further, a comprehensive screen of hybridomas should ideally involve the use of several plates under different sets of conditions. If the final application of the antibody is not considered then positive samples detected by the first screening should be assayed by a second directed to the final application, since the number of positive clones is much lower than the total number of clones.

g. **Time of assay**

The time of assay, in terms of clonal growth, should be soon after clones are microscopically visible and again a few days after when the clones are visible to the eye. Screening should continue for several weeks. The actual assay however, is influenced by the antibody concentration.

h. **pH of assay**

It is expected that the most antibodies have their optimal reaction with antigen at physiological pH. It is quite possible to fail to detect a good hybridoma by screening at a single pH. The pH of tissue culture fluid in which cells are growing can vary almost a whole pH unit. If a defined pH is required for the final application then the assay should be buffered accordingly during the incubation of antibody with antigen, particularly for *in vivo* uses. Where pH adjustments are necessary the nature of the buffer should be considered as the buffer components themselves may affect the assay.

i. **Temperature of assay**

Dissociation constant is the variable which is most sensitive to temperature. The best assay conditions for hybridoma favor incubation at 4 °C rather than at room temperature or higher but ideally both should be tried unless the final use precludes certain temperature.

j. **Ionic strength of assay**

There is no detailed information on the ionic strength variations in hybridoma assays. Non-specific binding is more likely to occur at low ionic strengths. If different pH buffers are used in the assay then all buffers used should have the same ionic strength.

11.4 Cell Culture Requirements for Hybridomas

1. **Media**

Two main types of media used for hybridoma production are Dulbecco's Modification of Eagles Medium (DMEM) and Rosewell Park Memorial Institute (RPMI) medium. Media are prepared in double distilled water, sterilized by filtration through 0.2 μm membranes, and usually stored in 500–1,000 ml aliquots for up to 6 weeks. Both the above media are bicarbonate buffered with phenol red indicator. The correct color for medium is bright orange indicating a pH of 7.2.

DMEM or RPMI 1640 supplemented with high glucose (4.58 g/l) are popularly used for culturing myeloma cells. Glutamine (2 mM final concentration), antibiotics such as penicillin (100 U/ml) or streptomycin (10 μg/ml) and 10 % fetal calf serum is also added. Myeloma cells should be grown in the presence of 8-azaguanine prior to fusion to ensure HGPRT negative character. For fusion, cells should be in logarithmic phase ($3–8 \times 10^6$/ml).

2. **Sera**

Fetal calf serum (FCS) is used in nearly all hybridoma work because of the low level of contaminating immunoglobulin. It is kept frozen at −20 °C. Horse and rabbit serum is also sometimes used.

3. **Antibiotics for prevention of contamination**

The main antibiotics used in hybridoma production are penicillin and streptomycin. Penicillin inhibits the growth of most gram-positive bacteria, whereas streptomycin inhibits the growth of most gram-negative bacteria. The antibiotic preparations are usually made up in 100 × stock solutions containing 10^7 units of sodium benzyl penicillin and 10 g of streptomycin sulfate per liter. This is filter sterilized through 0.2 μm membranes and stored in 20 ml aliquots at −20 °C. Some labs use fungizone (Amphotericin B) at 2.5 μg/ml medium.

Mycoplasma contamination may be prevented by kanamycin (100 μg/ml), tylocine (50 μg/ml) or lincomycin, and vanomycin to some extent. Exposure of cells at elevated temperatures has also been reported to curb mycoplasma contamination.

Viral contamination with Epstein Barr (EB) virus in a latent form is usually present if human lymphocytes are used. The virus transforms culture especially in the absence of cytotoxic T lymphocytes. Lymphocyte donors are usually tested for antibody to the viral capsid antigen by use of cell line P3HR1which secretes non-transforming EB.

Yeast contamination is rare. The use of fungizone or nystatin (50 μg/ml) is usually helpful in such situations.

4. **Feeder cells**

Feeder cells are absolutely essential for cloning of hybridomas. They feed or nourish the emerging hybridomas but may have a limited lifespan. If phagocytic cells such as macrophages and monocytes are used they can also be helpful in cleaning debris of dead cells found after aminopterin treatment. Spleen cells are convenient to use as feeders.

5. **Important chemicals**

Dimethyl sulphoxide (DMSO): DMSO is required for freezing and thawing of cells and some use it in fusion procedures. It is sterilized by autoclaving or filtration.

Polyethylene glycol (PEG): All fusions designed to produce hybridomas are performed with chemical fusogens and PEG is the main chemical used for this.

$$HO(CH_2CH_2O)nCH_2CH_2OH$$

Structure of polyethylene glycol (PEG)

PEG has a molecular weight range from 200 to 20,000 Daltons. It is toxic to cells; low molecular weight PEG is more toxic than high molecular weight PEG. Most successful fusions are performed with PEG of molecular weight 600–6000 Daltons. Most protocols use PEG at 37 °C. It has been also shown that room temperature is superior to 37 °C and that the optimum pH is 7.5.

The exact mechanism of fusion is not fully understood, but it is thought that hydrophilic PEG occupies the 'physical free space' leading to agglutination of the cells. This occurs at concentrations of PEG in the range of 40–50 % with some variations. Lower PEG concentrations such as 35 % can, however, be used with a longer exposure time. Mostly fusions are done using PEG in 15 % DMSO.

6. **Myeloma cells**

Myeloma cells are cancerous cells and impart immortality to hybridoma clones. Spleen cells are antibody-producing cells; hence provide the ability to generate the antibody. Most of the murine myeloma cell lines are available from American Type Culture Collection (ATCC), Maryland 20852 USA. A few commonly used murine myelomas are Sp 2/0 Ag.14, NS1/1, Ag 4.1 NSO, and P3X63 Ag8.

Rational for using HAT media for selecting hybridoma clones is that, myeloma cells are deficient in HGPRT and thymidine kinase (TK), hence lack the salvage pathway for nucleotide synthesis. Aminopterin is added to inhibit the de novo pathway of nucleotide synthesis. Thus myeloma cells do not grow in HAT medium. The spleen cells are primary cells and cease to grow after a few divisions. Only hybridoma cells which have acquired HGPRT and TK genes from spleen cells multiply and form clones. These cells secrete very high level of antibodies in the media.

7. **Subcloning of hybridoma clones**

To ensure monoclonality of the hybridoma clones, the cells should be re-cloned under limiting dilution conditions for a few cycles. Culture is diluted so that 1–2 cells per well are distributed. Limited dilution cloning technique requires presence of feeder cells in the cloning wells. This process eliminates non-secretors and contaminating hybridomas ensuring monoclonality of the antibody.

8. **Enzyme Linked Immunosorbent Assay for antibody detection**

The wells of the ELISA plate are coated with the antigen and the supernatant from the hybridoma wells is added. The unbound antibody is washed away. A second antibody conjugated to an enzyme which is specific to the Fc (fragment crystallizable) region of the first antibody is added and allowed to bind to the first antibody. After washing, the substrate for the enzyme is added and the reaction is allowed to proceed. If the first antibody binds to the antigen the second antibody will also bind (to the first antibody) and the enzyme will react with the substrate and colored product is formed which can be easily visualized.

11.5 Steps in Monoclonal Antibody Production

1. Balb/c mice are immunized with the antigen *via* suitable route. For most protein antigens, 2–3 immunizations are sufficient to evoke a strong immune response. Usually a group of 4–5 mice are immunized and the one exhibiting highest titer is used as the source of plasma cells. This mouse is immunized by IV (intravenous) route and after 3 days the mouse is sacrificed.
2. Spleen is removed from the mouse and single cell suspension is prepared.
3. Exponentially growing myeloma cells are mixed with splenocytes in the proportion of 1:10.
4. Fusion between these cells is accomplished in the presence of a fusion agent, usually poly ethylene glycol (MW 4,000). Both cells are centrifuged and the supernatant media is removed. The cell bottom is loosened by gentle tapping and 50 % PEG is slowly added drop-wise. The cell suspension is held at 37 °C for 2–3 min. After that, PEG is diluted with the serum free media and cells are centrifuged to remove the PEG.
5. The cell pellet is suspended in about 100 ml hypoxanthine, guanine, and aminopterin (HAT) medium. The HAT medium consists of RPMI 1,640 medium supplemented with HAT.
6. The suspension is distributed in ten 96-well microtiter plates.
7. The plates are incubated at 37 °C in a carbon dioxide incubator overnight (the incubator is operated at 37 °C, 95:5 air and CO_2 environment and 95 % humidity)

8. Next day, 0.1 ml HAT media is added to each well.
9. Plates are examined regularly for appearance of growth.
10. Between 10 and 14 days of incubation, some wells exhibit growth and medium turns slightly yellowish in color in the wells.
11. The supernatant from wells is screened for the presence of antibodies by Enzyme-Linked Immunosorbent Assay (ELISA) technique.
12. Hybridoma cells from positive wells are expanded by growing in 24-well plates followed by growth in 25 cm^3 culture flask.
13. Desirable hybridoma clones are preserved by cryopreservation in liquid nitrogen.

11.6 Large-Scale Production of Antibodies

Antibodies can be produced by two different techniques. In the first method the hybridoma cells are grown in serum free media. Growing cells secrete antibodies which can be purified by affinity chromatography technique. The antibody yield usually varies between 10 and 50 μg/ml. In exceptional cases one may get a yield of up to 100 μg/ml as well. The second method which involves production of ascitic fluid can produce up to 20 mg/ml antibody yield. For ascitic fluid production, the hybridoma cells are injected in the peritoneal cavity of pristine primed-immunocompromised Balb/c mice. The ascitic fluid develops in the peritoneal cavity and can be easily aspirated with the help of 18 gauge hypodermic needle. From ascitic fluid also antibodies can be purified by affinity chromatography using Protein A coupled Sepharose beads. Lately, due to objection from organizations for prevention of cruelty to animals, ascites production is becoming problematic. Antibodies can also be directly obtained from the fluid by ammonium sulfate precipitation.

The rapid progress being made in the commercialization of monoclonal antibodies has led to large-scale production of MAb. Commercial interests consider production scales of 0.1–10 g as small, 10–100 g as medium and over 100 g as large. Commercial-scale production is generally performed to produce MAb for three purposes: diagnosis, therapy and research, and development of new therapeutic agents. Monoclonal antibodies are being manufactured for clinical trials in large-scale suspension culture in fermenters. A completely automated pilot plant used for fermentation has been employed with direct digital control (DDC) technology for monitoring and regulating growth of human cells. A human hybridoma cell line (3D6) producing anti-human immunodeficiency virus (HIV)-1 antibodies was used as a model for large-scale production (300-liter airlift fermentor) in continuous culture. The production of anti-α-fetoprotein monoclonal antibodies for diagnostic use was carried out in a stirred tank fermenter equipped with a double membrane stirrer for bubble free aeration and continuous medium perfusion.

11.7 Applications of Monoclonal Antibodies

Monoclonal antibodies are widely used as diagnostic, research, and therapeutic agents for treatment of numerous human ailments. The applications can be divided in mainly three major groups:

1. Immunodiagnostic reagent
2. Research tools in molecular localization of antigens in intact cells and epitope analysis.
3. Immunotherapeutic and Immunoprophylactic agents.

11.7.1 Immunodiagnostic Reagent

Monoclonal antibodies are useful tools for detection of antigens in various body fluids and tissues. MAb-based commercial kits are available for detection of viral, bacterial, fungal, and parasitic pathogens. MAbs attached to latex beads or erythrocytes can be added to a body fluid. Presence of antigen causes agglutination. This is easily visible. It should be kept in mind that some antigens may not cause agglutination. Detection of antigens in intact tissue can be made with use of either a fluorochrome-labeled or enzyme-labeled or isotope-labeled antibody. Pathogens can be visualized by fixing a tissue slice on microscope slide or slide culture of the pathogen. Detection is made either with the help of a fluorescent microscope, use of an enzyme substrate which provide an insoluble colored product in case of fluorochrome-labeled or enzyme-labeled MAbs, respectively. A radioisotope can be detected by autoradiography. Many animal viruses (Koprowski and Wiktor 1980; Sonza et al. 1983), bacteria (Gustafson et al. 1982), and parasites (Wrisht et al. 1983; Scott 1983) have been diagnosed using MAbs.

A very large number of monoclonal antibodies have been produced to a wide range of viruses such as influenza, hepatitis, polio, Epstein Barr virus, and rabies. Their high specificity has led to accurate identification between similar stains of virus such as Herpes simplex Types I and II. They may also be used for early diagnosis of the IgM production in affected patients.

Antibodies have been designed for bacterial diseases as well though they are less common. They are helpful in analysis of bacterial spores which tend to show extensive cross-reactivity in conventional immunological analysis. Monoclonal antibodies to bacterial toxins have also been generated.

The potential of monoclonal antibodies in the study of parasitic diseases has already been widely exploited for diseases such as malaria, schistosomiasis, and leishmania. The advantage here is that the disease may be studied not only for diagnostic purpose but also in greater depth since many different surface antigens may be expressed at varying stages of the life cycle of the parasite.

Competitive or noncompetitive ELISA procedures have been developed for detecting antigens in body fluids. The MAb is usually attached to a microtiter well and body fluid (containing the antigen) is allowed to bind to the antibody. Detection can be made by using a different MAb coupled to an enzyme or radioisotope. Two MAbs used should be either for different epitopes or against a repetitive epitope. In competitive ELISA, the test sample is mixed with the known amount of labeled antigen and added to the antibody-coated microtiter well. Less labeled antigen binds to MAb with increasing concentrations of the test antigen.

Some precautions are required for diagnostic utility of the MAbs. Since there are multiple strains of the particular organism which can cause infection, a MAb may detect a single strain only. These strains may differ in their antigenic epitopes. Therefore, a monoclonal antibody against a common epitope must be selected for immunodiagnosis. MAbs are extremely useful for unambiguous diagnosis of diseases which produce closely related symptoms. As an example, MAb against Marek's disease does not bind to lymphoid leucosis virus though both cause similar symptoms. Similarly, MAbs can distinguish canine parvovirus infection from panleukopenia virus.

MAb are being used for cancer diagnosis too. MAbs to certain tumor markers are being used for detection in tissue fluid, tissue explants, or in intact organism through imaging. Precise information about the type of cancer is very helpful in its therapy. Tumor markers against prostate cancer, colorectal cancer, and ovarian tumors have been utilized for monitoring the tumor regression or progression as the marker levels fluctuate with its status. MAbs are also being extensively used for diagnosis of hematopoietic malignancies such as primary or acquired immunodeficiency syndrome (AIDS).

Immunoscintigraphy procedures have been developed for scanning human body for the location of cancerous tissue. The MAbs are labeled with radioisotopes such as ^{131}I (Iodine) or ^{99}Tc (Technetium) and injected intravenously into the patient. It binds to cancer cells in different locations which can be detected by imaging.

11.7.2 Basic Research

The potential of monoclonal antibody in basic research is substantial. In principle they can resolve a single protein from a complex mixture or a single epitope responsible for a specific function of a complex macromolecule. They have been widely used in basic enzymology, in nucleic acid structural studies, and in the analysis of hormone receptors. Another field in which monoclonal antibodies may prove of particular value is in the study of chromosomal proteins which are responsible for determining cell phenotype. For this they are ideal tools for the dissection of the complex mixture of proteins.

11.7.3 Therapeutic Applications

One of the earliest applications of MAb for humans came in the form of murine MAb OKT3 (orthoclone) which was used for preventing renal-transplant rejection. Human allografts are often rejected by the host and to sustain the allograft the patients are subjected to administration of highly toxic immunosuppressive drugs or high doses of glucocorticoids. OKT3 is an antibody which prevents the proliferation of cytotoxic T cells, hence prevents graft rejection.

Murine MAbs are not preferred for human therapeutic application, as this initiate immune response in human patients. In order to prevent or minimize deleterious immune responses murine MAbs have been humanized. Using genetic engineering techniques human Fc region is grafted onto the variable region of murine MAb. This reduces the immune response as the Fc region is strongly immunogenic. With further improvement in the technology, now even the framework regions of the MAbs have been grafted along with the Fc region from human immunoglobulins. Humanized murine MAbs which have only complementary determining region (CDR) from the murine MAb and rest of the molecule is of human immunoglobulin, are finding application mainly in cancer therapy.

HIV infects Th (T helper cell) lymphocytes *via* gp 120 envelope protein of the virus. Virus binds to Th lymphocytes on $CD4^+$ (Cluster of differentiation) receptors *via* gp 120 glycoprotein. Using genetic engineering techniques, Fc region of the antibody has been conjugated to a CD4 glycoprotein molecule. When this conjugate is administered to HIV patients, the HIV infected cells of the patient exhibiting gp 120 bind to $CD4^+$ Fc conjugate. This event triggers cell mediated immune response and the infected cell is killed. This strategy may prove as an effective therapy for HIV patients.

MAbs can also be used for targeting drugs to cancer cells or virally infected cells. The drug is conjugated onto the MAb which binds to only affected cell and it is selectively delivered to the cancer cell or virally infected cell. Not only drugs but also deadly toxins such as diphtheria toxin or ricin could be coupled to MAb molecules and can be used to selectively kill targeted cells. The normal cells are not killed by the drug or the toxin and prevent the side effects of chemotherapy.

Anti-idiotype MAbs can replace the antigen in detection assays. Anti-idiotype antibody (Ab2) has binding ability to antigen binding amino acid sequence (epitope) of the hypervariable region of the antibody (Ab1). The resulting Ab2 is capable of competing with the original antigen. Thus if the antigen availability is limiting or toxic, the MAb can easily be substituted for the antigen.

A substantial number of MAbs have been approved by U.S. Food and Drug Administration for use in humans. A few of these are listed here.

RiuxiMAb and Zevalin conjugated with isotopes of indium or yttrium have been approved for treatment of B cell lymphomas. Both these antibodies are targeted to CD20, a marker present on B cells. Administration of ^{131}I labeled TositumoMAb in B cell lymphoma patients have kept them free from disease for a substantial number of years. These antibodies also bind to CD40 receptors on B cells. CetruxiMAb and Herceptin have been used for successful treatment of breast cancer. CetruxiMAb is specific for HER1 (human epidermal growth factor) receptor present on breast cancer cells. Herceptin blocks the growth factor receptor HER2 which is also found in some breast cancers. MAb against CD30 conjugated with Vedotin blocks the proliferation of lymphomas. The preparation is called Adcetris. A number of MAb preparations are also available for treatment of leukemia. These are Lymphocide and AlemtuzuMAb. Lymphocide is against CD22 receptor and is used for treatment of B cell leukemias. AlemtuzuMAb is against CD52 receptor which is found on B as well as on T cells. Thus this antibody is useful for treatment of both B and T cell lymphomas. An antibody called Lym-1 has been shown to bind to HLA-DR which is expressed heavily on lymphoma cells. An antibody LipimuMAb has shown promise in suppressing all types of tumors acting as tumor suppressor agent.

Vitaxin is an antibody against integrin found on the blood vessels of tumor cells and has shown promise to be an inhibitor of angiogenesis selectively in tumor cells. This results in regression of solid tumors. BevaxizuMAb prevents clumping of platelets and prevents reclogging of arteries after angioplasty.

Autoimmune disorders are generally difficult to treat by using conventional drugs. Two MAbs namely InfliximAb and AdalimuMAb have been used for treatment of rheumatoid arthritis and Chrohan's disease. These antibodies bind to TNF-alpha (Tumor necrosis factor) and reduce the proliferation of Th1 cells which are active in enhancing cell-mediated immune response. OmalizuMAb binds to IgE receptors and is useful for treatment of allergic asthma. DaclizuMAb binds to IL-2 receptors on the surface of activated T cells preventing acute graft rejection of the transplanted kidney. This antibody has also shown promise against T cell lymphomas.

11.7.4 Monoclonal Antibodies and Infectious Diseases

A humanized MAb against respiratory syncytial virus has been approved for treatment of premature infants or infants suffering from bronchopulmonary dysplasia. The MAb called PalivizuMAb has shown great promise in such cases. Similarly, many other MAbs are under trial for combating viral infections. MAbs are also very useful tools for dissecting viral antigenic epitopes which is essential for developing efficacious vaccines for viral infections.

MAbs have also found applications in elucidating host–parasite relationships. In order to understand the relationship between the host and parasites, e.g., protozoa and helminthes in different stages of their life cycle, panels of antibodies have been produced against stage specific antigens. Antigens from some of the parasites have been fractionated using MAbs and the antigens providing protective immunity have been identified. These antigens could be used for vaccination against the parasite.

In case of bacterial pathogens, MAbs have been extremely useful tools for unambiguous diagnosis. Bacterial antigenic epitopes are being characterized with the objective of finding antigens with better immunogenic potential. Serotype specific MAbs have been developed for detection of various leptospira, causing infections in humans and animals.

11.7.5 Miscellaneous Uses of MAbs

MAbs can be effectively used for purification of antigens from a complex mixture. MAbs attached to solid particles can be used to prepare a series of columns. The complex mixture is then serially passed through different columns and from each column a single antigen can be obtained. Thus a panel of MAbs can be used for fractionation of antigens from a cell.

MAbs are useful reagent for mapping epitopes on complex antigen. They are also being used for isolating different cells of immune system. $CD4^+$ and $CD8^+$ MAbs are used for separating T helper and T cytotoxic cells. Further with appropriate MAbs TH1 and TH2 cells can also be fractionated.

Panel of MAbs used for generating protein microarrays are highly useful tools for protein characterization. MAbs are used for blood group analysis. Using chromosome Y specific MAbs, embryos can be sexed. This is being widely used in dairy industry for producing only female calves.

For passive immunization humanized MAbs are used as prophylactic agents. These can also be used for neutralizing pathogens in acute cases.

Monoclonal antibodies are also used in the detection of adulteration in meat. Poultry (chicken and turkey) tissue is a major source of protein and less expensive than red meat, which is consumed and imported throughout the world. Increasing use of mechanically separated poultry meat have a potential for the adulteration of red meat by poultry products. Martin (1989) produced and characterized MAbs against species-specific sarcoplasmic protein of chicken. Three MAbs are capable of distinguishing between muscle extracts of the most frequently, marketed avian (chicken and turkey) and mammalian (beef, pork, horse and lamb) species of meat animals. One of these antibodies has the added advantage of distinguishing between chicken and turkey extracts by ELISA. MAb technology will improve diagnosis and serotyping/pathotyping procedures in poultry disease management.

11.8 Conclusion

Monoclonal antibodies are antibodies having a single specificity and are continuously secreted by "immortalized" hybridoma cell, which is a biologically constructed hybrid between an antibody-producing, mortal, lymphoid cell, and a malignant, or "immortal", myeloma cell. MAb are more pure than the polyclonal antibodies and because of their specificity, are used in biomedical research, in diagnosis of diseases, and in treatment of infections and cancer. These have also been used for animal disease diagnosis of poultry diseases, infectious bronchitis, and avian influenza. Commercialization of MAb has led to their large-scale production in fermentors.

References

Burnet FM (1959) The clonal selection theory of acquired immunity. Cambridge University press, Cambridge

Gustafson B, Rosen A, Holmes T (1982) MAbs against *Vibrio cholera* lipopolysaccharide. Infect Immun 38:91–99

Kohler G, Milstein C (1975) Continuous cultures of fused cells secreting antibody of predefined specificity. Nature London 256:495–497

Koprowsik H, Wiktor T (1980) Monoclonal antibodies against rabies virus. In: Kemet RH, McKearn IJ, Bechtol KB (eds) Monoclonal antibodies. Plentum Press, NY, pp 335–351

Martin M (1989) Production and characterization of monoclonal antibodies to species-specific sarcoplasmic protein of chickens. Meat Sci 25:199–207

Scott AL (1983) MAbs against somatic antigens of *Dirofilaria immitis*. Fed Proc 42:1089

Sinkovics JG, Shirato E, Gyroky F, Cabiness JR, Howe CD (1970) In leukemia lymphoma. Year book Med Publication, Chicago

Sonza S, Breschkin AM, Holmes IH (1983) Derivation of neutralizing MAbs against rotavirus. J Virol 45:1143–1146

Wright IG, White M, Tracey-Patte PD, Donaldson RA, Goodger BV, Waltibuhl DJ, Mahoney DF (1983) *Babesia bovis:* isolation of protective antigen by using monoclonal antibodies. Infect Immun 41:244–250

12 Edible Vaccines

Jyoti Saxena and Shweta Rawat

Abstract

In recent years edible vaccine emerged as a new concept developed by biotechnologists. Edible vaccines are subunit vaccines where the selected genes are introduced into the plants and the transgenic plant is then induced to manufacture the encoded protein. Foods under such application include potato, banana, lettuce, corn, soybean, rice, and legumes. They are easy to administer, easy to store and readily acceptable delivery system for different age group patients yet cost effective. Edible vaccines present exciting possibilities for significantly reducing various diseases such as measles, hepatitis B, cholera, diarrhea, etc., mainly in developing countries. However, various technical and regulatory challenges need to overcome in the path of this emerging vaccine technology to make edible vaccine more efficient and applicable. This chapter attempts to discuss key aspects of edible vaccines like host plants, production, mechanism of action, advantages and limitations, applications, and different regulatory issues concerned to edible vaccines.

12.1 Introduction

Vaccine is a biological preparation intended to produce immunity to a disease by stimulating the production of antibodies. Dead or attenuated organisms or purified products derived from them are generally used to produce various vaccines. Over the past decade, scientific advances in genetics, molecular biology, and plant biotechnology have improved the understanding of many infectious diseases and led to the development of vaccination programs. The most common method of administering vaccines is by injection but some are given by mouth or nasal spray. Though immunization is the safest method to combat the diseases worldwide but there are many constraints regarding its mode of production, distribution, delivery, cost, and lack of enough research. Hence it is desirable to look for an effectual and powerful yet cost effective, easy for storage and distribution yet safe method of immunization. It should also be readily

J. Saxena (✉) · S. Rawat
Biochemical Engineering Department, BT Kumaon Institute of Technology, Dwarahat, Uttarakhand 263653, India
e-mail: saxenajyoti30@gmail.com

S. Rawat
e-mail: shweta.biotech24@gmail.com

I. Ravi et al. (eds.), *Advances in Biotechnology*,
DOI: 10.1007/978-81-322-1554-7_12, © Springer India 2014

acceptable to all sociocultural groups around the globe. Research underway is dedicated to solving these problems by finding ways to produce edible vaccines in the form of transgenic plants which have been investigated as an alternative means to produce and deliver vaccine.

Edible vaccines are called by several alternative names such as food vaccines, oral vaccines, subunit vaccines, and green vaccines. They seem to be a viable alternative especially for the poor and developing countries. They have come up as great boon in medicinal science for which biotechnologists should be given all credit. The concept of edible vaccines lies in converting the edible food into potential vaccines to prevent infectious diseases. It involves introduction of selected desired genes into plants and then inducing these altered plants to manufacture the encoded proteins. It has also found application in prevention of autoimmune diseases, birth control, cancer therapy, etc. Edible vaccines are currently being developed for a number of human and animal diseases. This new technology hopefully will contribute positively toward the global vaccine programs and have a dramatic impact on health care in developing countries.

12.2 Historical Background

Many people in developing countries do not have access to the vaccines they need, as the traditional vaccines are costly and require skilled medical people for administration and are less effective in inducing mucosal immune response. It was these needs which inspired Hiatt et al. (1989) who attempted to produce antibodies in plants which could serve the purpose of passive immunization. The first report of edible vaccine (a surface protein from *Streptococcus*) in tobacco, at 0.02 % of total leaf protein level, appeared in 1990 in the form of a patent application published under the international patent cooperation treaty. By conceiving the idea of edible vaccine Dr. Charles Arntzen tried to realize it (Arntzen 1997). In 1992, Arntzen and coworkers introduced the concept of transgenic plants as a production and delivery system for subunit vaccines in which edible tissues of transgenic crop plants were used (Mor et al. 1998). They found that this concept could overcome the limitations of traditional vaccines, thereby triggering the research on edible vaccine. In 1990s, *Streptococcus mutans* surface protein antigen A was expressed for the first time in tobacco. The same group also pioneered the field with work on hepatitis B and heat-labile toxin, B subunit in tobacco plants and potato tubers. In the same year, the successful expression of hepatitis B surface antigen (HBsAg) in tobacco plants was also achieved (Mason et al. 1992). To prove that plant-derived HBsAg could stimulate mucosal immune responses *via* oral route, potato tubers were used as an expression system and were optimized to increase the accumulation of the protein in plant tubers (Richter et al. 2000).

Parallel to the evaluation of plant-derived HBsAg, Mason and Arntzen explored plant expression of other vaccine candidates including the labile toxin B subunit (LT-B) of enterotoxigenic *Escherichia coli* (ETEC) and the capsid protein of Norwalk virus. The plant-derived proteins correctly assembled into functional oligomers that could elicit the expected immune responses when given orally to animals (Mason et al. 1998).

In 1998 a new era was opened in vaccine delivery when researchers supported by the National Institute of allergy and infectious diseases (NIAID) have shown for the first time that an edible vaccine can safely generate significant immune responses in people. The report by collaborators from the University of Maryland in Baltimore, the Boyce Thompson Institute for Plant Research in Ithaca, N.Y., and Tulane University in New Orleans appeared in the May issue of Nature Medicine. According to the then Director of NIAID "Edible vaccines offer exciting possibilities for significantly reducing the burden of diseases like hepatitis and diarrhea, particularly in the developing world where storing and administering vaccines are often major problems," Mor et al. (1999) also discussed the rapid increase of research in the edible-vaccine field and pointed out that plants could be used to create multicomponent vaccines that can protect against several pathogens at once. This is an aspect of the edible-vaccine approach that further strengthens its impact. Later, in 2003 Sala and research group reported that proteins produced in these plants induced the mucosal immune response which was the main aim behind this concept.

Research into edible vaccine is still at a very early stage and scientists have a long way to go before it will become a major part of immunization program world wide.

12.3 Choice of Host Plant for Edible Vaccine

To date, many plant species have been used for vaccine production. The choice of the plant species is important. An edible, palatable plant is necessary if the vaccine is planned for raw consumption. In case of vaccine for animal use, the plant should preferentially be selected among those consumed as normal component of the animal's diet. Some food vehicles are discussed below:

12.3.1 Tobacco

The concept of edible vaccine got impetus after Arntzen and coworkers expressed HBsAg in tobacco. The first edible vaccine was produced in tobacco in 1990 in which 0.02 % recombinant protein (a surface protein from *Streptococcus*) of the total soluble leaf proteins was found. It appeared in the form of a patent application published under the International Patent Cooperation Treaty. Transgenic tobacco is successfully engineered for the production of edible vaccines against hepatitis B antigen using 's' gene of hepatitis B virus (HBV). The optimum level of recombinant protein was obtained in leaves and seeds. Since acute watery diarrhea is caused by enterotoxigenic *E. coli* and Vibrio cholerae that colonize the small intestine and produce one or more enterotoxin, an attempt was made toward the production of edible vaccine by expressing heat-labile enterotoxin (LT-B) in tobacco. Besides, antibodies against dental caries, expressed in tobacco, are already in preclinical human trials. Italian researchers have now developed an immunologically active, cost-efficient vaccine against human papilloma viruses (HPV). HPV are the causative agents for cervical cancer, and are also involved in skin, head, and neck tumors. Cervical cancer is one of the main causes of cancer-related deaths.

12.3.2 Potato

Genetically modified potatoes are also a viable option and seem to be the desired vector. Many of the first edible vaccines were synthesized in potato plants. The transgenic potatoes were developed and grown by Arntzen and Mason and their research group at the Boyce Thompson Institute for Plant Research, Cornell University. Previously, NIAID supported *in vitro* and preclinical studies by John Clements and colleagues at Tulane University School of Medicine, in which 14 volunteers ate bite-sized pieces of raw potato that had been genetically engineered to produce part of the toxin secreted by *E. coli* causing diarrhea. The investigators periodically collected blood and stool samples from the volunteers to evaluate the vaccine's ability to stimulate both systemic and intestinal immune responses. Ten of the 11 volunteers (91 %) who ingested the transgenic potatoes had fourfold rise in serum antibodies at some point after immunization, and 6 of the 11 (55 %) also showed fourfold mount in intestinal antibodies. The potatoes were well tolerated and no one experienced serious adverse side effects. Vaccine development has successfully tested a potato-based vaccine to combat the Norwalk Virus, which is spread by contaminated food and water. The virus causes severe abdominal pain and diarrhea.

A research team led by William Langridge of the Loma Linda University in California has reported that transgenic potatoes engineered with a cholera antigen, CTB can effectively immunize mice. Mice fed transgenic potatoes produce cholera-specific antibodies in their serum and intestine; IgA and IgG antibodies reach their highest levels after the fourth feeding. In yet another experiment genetically engineered potatoes containing a hepatitis B vaccine have successfully boosted immunity in their first human trials.

Attempts have also been made to boil the potatoes as raw potatoes are not very appetizing but unfortunately the cooking process breaks down about 50 % of the proteins in the vaccine. While some proteins are more tolerant to heat, for most proteins it will be necessary to amplify the amount of protein in the engineered foods if they are to be cooked before consumption.

12.3.3 Tomato

Tomatoes are an excellent candidate because they are easy to manipulate genetically and new crops can be grown quickly. Moreover, they are palatable and can be eaten raw. While tomatoes do not grow well in the regions in which the edible vaccines are most needed, the engineered tomatoes can be dried or made into a paste to facilitate their delivery.

The anti-malaria edible vaccines in different transgenic tomato plants expressing antigenic type(s) have been proposed by Chowdhury and Bagasara in 2007. They hypothesized that immunizing individuals against 2–3 antigens and against each stage of the life cycle of the multistage parasites would be an efficient, inexpensive and safe way of vaccination. Tomatoes with varying sizes, shapes, and colors carrying different antigens would make the vaccines easily identifiable by lay individuals.

Tomatoes serve as an ideal candidate for the HIV antigen because they unlike other transgenic plants that carry the protein, are edible and immune to any thermal process, which help to retain their healing capabilities. Scientists have claimed that tomatoes could be used as a vaccine against Alzheimer's disease. The work is in progress to genetically modify the fruit to create an edible vaccine that fires up the immune system to tackle the disease by attacking the toxic beta-amyloid protein that destroys vital connections between brain cells, causing Alzheimer's.

Researchers have engineered tomato plants (*Lycopersicon esculentum* Mill var. UC82b) to express a gene for the glycoprotein (G-protein), that coats the outer surface of the rabies virus. The recombinant constructs contained the G-protein gene from the ERA strain of rabies virus, including the signal peptide, under the control of the 35S promoter of cauliflower mosaic virus (CaMV).

12.3.4 Banana

A common fruit—the banana—is currently being considered as a potential vehicle for vaccines against serious as well as too common diseases. The advantage of bananas is that they can be eaten raw as compared to potatoes or rice that need to be cooked and can also be consumed in a pure form. Furthermore, children tend to like banana and the plants grow well in the tropical areas in which the vaccines are needed the most. Hence, the research is leaning toward the use of banana as the vector since a large number of third-world countries, who would benefit the most from edible vaccines have tropical climates. On the negative side, a new crop of banana plants takes about 12 months to bear fruit. After fruiting, the plants are cut down and a new crop of vaccine-bearing plants must be planted.

Researchers have also developed bananas that deliver a vaccine for HBV. The banana vaccine is expected to cost just 2 cents a dose, as compared to the $125 for the currently available injectable vaccine.

12.3.5 Maize

Maize has also been used as a vector for various edible vaccines. Egyptian scientists have genetically engineered the maize plants to produce a protein known as HbsAg which elicits an immune response against the hepatitis B virus and could be used as a vaccine. If human trials are successful more than 2 billion people infected with hepatitis B, and about 350 million of these at high risk of serious illness and death from liver damage and liver cancer would be benefited.

Researches are in offing at Iowa State University with the aim to allow pigs and humans to get a flu vaccination simply by eating corn or corn products. It is quite likely that corn vaccine would work in humans when they eat corn or even corn flakes, corn chips, tortillas, or anything that contains corn.

Genetically modified maize could provide protection to chickens against a highly contagious and fatal viral disease affecting most species of birds. Mexican researcher Octavio Guerrero-Andrade and his colleagues at the Centre for Research and Advanced Studies in Guanajuato, Central Mexico, genetically modified maize to create an edible vaccine against Newcastle disease virus (NDV). They inserted a gene from the NDV, a major killer of poultry in developing countries, into the maize DNA and found antibodies against the virus in chickens that ate the genetically modified maize. One pig vaccine has also been produced in corn successfully.

Efforts are being made by US company ProdiGene to genetically modify maize to contain a key protein found on the surface of the monkey form of HIV. According to US National Institute of Health this development brings an edible, more effective, HIV vaccine for people a step closer.

Transgenic maize expressing the rabies virus glycoprotein (G) of the Vnukovo strain has also been produced using ubiquitin maize promoter fused to the whole coding region of the rabies virus G gene, and a constitutive promoter from CaMV. Maize embryogenic callus were transformed with the above construct by biolistics. Regenerated maize plants were recovered and grown in a greenhouse. The amount of G-protein detected in the grains was approximately 1 % of the total soluble plant protein.

12.3.6 Rice

Rice is another potential crop which has been used for developing vaccines. It offers several advantages over traditional vaccines; it does not require refrigeration. In fact, the rice proved just as potent after 18 months of storage at room temperature and the vaccine did not dissolve when exposed to stomach acids. In an attempt, predominant T cell epitope peptides, which were derived from Japanese cedar pollen allergens, were specifically expressed in rice seeds and delivered to the mucosal immune system (MIS); the development of an allergic immune response of the allergen-specific Th2 cell was suppressed. Furthermore, not only the specific IgE production and release of histamine from mast cells were suppressed, but the inflammatory symptoms of pollinosis, such as sneezing, were also suppressed. These results suggest the feasibility of using an oral immunotherapy agent derived from transgenic plants that accumulate T cell epitope peptides of allergens for allergy treatment.

The transfer of genetic material from the microbe responsible for producing cholera toxin into a rice plant has been achieved. The plants produced the toxin and when the rice grains were fed to mice they provoked immunity from the diarrhea-causing bacterium.

12.3.7 Spinach

Genetically modified spinach has also been considered for the development of edible vaccine. Spinach is being investigated as a plant-derived, edible vehicle for anthrax vaccine, as well as a vehicle for the HIV-1 Tat protein (a prospective vaccine candidate). In an experiment a fragment of protective antigen (PA) that represents most of the receptor-binding domain was expressed as a translational fusion with a capsid protein on the outer surface of tobacco mosaic virus, and spinach was inoculated with the recombinant virus. The plant-expressed PA is highly immunogenic in laboratory animals.

Among other food crops with potential to be developed as edible vaccine; sweet potato, peanuts, lettuce, watermelon, and carrots are on the top priority. The development of plant-based vaccines to protect against many other diseases, such as HIV-1, hepatitis B, rabies, and non-Hodgkin's lymphoma are ongoing throughout the globe using one of these edible plants.

The advantages and disadvantages of various plant host systems are given in Table 12.1.

12.4 Advantages

Conventional subunit vaccines are expensive and technology-intensive, need purification, require refrigeration, and produce poor mucosal response. In contrast, edible vaccines would

Table 12.1 Features of different plant host systems

Plant	Advantages	Disadvantages
Tobacco	Facile and efficient transformation system; abundant material for protein characterization	Toxic alkaloids incompatible with oral delivery; potential for outcrossing in field
Banana	Cultivated widely in developing countries where vaccines are needed; eaten raw by infants and adults; clonally propagated; low potential for outcrossing in field; once established, plentiful and inexpensive fruits are available on a 10–12 month cycle	Inefficient transformation system; little data available on gene expression, especially for fruit specific promoters; high cultivation space requirement; very expensive in greenhouse
Potato	Facile and efficient transformation system; tuber is edible raw though not palatable; tuber-specific promoters available; microtuber production for quick assay; clonally propagated, low potential for outcrossing in field; Industrial tuber processing well established	Relatively low tuber protein content; unpalatable in raw form; cooking might cause denaturation and poor immunogenicity of vaccine
Tomato	Relatively efficient transformation system; fruit is edible raw; fruit specific promoters available; crossing possible to stack antigen genes; industrial greenhouse culture and industrial fruit processing well established	Relatively low fruit protein content; acidic fruit may be incompatible with some antigens or for delivery to infants; no *in vitro* system to test fruit expression
Legumes	Production technology widely established; high protein content in seeds; stable protein in stored seeds; well suited for animal vaccines; industrial seed processing well established	Inefficient transformation systems; heating or cooking for human use might cause denaturation and poor immunogenicity of vaccine; potential for outcrossing in field for some species
Alfalfa	Relatively efficient transformation system; high protein content in leaves; leaves edible uncooked	Potential for outcrossing in field; deep root system problematic for cleaning field

(Mason et al. 2002)

enhance compliance, especially in children, and because of oral administration would eliminate the need for trained medical personnel. Their production is highly efficient and can be easily scaled up. For example, hepatitis B antigen required to vaccinate whole of China annually, could be grown on a 40-acre plot and all babies in the world each year on just 200 acres of land. They are cheaper, sidestepping demands for purification (single dose of hepatitis B vaccine would cost approximately 0.43 cents), grown locally using standard methods and do not require capital-intensive pharmaceutical manufacturing facilities. Mass-indefinite production would also decrease dependence on foreign supply. Fear of contamination with animal viruses—like the mad cow disease, which is a threat in vaccines manufactured from cultured mammalian cells, is eliminated as plant viruses do not infect humans.

Edible vaccines activate both mucosal and systemic immunity, as they come in contact with the digestive tract lining which is not possible with subunit vaccines which provide poor mucosal response. This dual effect of edible vaccines provides first-line defense against pathogens invading through mucosa, such as *Mycobacterium tuberculosis* and agents causing diarrhea, pneumonia, STDs, HIV, etc.

The specific advantages are stated below:

1. Edible means of administration.
2. No need of medical personnel and syringes.
3. Sterile injection conditions are no more required.
4. Economical in mass production by breeding compared to an animal system.
5. Easy for administration and transportation.
6. Effective maintenance of vaccine activity by controlling the temperature in plant cultivation.
7. Therapeutic proteins are free of pathogens and toxins.
8. Storage near the site of use.
9. Heat stable, thus eliminating the need of refrigeration.

10. Antigen protection through bioencapsulation.
11. Subunit vaccine (not attenuated vaccine) means improved safety.
12. Seroconversion in the presence of maternal antibodies.
13. Generation of systemic and mucosal immunity.
14. Enhanced compliance (especially in children).
15. Delivery of multiple antigens.
16. Integration with other vaccine approaches.
17. Plant-derived antigens assemble spontaneously into oligomers and into virus like particles.
18. No serious side effect problems have been noticed until now.
19. Reduced risk of anaphylactic side effects from edible vaccine over injection system is one benefit reported by the Bio-Medicine.org. They reported that the edible vaccine carries only part of the allergen compared to injection methods which reduce anaphylactic risk.
20. Administration of edible vaccines to mothers to immunize the *fetus*-in utero by transplacental transfer of maternal antibodies or the infant through breast milk. Edible vaccines have a potential role in protecting infants against diseases like group-B *Streptococcus*, respiratory syncytial virus (RSV), etc., which is under investigation.
21. Edible vaccines would also be suitable against neglected/less common diseases like dengue, hookworm, rabies, etc. They may be integrated with other vaccine approaches and multiple antigens may also be delivered.

12.5 Limitations and Challenges

With advancement come many hurdles and problems, so is true for edible vaccines. Like, one could develop immunotolerance to the vaccine peptide or protein, though a little research has been done on it. One of the key goals of the edible-vaccine pioneers was to reduce immunization costs but later many limitations were reported as given below:

1. Consistency of dosage from fruit to fruit, plant to plant, lot to lot, and generation to generation is not similar.
2. Stability of vaccine in fruit is not known.
3. Evaluation of dosage requirement is tedious.
4. Selection of best plant is difficult.
5. Certain foods like potatoes are generally not eaten raw and cooking the food might weaken the medicine present in it.
6. Not convenient for infants as they might spit it, eat a part or eat it all, and throw it up later. Concentrating the vaccine into a teaspoon of baby food may be more practical than administering it in a whole fruit.
7. There is always possibility of sideeffects due to the interaction between the vaccine and the vehicle.
8. People could ingest too much of the vaccine, which could be toxic, or too little, which could lead to disease outbreaks among populations believed to be immune.
9. A concern with oral vaccines is the degradation of protein components in the stomach due to low pH and gastric enzymes. However, the degradation can be compensated by repeating the exposure of the antigen until immunological tolerance is accomplished (Mason et al. 2002).
10. Potential risk of spreading the disease by edible vaccine delivery is a concern of many. Potential contamination of the oral delivery system is a possible danger.

Foreign proteins in plants accumulate in low amounts (0.01–2 % of total protein) and are less immunogenic, therefore the oral dose far exceeds the intranasal/parenteral dose. For example oral hepatitis B dose is 10–100 times the parenteral dose and 100 g potato expressing B subunit of labile toxin of ETEC (LT-B) is required in three different doses to be immunogenic. Attempts at boosting the amount of antigens often lead to stunted growth of plants and reduced tuber/fruit formation as too much mRNA from the transgene causes gene silencing in plant genome.

Techniques to overcome these limitations are given below:

1. Optimization of coding sequence of bacterial/viral genes for expression as plant nuclear genes
2. Expression in plasmids
3. Plant viruses expressing foreign genes
4. Coat-protein fusions
5. Viral-assisted expression in transgenic plants
6. Promoter elements of bean yellow dwarf virus with reporter genes GUS (β -glucuronidase) and green fluorescent protein (GFP), substituted later with target antigen genes.
7. Antigen genes may be linked with regulatory elements which switch on the genes more readily or do so only at selected times (after the plant is nearly fully grown) or only in its edible regions. Exposure to some outside activator molecule may also be tried.

12.6 Side Effects

Development of edible vaccines is a possible high-volume, low-cost delivery system for third-world countries to fight against fatal maladies like AIDS, hepatitis, and diarrhea. Researches by the NIAID and the University of Maryland showed no significant side effects in a small study using genetically engineered potatoes to make toxin of the *E. coli*, a diarrhea-causing bacterium. Volunteers reported no serious adverse reactions to genetically altered potatoes used to deliver edible vaccine toxin, according to the National Institutes of health (NIH). The NIH reported that 10–11 volunteers who ate the raw potato bites developed four times the antibodies against *E. coli* without obvious side effects.

Long-term reactions to edible vaccines are yet to be determined and possible delayed reactions have not yet been discovered. An organized large scale study is required before edible vaccines are put into large scale production.

12.7 Production of Edible Vaccines

Creating edible vaccines involves the genetic engineering approach for the introduction of selected desired genes into plants and then inducing these altered plants to manufacture the encoded proteins. This process is known as "transformation" and the plants altered with new characteristics are called "transgenic plants". Like conventional subunit vaccines, edible vaccines are composed of antigenic proteins and are devoid of genes responsible for pathogenicity. Thus, they have no way of establishing infection, assuring its safety, especially in immune-compromised patients.

The gene which codes the active antigenic protein is first isolated from the pathogen and is incorporated in a suitable "gene vehicle". This gene vehicle is integrated into the genome of the plant and is allowed to express the corresponding antigen. Then these plant parts are fed to animals and humans to run their course.

Two main strategies are used for the production of candidate vaccine antigen in plant tissues (Fig. 12.1 a, b).

1. Stable genomic integration: This is the most popular method used for the published edible vaccine clinical trials to date. Under this method the genetic line is produced that can be propagated either by vegetative (stem cuttings) or sexual (seeds) reproduction methods. The stable expression strategy provides an opportunity to introduce more than one gene for possible multicomponent vaccine production. Furthermore, the choice of genetic regulatory elements allows organ and tissue-specific expression of foreign antigens. Stable transformation causes the desired gene to be incorporated either in nucleus or chloroplast. *Agrobacterium* mediated gene transfer is used for transforming the plants in which the gene is integrated in nucleus. Besides, direct delivery of DNA into the tissue can also be applied, biolistic being the most popular method. However, chloroplast

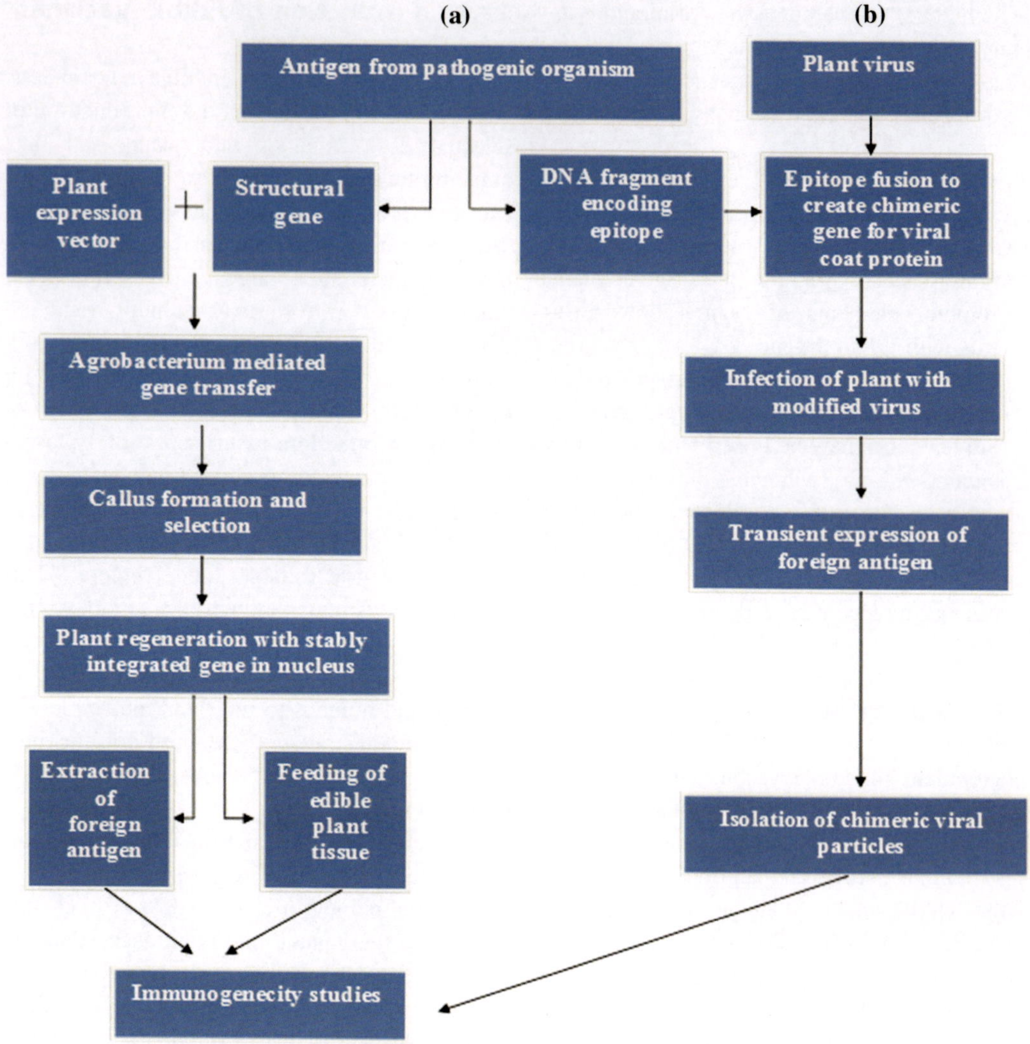

Fig. 12.1 (**a**) Stable and (**b**) transient strategies for the production of candidate vaccine in potato and tobacco plant tissue, respectively

engineering has also got impetus in last decade due to its various advantages over nuclear engineering.

2. Transient expression using viral vectors: In this method viral vectors are used as a tool to deliver genetic material into cells. A recombinant plant virus is selected that can carry the vaccine gene and can cause the plant to express the antigen by systemic infection. As compared to stable expression, transient expression is difficult to initiate, because the viral vectors must be inoculated into individual host plants, but gives higher level of expression as it allows the virus to replicate and amplify the gene copy number.

Some of the most popularly used techniques are described below:

12.7.1 *Agrobacterium tumefaciens* Mediated Gene Transfer Method

Plant transformation mediated by the plant pathogen, *A. tumefaciens* has become the most popular method lately. It is a naturally occurring gram-negative soil bacterium, which infects the wound sites in dicot plants causing the formation of the crown gall tumor. This bacterium is capable of transferring a particular DNA segment (T-DNA) of the tumor-inducing (Ti) plasmid into the nucleus of infected cells where it is subsequently integrated into the host genome and transcribed. The T-DNA usually contains cancer-causing oncogenic genes and genes that synthesize opines which are excreted by infected crown gall cells and are a food source for bacterium. During the genetic manipulation, the Ti plasmid is engineered to carry the desired gene for vaccine and the virulent genes that cause tumor growth in plants are deleted. The transgene is integrated, expressed, and inherited in mendelian fashion. The whole plant can be then regenerated from individual transformed plant cell. It has been studied that genes are successfully expressed in experimental model plants and when given orally to animals, the extract of transgenic plant containing the antigen induced serum antibodies, thus can be used to produce the edible vaccine.

The application of *Agrobacterium* mediated transformation is at present possible to most species of agronomic interest, including members of family Graminae and Leguminosae. This opens interesting new aspect for the development of edible vaccines for both human and veterinary uses.

12.7.2 Biolistic Method

The second approach for nuclear transformation is based on the microprojectile bombardment method, also known as the gene gun or biolistic method. This method is especially beneficial for those plants which can not be transformed by *A. tumefaciens* mediated gene transfer method. Selected DNA sequences are precipitated onto metal microparticles and bombarded with a particle gun at an accelerated speed in a partial vacuum against the plant tissue placed within the acceleration path. Microparticles penetrate the walls and release the exogenous DNA inside the cell where it will be integrated in the nuclear genome. Thus, this method effectively introduces DNA. The cells that take up the desired DNA, are identified through the use of a marker gene (in plants the use of GUS is most common), and then cultured to replicate the gene and possibly cloned. This method has various advantages including (1) thousands of particles are accelerated at the same time causing multiple hits resulting in transferring of genes into many cells simultaneously, (2) since intact cells can be used, the difficulties encountered with the use of protoplast are automatically circumvented, and (3) the method is universal in its application so that cell type, size, and shape or the presence/absence of cell wall do not significantly alter its effectiveness.

Another important use of the gene gun involves the transformation of organelles such as chloroplasts, and yeast mitochondria. The biolistic particle delivery system "shoots" adequately processed DNA particles, which penetrate into the chloroplast and integrate with its genome.

12.7.3 Chloroplast Transformation

The chloroplast's transformation is an interesting alternative to nuclear transformation which has come up in recent past. All plant cells have chloroplasts that capture light energy from the sun to produce free energy through a process called photosynthesis. In chloroplast genetic engineering, the recombinant DNA plasmid is bound to small gold nanoparticles that are injected into the chloroplasts of the leaf using a gene gun as described above. This device uses high pressure to insert the plasmid coated particles into the cells. These plasmids contain multiple genes of importance such as the therapeutic gene, a marker gene (may or may not be for antibiotic resistance), a gene that enhances the translation of therapeutic gene and two targeting sequences that flank the foreign gene. The foreign genes are inserted through homologous recombination *via* flanking sequences at a precise and predetermined location in the organelle genome. The gene expression level in plastids is predominately determined by promoter and 5′-untranslated regions (5′-UTR elements) (Gruissem and Tonkyn 1993). Therefore, suitable 5′-UTRs including a ribosomal binding site (RBS) are important elements of plastid expression vectors (Eibl et al. 1999). In order to obtain high-level protein accumulation from expression of the transgene, the first requirement is a strong promoter to ensure high levels of mRNA. Most laboratories use the strong plastid rRNA operon (*rrn*) promoter (*Prrn*).

Besides gene gun, PEG mediated transformation and Galistan Expansion Femto Syringe microinjection techniques are also used for gene delivery in chloroplast. Some of the advantages of chloroplast transformation technology are its low cost, natural gene containment, site specific insertion, very high level of stable expression, generation of production lines with a competitive timeline, elimination of gene silencing, and high accumulation of the recombinant protein. Precise steps are given below:

Step 1:

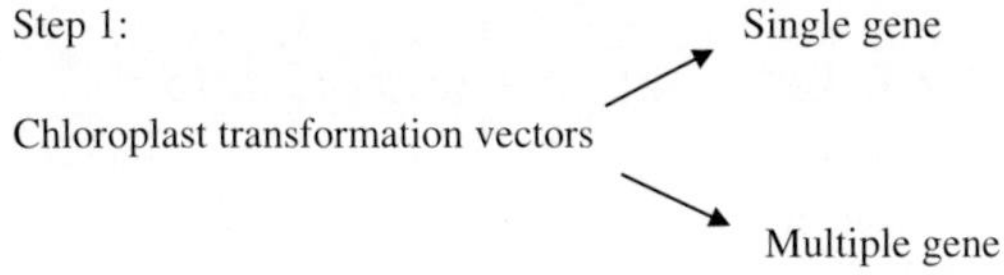

Step 3:

A heteroplasmic diploid plant cell (First round of selection)

A homoplasmic diploid plant cell (Second round of selection)

Step 4:

Multiple gene expression

Step 5:

Reproductive organs

Disintegration of paternal plastids

Step 6:

Maternal inheritance of transgenic traits.

12.8 Mechanism of Action

The most common entry point for pathogens is at mucosal epithelia lining the gastrointestinal, respiratory, and urino-reproductive tracts, which are collectively the largest immunologically active tissue in body. The MIS is the first line of defense and the most effective site for vaccination against those pathogens; nasal, and oral vaccines being the most effective. The goal of oral vaccine is to stimulate both mucosal and humoral immunity against pathogens.

Edible vaccines have plant parts which are fed directly and the outer tough wall of plant cell acts to protect the antigens against attack by

Step 2:

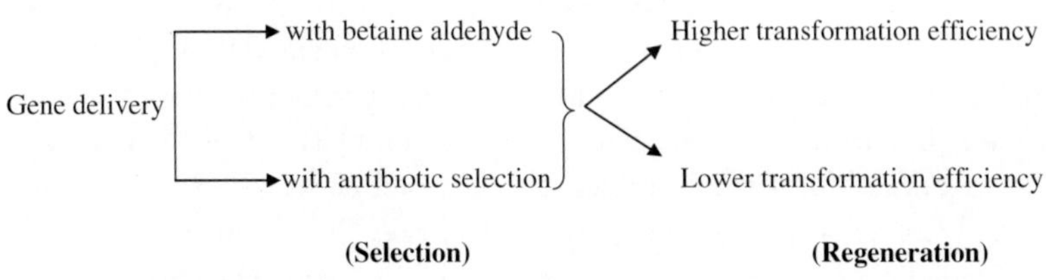

enzymes, and gastric and intestinal secretions. This method is known as bioencapsulation. Therefore, the majority of the plant cell degradation occurs in the intestine as a result of action on digestive or bacterial enzymes. The antigens thus released are taken up by M cells in the intestinal lining that are present over the Peyer's patches (PP) in the ileum and the gut-associated lymphoid tissue. PP are an enriched source of IgA producing plasma cells and populate mucosal tissue and serves as mucosal immune effectors sites. The breakdown of edible vaccine occurs near PP, consisting of 30–40 lymphoid nodules on the outer surface of the intestine and contain follicles from which germinal center develops upon antigenic stimulation. These follicles act as the sites from which antigen penetrates the intestinal epithelium, thereby accumulating antigen within organized lymphoid structure. The antigens then come in contact with M cells which in turn express class II MHC molecules. Antigens transported across the mucous membrane by M cells can activate B-cells within these lymphoid follicles. The activated B-cells leave the lymphoid follicles and migrate to diffuse mucosal associated lymphoid tissue (MALT) where they differentiate into plasma cells that secrete the IgA class of antibodies. These IgA antibodies are transported across the epithelial cells into secretions of the lumen where they can interact with antigens present in the lumen and immediately neutralize the infectious agent. The induction of mucosal immunity by edible vaccine is depicted in the flow diagram (Fig. 12.2).

12.8.1 How Edible Vaccine Provides Protection?

An antigen in a food vaccine is taken up by M cells in the intestine and passed to various immune system cells, which then start a defensive attack, as antigen is a true infectious agent, not just part of one. The response leaves long-lasting "memory cells" able to promptly neutralize the real infectious agent. The whole procedure can be explained in two stages (Fig. 12.3 a, b) Antibodies and antibody fragments produced against specific antigens are given in Table 12.2.

12.9 Applications

There are numerous therapeutic and diagnostic applications of edible vaccines which are summarized in Table 12.3. Some of the diseases on which the work is going on are described below:

12.9.1 Malaria

Malaria is a disease of humans transmitted by the bite of an infected mosquito. It remains one of the most significant causes of human morbidity and mortality worldwide. According to WHO's 2010 world malaria report there are more than 225 million cases of malaria killing around 781,000 people. Three antigens are currently being investigated for the development of a plant-based malaria vaccine, merozoite surface protein (MSP) 4, MSP 5 from *Plasmodium falciparum* and MSP 4/5 from *P. yoelli*. Wang et al. (2004) have demonstrated that oral immunization of mice with recombinant MSP 4, MSP 4/5, and MSP 1, co-administered with CTB as a mucosal adjuvant, induces antibody responses effective against blood stage parasite.

12.9.2 Measles

Measles is an infection of the respiratory system caused by a virus. In an experiment mice fed with tobacco expressing MV-H (measles virus haemagglutinin from Edmonston strain) antibody titers five times the level considered protective for humans could be attained and secretory IgA was found in their feces. Prime boost strategy by combining parenteral and subsequent oral MV-H boosters could induce titers 20 times the human protective levels.

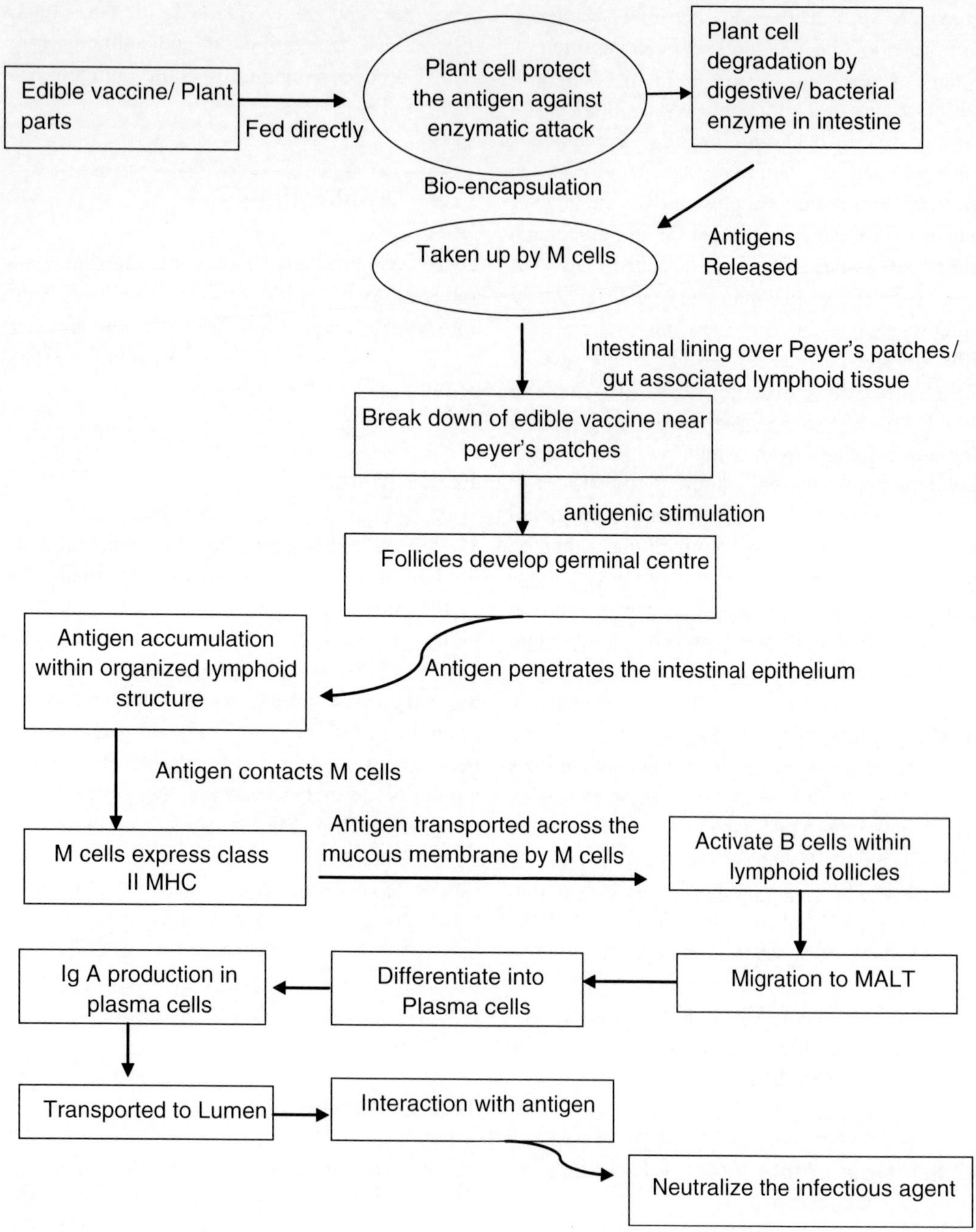

Fig. 12.2 Induction of mucosal immunity

These titers were significantly greater than with either of the vaccine administered alone. MV-H edible vaccine does not cause atypical measles, which may be occasionally seen with the current vaccine. Thus, it may prove better for achieving its eradication. The success in mice has prompted similar experiments in primates. Transgenic rice, and lettuce, and baby food against measles are also being developed. When given with CTB (adjuvant), 35–50 g MV-H lettuce is enough; however, an increased dose would be required if given alone.

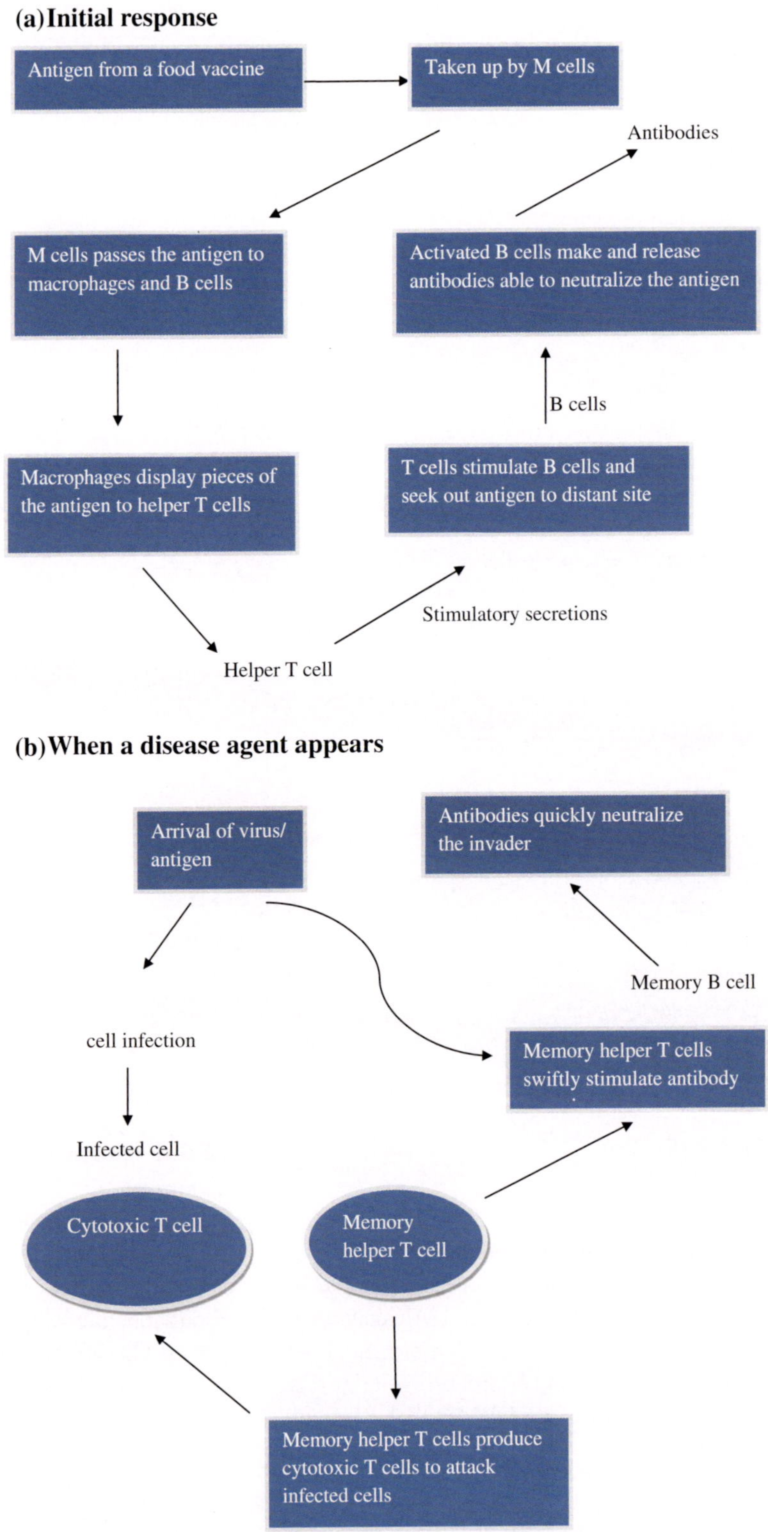

Fig. 12.3 **a** Production of antibodies. **b** Neutralization of invader

Table 12.2 Details of antigens produced by host plants and antibodies produced against them

Antibody	Antigen	Plant
IgG	Transition stage analog	Tobacco
IgM	NP(4-hydroxy-3-nitrophenyl) acetyl hapten	Tobacco
Single domain (dAb)	Substance B	Tobacco
Single chain Fv	Phytochrome	Tobacco
Single chain Fv	Artichoke mottled virus coat protein	Tobacco
Fab, IgG	Human creatine kinase	*Arabidopsis*
IgG	Fungal cutinase	Tobacco
IgG(k) and SIgG/A hybrid	*S. mutagens* adhesion	Tobacco
Single chain Fv	Abscisic acid	Tobacco
Single chain Fv	Nematode antigen	Tobacco
Single chain Fv	Alpha-glucuronidase	Tobacco
IgG	Glycoprotein B of herpes simplex virus	Soybean

(Das 2009)

Table 12.3 Therapeutic and diagnostic application of edible vaccines

Name of the vaccine	Vector	Pathological condition
Rabies virus	Tobacco, spinach	Rabies
Hepatitis B	Potato, Tobacco, Banana	Hepatitis B
HIV	Tomato	AIDS
Vibrio cholerae	Potato	Cholera
Cancer	Wheat, Rice	Cancer
Norwalk virus	Tobacco, potato	Hepatitis B
Rabbit hemorrhagic disease virus	Potato	Hemorrhage
Transmissible gastroenteritis corona virus	Tobacco	Gastroenteritis
Alzheimer's disease	Tomato	Alzheimer's disease
Colon cancer	Tobacco and potato	Colon cancer
Paramyxovirus	Banana, rice, lettuce	Measles
Plasmodium falciparum	Tobacco	Malaria
Type-I Diabetes	Potato	Type-I diabetes
Cysticercosis	Arabidopsis	Cysticercosis, foot and mouth disease

12.9.3 Rabies

Rabies is a deadly viral infection that is transmitted to humans from animals. Tomato plants expressing rabies antigens could induce antibodies in mice. Alternatively, TMV may also be used. Transformed tomato plants using CaMV with the glycoprotein (G-protein) gene of rabies virus (ERA strain) was shown to be immunogenic in animals.

12.9.4 Hepatitis B

Hepatitis B is a potentially life-threatening liver infection caused by the *hepatitis B* virus. It is estimated to have infected 400 million people throughout the globe, making the virus one of the most common human pathogen. First human trials of a potato-based vaccine against hepatitis B have reported encouraging results. Since immunization is the only known method to prevent the disease of the hepatitis B virus, any attempt to reduce its infection requires the availability of large quantities of vaccine HBsAg. The amount of HBsAg needed for one dose could be achieved in a single potato. Levels of specific antibodies significantly exceeded the protective level of 10 mIU/mL in humans. When cloned into CaMV, the pCMV-S plasmid encoding the HBsAg subtype ayw showed higher expression in roots as compared to leaf tissue of the transgenic potato. Furthermore, expression of the antigen was found to be higher in roots of transgenic potato than in leaf tissues. However, the expression of HBsAg in transgenic potatoes is not sufficient for using as oral vaccine. Further studies are underway to increase the level of HBsAg by using different promoters e.g., patatin promoter, and different transcription regulating elements.

12.9.5 Cholera

Cholera is an infection of the small intestine that causes a large amount of watery diarrhea. It causes up to 10 million deaths per year in the developing world, primarily among children. Studies supported by WHO have demonstrated possibility of an effective vaccine for cholera, which provides cross protection against enterotoxic *E. coli*. To address this limitation, plants were transformed with the gene encoding B subunit of the *E. coli* heat liable enterotoxin (LT-B). Transgenic potatoes expressing LT-B were found to induce both serum and secretory antibodies when fed to mice; these antibodies were protective in bacterial toxin assay *in vitro*. This is the first "proof of concept" for the edible vaccine.

Since people eat only cooked potatoes, the effect of boiling on the properties of CTB expressed in transgenic potatoes was examined. After boiling for 5 min, over half of the vaccine protein survived in its biologically active form, providing evidence that cooking does not always inactivate edible vaccines. Thus, the spectrum of plants for producing edible vaccines may be expanded beyond raw food plants such as fruits. Co-expression of mutant cholera toxin subunit (mCT-A) and LT-B in crop seeds has been shown to be effective by nasal administration and is extremely practical.

12.9.6 Diabetes

The prevalence of *diabetes* is increasing globally and India is no exception. More than 100 million people are affected with diabetes worldwide. Type-I diabetes, also known as insulin-dependent diabetes mellitus (IDDM) or juvenile-onset diabetes, primarily affects children and young adults and accounts for 5–10 % of the diagnosed diabetes in North America. Research by Ma and Hein (1995) at the University of Western Ontario showed that diabetes can be prevented in mice by feeding them with plants engineered to produce a diabetes related-protein. The idea is based on 'oral tolerance' where the autoimmune system is selectively turned off early by teaching the body to tolerate the "antigenic proteins". The pancreatic protein, glutamic acid decarboxylase (GAD67) is linked to the onset of IDDM, and when injected into mice it is known to prevent diabetes.

The Canadian group developed transgenic potato and tobacco plants with the gene for GAD67, fed them to nonobese diabetic mice, which developed insulin-dependent diabetes spontaneously. The results were intriguing, only 20 % of the prediabetic mice fed with transgenic plants developed the diabetes, while 70 % non-treated mice developed the disease. The treated mice also showed increased levels of IG1, an antibody associated with cytokines, which suppresses harmful immune responses. Thus, the antigen produced in plants appears to retain immunogenicity and prevent diabetes in an animal model. According to Canadian scientists, this is the first proof of principle for the use of edible vaccines in the treatment of the autoimmune diseases.

12.9.7 HIV

Human immunodeficiency virus (HIV) is a retroviral that causes acquired immunodeficiency syndrome (AIDS), a condition in humans in which progressive failure of the immune system allows life-threatening opportunistic infections and cancer to thrive. In order to produce edible vaccine initial success in splicing HIV protein into CPMV has been achieved. Two HIV protein genes and CaMV as promoter were successfully injected into tomatoes with a needle, and the expressed protein was demonstrable by polymerase chain reaction (PCR) in different parts of the plant, including the ripe fruit, as well as in the second generation plants. Recently, spinach has been successfully inoculated for Tat protein expression cloned into TMV. Each gram of leaf tissue of spinach was shown to contain up to 300–500 μg of Tat antigen. Mice fed with this spinach followed by DNA vaccinations resulted in higher antibody titers than the controls, with the levels peaking at 4 weeks post-vaccination.

12.10 Regulatory Issues

It is still unclear whether the edible vaccines would be regulated under food, drugs. or agricultural products and what vaccine component would be licensed—antigen itself, genetically engineered fruit or transgenic seeds. They would be subjected to a very close scrutiny by the regulatory bodies in order to ensure that they never enter the food supply. This would include greenhouse segregation of medicinal plants from food crops to prevent outcrossing and would necessitate separate storage and processing facilities. Although edible vaccines fall under "Genetically modified" plants, it is hoped that these vaccines will avoid serious controversy, because they are intended to save lives.

12.10.1 Clinical Trials

Edible vaccines are future vaccines and some challenges are yet to be overcome before these can become a reality. Like all products regulated by Food and Drug Administration, edible vaccines undergo a rigorous review of laboratory, and clinical testing that are conducted to get information regarding safety, efficacy, purity, and potency of these products. These trials can take place only after satisfactory information has been collected on the quality of the nonclinical safety.

Successful expression of antigens in plants has been demonstrated in the past. The vaccines have also been checked for their efficacy in humans. Results from the primary phase of the first-ever human clinical trial of an edible vaccine were published in the journal Nature Medicine in 1998 (Blaine P. Friedlander, Boyce Thomson Institute of Plant Research), which indicated that consumption of servings of raw potatoes resulted in immunity to specific diseases. The human clinical study was conducted under the direction of Dr. Carol Tacket at the Center for Vaccine Development, University of Maryland School of Medicine in Baltimore. In the first phase of human testing, the potatoes eaten by volunteers contained a vaccine against travelers' diarrhea, a common condition resulting from intestinal infection by the bacterium *E. coli*, which contaminates food or water supplies. The clinical trials were approved in advance by the Food and Drug Administration.

Encouraged by the results of this study, scientists started exploring the use of this technique for administering other antigens. In 2005 Thanavala's group has developed a potato vaccine booster for use in conjunction with injected hepatitis B vaccine. It is currently in phase II clinical trial and phase I for patients who have previously been vaccinated. In 2000, Tacket and his team mates studied the human immune response to the Norwalk virus capsid protein expressed in potatoes. Overall, 95 % (19 out of 20 volunteers) developed some kind of immune response, although the antibody increase in some cases was modest. In same year, Pogrebnyak's lab developed an effective vaccine against the coronavirus which causes severe acute respiratory syndrome (SARS). Tomato and Tobacco plants are used for high expression of the coronavirus spike protein (S1). First, lyophilized tomato fruit was fed to mice and then boosting occurred with S1 protein expressed in tobacco roots; high IgG1 immune responses and significant IgG2a and IgG2b responses were observed in their sera. Research is also going in the direction to engineer the plants to produce a variety of functional monoclonal antibody (Ma et al. 2005).

In the first human study of transgenic plant vaccine designed to induce active immunity, 14 adult volunteers were given either 100 g of transgenic potato, 50 g of transgenic potato or 50 g of wild type potato, each transgenic potatoes containing from 3.7 to 15.7 μg/g of LT-B. The variable dose per gram of potato was due to the tissue specificity of the promoter, therefore, that LT-B was expressed to a different degree in the different tissues of the potatoes. The potatoes in this study were ingested raw; however, subsequent studies have shown that transgenic potatoes expressing the B subunit of cholera toxin could be boiled for 3 min until the tissue becomes soft with loss of only about 50 % of the CT- B pentameric GM1-binding form. Serologic responses were also detected after vaccination. Totally 10 out of the 11 volunteers (91 %) who ingested transgenic potatoes developed IgG anti-LT and in half of them responses occurred after the first dose. There are 6 of the 11 (55 %) volunteers developed fourfold rise in serum IgA anti-LT.

Researchers supported by the NIAID have shown for the first time that an edible vaccine can safely trigger significant immune responses in people. The goal of the Phase 1 proof-of-concept trial study was to demonstrate that an edible vaccine could stimulate an immune response in humans. Volunteers ate bite-sized pieces of raw potato that had been genetically engineered to produce part of the toxin secreted by *E. coli,* which causes diarrhea. The trial enrolled 14 healthy adults, 11 were chosen at random to receive the genetically engineered potatoes and 3 received pieces of ordinary potatoes. The investigator periodically collected blood and stool samples from the volunteers to evaluate the vaccine's ability to stimulate both systemic and intestinal immune responses. The potatoes were well tolerated and no one experienced serious adverse side effects.

NIAID supported scientists are exploring the use of this technique for administering other antigens. Edible vaccines for other intestinal pathogens are already in the pipeline. Potatoes and bananas that might protect against Norwalk virus, a common cause of diarrhea, and potatoes and tomatoes that might protect against hepatitis B are being developed.

12.11 Future Perspective and Conclusion

Thirty million children throughout the world do not receive even the most basic immunizations each year. As a result, at least three million of these children die from diseases that are fully vaccine-preventable. The solution to vaccinate these children might seem simple with the idea of large scale production of edible vaccines for various diseases.

As a recent progress, the first human clinical trials for plant-based vaccine have been performed; it brings many challenges like optimization of expression levels, stabilization during post harvest storage, etc. Long-term reactions to edible vaccines are yet to be determined. Possible delayed reactions not yet discovered may be the point of consideration. In addition to that, edible vaccines can be further improved for their oral immunogenicity by the use of specific adjuvant which can be applied either as a fusion to the candidate gene or as an independent gene. Some of the diseases to which edible vaccines have shown promising application may be elaborated in the veterinary as well as human spectrum. These studies conclude plant-derived vaccines as a new hope and promise for more immunogenic, more effective, and less expensive vaccination strategies against both respiratory as well as intestinal mucosal pathogens.

Research in the field of edible vaccines holds immense potential for the future and every advancement made in this direction is bringing the dream of edible vaccine one step closer. There is hope that in coming future edible vaccines will conquer all serious diseases and make the planet beautiful to live in.

References

Arntzen CJ (1997) Edible vaccines. Public Health Rep 112(3):190–197

Chowdhury K, Bagasra O (2007) An edible vaccine for malaria using transgenic tomatoes of varying sizes, shapes and colors to carry different antigens. Med Hypo 68:22–30

Das DK (2009) Plant derived edible vaccines. Curr Tren Biotech Pharm 3(2):113–127

Eibl C, Zou Z, Beck A, Kim M, Mullet J, Koop HU (1999) *In vivo* analysis of plastid psbA, rbcL and rpl32 UTR elements by chloroplast transformation: tobacco plastid gene expression is controlled by modulation of transcript levels and translation efficiency. Plant J 19:333–345

Gruissem W, Tonkyn J (1993) Control mechanisms of plastid gene expression. CRC Critical Rev Plant Biol 12:19–55

Hiatt A, Cafferkey R, Bowdish K (1989) Production of antibodies in transgenic plants. Nature 342(6245):76–78

Ma JKC, Hein MB (1995) Immunotherapeutic potential of antibodies produced in plants. Trends Biotechnology 13:522–527

Ma S, Huang Y, Davis A, Yin Z, Mi Q, Menassa R, Brandle JE, Jevnikar AM (2005) Production of biologically active human interleukin-4 in transgenic tomato and potato. Plant Biotechnol J 3(3):309–318

Mason HS, Arntzen CJ (1995) Transgenic plants as vaccine production systems. Trends Biotechnol 13:388–392

Mason HS, Lam DMK, Arntzen CJ (1992) Expression of hepatitis B surface antigen in transgenic plants. Proc Natl Acad Sci USA 89:11745–11749

Mason HS, Haq TA, Clements JD, Arntzen CJ (1998) Edible vaccine protects mice against *Escherichia coli* heat-labile enterotoxin (LT): potatoes expressing a synthetic LT-B gene. Vaccine 16:1336–1343

Mason HS, Warzecha H, Tsafir MS, Arntzen CJ (2002) Edible plant vaccines: applications for prophylactic and therapeutic molecular medicine. Trends Mol Med 8:324–329

Mor TS, Gomez Lim MA, Palmer KE (1998) Edible vaccines: a concept comes of age. Trends Microbiol 6:449–453

Mor TS, Richter L, Mason HS (1999) Expression of rotavirus proteins in transgenic plants. In: Altman A, Ziv M, Izhar S (eds) Plant biotechnology and *in vitro* biology in the 21st century. Kluwer Academic publishers, Dordrecht, pp 521–524

Richter LJ, Thanavala Y, Arntzen CJ, Mason HS (2000) Production of hepatitis B surface antigen in transgenic plants for oral immunization. Nat Biotechnol 18(11):1167–1171

Tacket CO, Mason HS, Losonsky G, Estes MK, Levine MM, Arntzen CJ (2000) Human immune responses to a novel Norwalk virus vaccine delivered in transgenic potatoes. J Infect Dis 182:302–305

Thanavala Y, Mahoney M, Pal S, Scott A, Richter L, Natarajan N, Goodwin P, Arntzen CJ, Mason HS (2005) Immunogenicity in humans of an edible vaccine for hepatitis B. Proc Natl Acad Sci USA 102:3378–3382

Wang L, Goschnick MW, Coppel RL (2004) Oral immunization with a combination of *Plasmodium yoelii* merozoite surface protein 1 and 4/5 enhances protection against lethal malarial challenge. Infect Immunol 72:6172–6175

13 Engineering Plants for Phytoremediation

Mamta Baunthiyal

Abstract

In order to restore environmental balance, the utility of phytoremediation to remediate environmental contamination has received much attention in the last few years. Considerable effort has been devoted to making the transition from the laboratory to commercialization. Although plants have the inherent ability to detoxify contaminants, they generally lack the catabolic pathway for the complete degradation of these compounds as compared to microorganisms. There are also concerns over the potential for the introduction of contaminants into the food chain and how to dispose of plants that accumulate them in high quantities is also a serious concern. Hence, the utility of phytoremediation to remediate environmental contamination is still somewhat in question. For these reasons, researchers have endeavored to engineer plants with genes that can bestow superior degradation abilities. Genes from microbes, plants, and animals are being used successfully to enhance the ability of plants to tolerate, remove, and degrade pollutants. Although improvement of plants by genetic engineering opens up new possibilities for phytoremediation, it is still in its research and development phase, with many technical issues needing to be addressed.

13.1 Introduction

Elemental pollutants such as arsenic (As), copper (Cu), cadmium (Cd), mercury (Hg), zinc (Zn), and lead (Pb) are difficult to remediate from air, water, and soil because these toxic elements are immutable by all biochemical reactions, hence remain in the ecosystem. They seep into surface water, groundwater, or soil and may channel into the food chain through crops growing on such soil and water. Apart from these metals, small organic contaminants and volatile organic compounds are other prevalent contaminants of the environment. These include petroleum hydrocarbons (PHC), halogenated hydrocarbons, polycyclic aromatic hydrocarbons (PAHs), polychlorinated biphenyls (PCBs) and pesticides, and solvents like trichloroethylene (TCE), carbon tetrachloride, and salts. PAHs, in

M. Baunthiyal (✉)
Govind Ballabh Pant Engineering College,
Ghurdauri, Pauri, Uttarakhand 246194, India
e-mail: mamtabaunthiyal@yahoo.co.in

I. Ravi et al. (eds.), *Advances in Biotechnology*,
DOI: 10.1007/978-81-322-1554-7_13, © Springer India 2014

general, are ubiquitous environmental pollutants and are formed from both natural and anthropogenic sources. Anthropogenic sources are by far the major contributors and include the burning of coal refuse banks, coke production, automobiles, commercial incinerators, and wood gasifers. PAHs are an environmental concern because they are not only toxic to aquatic life but also several are suspected human carcinogens

The term phytoremediation ("phyto" meaning plant, and the Latin suffix "remedium" meaning to restore) refers to a diverse collection of plant-based technologies that use plants (either naturally occurring or genetically engineered) to clean contaminated environments. Phytoremediation is popular because of its cost-effectiveness, environmentally friendly aesthetic advantages, long-term applicability, and good influence on the physical, chemical, and biological aspects of the environment. The technology can be used at hazardous waste sites where other methods of treatment are too expensive or impractical; it may also be used at low-level contaminated sites where only "polishing treatment" is required over a long period of time. Compared to the use of other organisms, the use of plants costs less as they capture solar energy to synthesize the proteins and other structures that are necessary for the remediation. Phytoremediation can be used in conjunction with other technologies as a final cap. Phytoextraction, a type of phytoremediation, is an effective means of remediating a site because it reduces the overall mass to be treated from tons of widespread contaminated soil to plant tissue that can be dried to a small volume. Unlike other engineering methods that might remove the fertile topsoil, phytoremediation does not reduce the fertility of the site (Chaney et al. 1997).

To be effective, the concentration of the metal in the harvestable part of the plants must be higher than its concentration in the soil so that the volume of the hazardous plant waste is less than the volume of the contaminated soil. Plants that are especially good at concentrating the pollutants are termed as hyperaccumulators. The 'bioconcentration factor' (BCF) refers to the metal concentration in plant tissues relative to the metal concentration in the substrate, and a value >1 indicates that the plant actively concentrates the metal. A metal hyperaccumulator is defined as a plant that can concentrate the metals to a level of 0.1 % for nickel (Ni), cobalt (Co), Cu, and Pb, 1 % for Zn and 0.01 % for Cd (Baker et al. 2000). For example, *Pteris vittata* (Chinese brake fern) efficiently hyperaccumulates arsenic in its fronds. Another hyperaccumulator is *Thlaspi caerulescens*, which can concentrate cadmium, a highly toxic and probably carcinogenic metal, in the above-ground tissues at concentrations 1,000 times higher than the normal toxic concentration of only 1 ppm.

Although plants have innate capabilities of remedying hazardous contaminants from the environment, the rate is directly proportional to plant growth rate and the total amount of remediation is correlated with the plant total biomass, making the process very slow. Further, phytoremediation using natural plants has other limitations that include the potential for introducing the contaminant or its metabolites into the food chain, long cleanup times required to achieve regulatory action levels, and toxicity encountered in establishing and maintaining vegetation at waste sites. The plants may also suffer toxicity symptoms.

Keeping these limitations in view, it would be beneficial to enhance plants' capacity for phytoremediation. This necessitates the identification of a fast growing (largest potential biomass and greatest nutrient responses) and more strongly metal accumulating genotype. Genetic engineering provides an excellent means to achieve this goal. It is possible to generate transgenic plants that can overexpress their genes or express exogenous genes, and thereby acquire new or improved phenotypes. One major advantage of genetically engineered plants is that the ability of specific enzymes for degradation of a contaminant could be transferred to a plant species that is either indigenous to an ecosystem or has other desirable remediation properties such as rapid growth, deep root structures, or high water uptake. Once an individual plant with suitable phenotype is produced, it is easy to multiply it by cloning methods, such as cuttings and somatic embryogenesis. Moreover, many trees continue to grow for many years as long as the conditions are suitable.

13.2 Genetic Engineering Strategies for Plant Modification to Enhance Phytoremediation

Many transgenic plants with new characteristics due to expression of genes from microorganisms, yeasts, mouse, or even human have been reported in scientific journals. Till date, all commercial and research activities have used naturally occurring plant species. Many of these are species that can be genetically engineered. In general, any dicotyledonous plant species can be genetically engineered using the *Agrobacterium* vector system, while most monocotyledonous plants can be transformed using particle gun or electroporation techniques.

Recent studies using transgenic aspen over-expressed with a nitroreductase, *pseudomonas nitroreductase A* (*pnrA*), isolated from the bacterium *Pseudomonas putida* have shown higher accumulation of TNT (2, 4, 6-trinitrotoluene) from liquid culture and soil, compared to nontransgenic plants. Furthermore, the tolerance limit toward TNT was also significantly higher than for nontransgenic plants.

The genetic engineering approach has successfully facilitated to alter the biological functions of plants through modification of primary and secondary metabolism and by adding new phenotypic and genotypic characters to plants with the aim of understanding and improving their phytoremediation properties (Fig. 13.1). Transgenic plants produced for metabolizing herbicides and long-persisting pollutants can be used for phytoremediation of foreign chemicals in contaminated soil and water (Dowling and Doty 2009).

Metal-hyperaccumulating plants and microbes with unique abilities to tolerate, accumulate, and detoxify metals and metalloids represent an important reservoir of unique genes. These genes could be transferred to fast-growing plant species to improve phytoremediation (Fulekar et al. 2009). It has been established that the adaptive metal tolerance is governed by a small number of major genes and some minor modifier genes. Probably, it is this adaptive metal tolerance that gears a plant species for hyperaccumulation.

Use of tissue culture techniques to select genes having enhanced bio-degradative properties (for organics) or the enhanced ability to

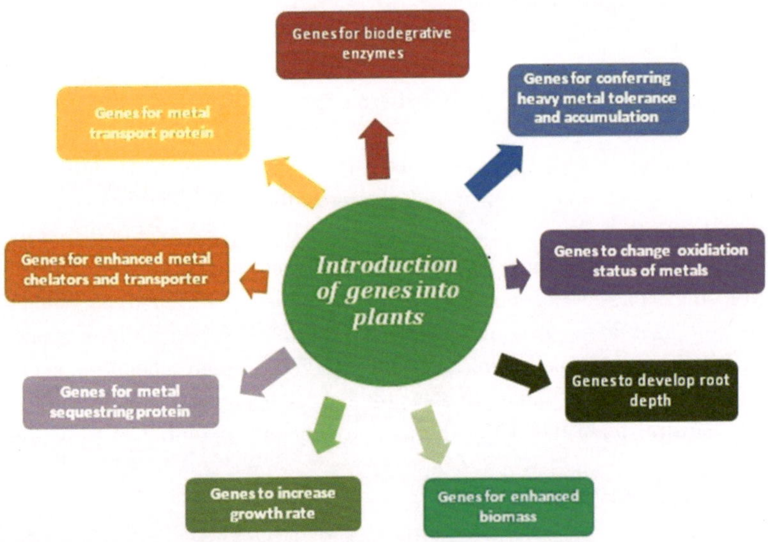

Fig. 13.1 Genetic engineering in phytoremediation

assimilate metals helps to select plants with the desired characters. Molecular techniques such as the analysis of molecular variance of the random amplified polymorphic DNA (RAPD) markers are useful to investigate the genetic diversity and heavy metal tolerance in plant populations, providing the opportunity to investigate the first steps in the differentiation of plant populations under severe selection pressure and to select plants for phytoremediation.

The genetic and biochemical basis is becoming an interesting target for genetic engineering. A fundamental understanding of both uptake and translocation processes in normal plants and metal hyperaccumulators, regulatory control of these activities, and the use of tissue specific promoters offers the ability to develop effective and economic phytoremediation plants for soil metals. Examples include genes controlling the synthesis of peptides that sequester metals, like phytochelatins, genes encoding transport proteins, or genes encoding enzymes that change the oxidation state of heavy metals. The genes involved in the metabolism of chemical compounds can be isolated from various organisms, including bacteria, fungi, plants, and animals, and these genes are then introduced into candidate plants. The desired characters for phytoremediation can be improved by identifying candidate protein, metal chelators, and transporter genes for transfer and/or overexpression of a particular gene.

13.2.1 Plant Metal Transporters

One of the promising approaches to enhance the ability of metal ions to enter plant cells is the identification of the metal transporter proteins and introducing genes encoding transporter molecules. These are generally proteins found in the cell membrane, that either have an affinity for metal ions, or that create favorable energetic conditions to allow metals to enter the cell. Till date, several plant metal transporters have been reported and more remain to be recognized. Some of the transporters identified so far include the *Arabidopsis* IRT1 gene that encodes a protein that regulates the uptake of iron and other metals (Eide et al. 1996) and the MRP1 gene encoding, an Mg-ATPase transporter, also from *Arabidopsis* (Lu et al. 1997).

Further, success in this approach is achieved by identifying proteins such as ZNT1, ZIP1-4, IRT1, COPT1, LCT1, and tVramp-1/3/4 on the plasma membrane-cytosol interface; ZAT, ABC type, HMT1, AtMRP, CAX2 seen in vacuoles; and RAN1 in Golgi bodies. Manipulations of these transporters to achieve the removal of metal ions from the cell hold great potential (Tong et al. 2004). The natural resistance-associated macrophage proteins (Nramp) family of transporters has been recently characterized from rice and *Arabidopsis*.

13.2.2 Phytochelatins for Metal Sequestering

The principal classes of metal chelators include phytochelatins, metallothioneins, organic acids, and amino acids. Phytochelatins (PCs) are small metal-binding peptides found in plants. Iso-PCs, a series of PC-like homologous chelating peptides, are reported with varying terminal amino acids and have a C-terminal modified residue other than glycine. The PC and iso-PC molecules form complexes with heavy metals like Cd. In addition to PC-Cd complex other PC-metal complexes include Ag, Cu, and As (Shah and Nongkynrih 2007).

In vitro experiments have shown that a series of metal-sensitive plant enzymes can tolerate a 10- to 1,000-fold concentration of Cd in the form of a PC complex than as free radical ion. PC reactivate metal poisoned plant enzymes such as nitrate reductase up to 1,000-fold better than chelators such as glutathione (GSH) or citrate, signifying the extraordinary sequestering potential of these peptides.

13.2.3 Metallothioneins

Metallothioneins (MTs) are metal-binding proteins that confer heavy metal tolerance and accumulation (Hamer 1986; Malin and Bülow 2001). These are low molecular mass cysteine rich proteins that were originally isolated as Cu, Cd, and Zn binding proteins in mammals. MTs are able to bind a variety of metals by the formation of mercaptide bonds between the numerous cysteine (Cys) residues present in the proteins and the metal, and it is the arrangement of these residues that in part determines the metal-binding properties of the MT proteins. Researchers successfully reported more than 50 MTs in different plants categorized into four types, types 1–4. These are based on the Cys arrangement. Although MTs are expressed throughout the plant, some have been found to be expressed in a tissue-specific manner. Transgenic plants expressing MTs have been created, and although these plants exhibited enhanced tolerance to high metal concentrations, the uptake of metals was not enhanced. To enhance higher plant metal sequestration, the yeast MT CUP1 was introduced into tobacco plants, and the *cup 1* gene expression and Cu and Cd phytoextraction were determined. Overexpression of Cu inducible MT *cup 1* also enhanced Cu tolerance in plants. In plants, a wide range of MT genes from various sources have been overexpressed including those from human, mouse, Chinese hamster, and yeast.

The vacuole is generally considered to be the main storage site for metals in yeast and plant cells and there is evidence that phytochelatin-metal complexes are pumped into the vacuole. The best characterized of the known vacuolar transporters and channels involved in metal tolerance are YCF1 from yeast, *Saccharomyces cerevisiae*. YCF1 is an MgATP-energized glutathione *S*-conjugate transporter responsible for vacuolar sequestration of compounds after their *S*-conjugation with glutathione. Overexpression of the *YCF1* gene in *Arbidobsis thaliana* and the YCF1 proteins were found to be associated with the tonoplast and the plasma membrane. The vacuoles of the YCF1-transgenic plants exhibited a 4-fold higher rate of glutathione-Cd uptake than those of wild-type plants, indicating that expression of *YCF1* - strongly increases Cd transport activity. The transgenic plants showed improved resistance to both Cd and Pb and elevated metal content, characteristics desirable for phytoremediation.

13.2.4 Genes to Change the Oxidation State of Heavy Metals

Genes may be introduced that code for enzymes to change the oxidation state of heavy metals like Hg and selenium (Se). For instance, introduction of the bacterial merA gene encoding mercuric oxide reductase (Rugh et al. 1996), or that which converts metals into less toxic forms, such as enzymes that can methylate Se into dimethylselenate (Hansen et al. 1998). In both these cases, the resulting form of the metal is volatile, so that one can create a plant capable of metal remediation by phytovolatilization, another type of phytoremediation.

13.3 Genetically Modified Plants for Metal Uptake, Tolerance, and Detoxification

Several researchers have already reported encouraging results using plants bioengineered with increased heavy metal tolerance and uptake of heavy metals for the purpose of phytoremediation. The majority of these novel plants have only been tested under limited laboratory conditions and very few have been grown in the field (Table 13.1).

13.3.1 *Arabidopsis*

Arsenic is an extremely toxic metalloid pollutant and the decontamination of polluted sites can be environmentally destructive. It is a lethal poison that is released into the environment from

Table 13.1 Genetically modified organisms used for enhanced phytoremediation of organic contaminants

Transgenic species	Gene source	Gene(s)	Enzyme(s)	Transgene effects	References
Arabidopsis thaliana	Cotton	*LACI*	Root-specific laccase	Cotton secretes laccase to the rhizosphere and has shown enhanced resistance to phenolic alleleochemicals and enhanced tolerance to 2, 4, 6,-trichlorophenol	Wang et al. (2004)
A. thaliana	*A. thaliana*	*743B4, 73C1*	Glycosyltransferases (UGTs)	Overexpression of UGTs genes resulted in the enhanced detoxification of TNT and enhanced root growth	Gandia-Herrero et al. (2008)
Potato (*Solanum tuberosum*)	Rat	*CYP1A1*	Cytochrome P450 monooxygenase	Increased tolerance to atrazine and chlortoluron, assumed to be *via* metabolism to less-toxic derivatives	Yamada et al. (2002)
Tobacco (*Nicotiana tabacum*)	Human	*CYP2E1*	Cytochrome P450 monooxygenase	Dramatically enhanced metabolism of trichloroethylene; increased uptake and debromination of ethylene dibromide	Doty et al. (2000)
Tobacco (*N. tabacum*)	*Enterobacter cloacae*	*onr*	Pentaerythritol tetranitrate	Enhanced denitration of glycerol trinitrate	French et al. (1999)
Tomato (*Lycopersicon esculentum*)	Roots of *L. esculentum*	*tpxI*	Peroxidases	The overexpression of tpxI gene in transgenic tomato hairy roots resulted in the enhanced removal of phenol	Oller et al. (2005)
Hybrid poplar (*Populus tremula* × *Populus alba*)	Rabbit	*CYP2E1*	Cytochrome P450 monooxygenase	Enhanced removal and degradation of trichloroethylene, vinyl chloride, carbon tetrachloride, benzene, and chloroform; enhanced removal of gaseous trichloroethylene, chloroform, and benzene	Doty et al. (2007)
Brassica juncea	*Brassica juncea*	*γ-ECS, GS*	γ-Glutamycysteine synthetase; Glutathione synthetase	Overexpression of ECS and GS resulted in enhanced tolerance to atrazine, 1-chloro-2, 4-dinitrobenzene, phenanthrene, metolachlor	Flocco et al. (2004)
Rice (*Oryza sativa*)	Human	*CYP1A1*	Cytochrome P450 monooxygenase	Enhanced metabolism of atrazine, norflurazon, and chlortoluron (should also metabolize PAHs)	Kawahigashi et al. (2003)
Rice (*O. sativa*)	Human	*CYP1A1, CYP2B6, CYP2C19*	Cytochrome P450 monooxygenases	Enhanced metabolism of atrazine, norflurazon, and metolachlor from soil (should also metabolize PAHs)	Kawahigashi et al. (2006)
Rice (*O. sativa*)	Human	CYP2C9	Cytochrome P450 monooxygenases	Tolerance to sulfonylurea	Hirose et al. (2005)
Pseudomonas fluorescens	*Burkholderia* sp.	*bph operon*	Suite of enzymes for the complete PCB degradation pathway	Enhanced rate of degradation of numerous PCBs (resting cell assays)	Villacieros et al. (2005)
P. fluorescens (psychrotolerant strain)	*Burkholderia* sp.	*dnt* genes	Suite of enzymes for degradation of 2, 4-dinitrotoluene to pyruvate and propionyl-CoA	Complete degradation of 2, 4-dinitrotoluene as a co-substrate at temperatures as low as 10 °C	Monti et al. (2005)
P. fluorescens	*Burkholderia cepacia*	*tomA*	Toluene-o-mono-oxygenase	Sixty-three percent degradation of trichloroethylene after 4 days in wheat rhizosphere	Yee et al. (1998)

The results are obtained from laboratory and greenhouse experiments. None of the transgenic species were field tested (Modified from Gerhardt et al. 2009)

natural processes and *via* arsenic-based chemicals. Genetically engineered *Arabidopsis* plants can sequester As from the soil. These plants can transport As aboveground, reduce it to arsenite, and sequester it in thiol-peptide complexes (Dhankher et al. 2002). By coexpressing two bacterial genes, arsenate reductase (ArsC) and γ-glutamylcysteine synthetase (γ-ECS) in *Arabidopsis* plants, it was observed that plants expressing SRS1p/ArsC and ACT2p/γ-ECS together showed substantially greater As tolerance than wild-type plants or plants expressing γ-ECS alone. In addition, when grown on As, these plants accumulated 4- to 17-fold greater fresh shoot weight and accumulated 2- to 3-fold more As per gram of tissue than wild-type plants or plants expressing γ-ECS or ArsC alone.

Extensive progress has also been achieved in identifying genes and proteins involved in uptake of iron (Fe) by yeast and plants. As described earlier, the utility of the yeast protein YCF1, a protein which detoxifies Cd by transporting it into vacuoles, has been implemented for the remediation of Cd and Pb. Transgenic *A. thaliana* plants overexpressing YCF1 showed an enhanced tolerance and accumulated greater amounts of Cd and Pb.

The close relationship between *Arabidopsis halleri,* a metal tolerant and hyperaccumulating relative of the biological model species *A. thaliana*, has recently allowed the use of *A. thaliana* GeneChips to compare gene expression levels between *A. halleri* and the nontolerant *A. thaliana* and, consequently, permitted the identification of genes potentially involved in metal tolerance and/or hyperaccumulation. The complete annotation of the *A. thaliana* genome sequence provides a solid foundation for comparative mapping studies within the Brassicaceae family. The genome organization of *A. thaliana* has already been compared with those of several species like *Arabidopsis lyrata*, *Arabidopsis petraea* and *Capsella rubella*. Moreover, *A. halleri* is a species that has undergone natural selection for Zn tolerance. Isolation of the quantitative trait loci (QTL) associated with this trait holds great promise for the identification of the main genes responsible for this adaptation.

13.3.2 *Brassica*

Brassica juncea was genetically engineered to investigate rate-limiting factors for glutathione and phytochelatin production. To achieve this *Escherichia coli gshII* gene was introduced. Cd-treated GS plants had higher concentrations of glutathione, phytochelatin, thiol, sulphur, and calcium than wild-type plants (Zhu et al. 1999). A study showed that γ-glutamylcysteine synthetase inhibitor, L-buthionine-[S,R]-sulphoximine (BSO), dramatically increased As sensitivity both in nonadapted and As-hypertolerant plants, showing that phytochelatin-based sequestration is essential for both normal constitutive tolerance and adaptative hypertolerance to this metalloid (Schat et al. 2002).

Astragalus bisulcatus is a native plant that has the capacity to grow on Se containing soils and accumulate Se to high concentrations but it has a slow growth rate. It has been proposed that in *A. bisulcatus* selenocysteine methyltransferase (SMT) converts the amino acid SeCys into the nonprotein amino acid (MetSeCys). By incorporating gene for SMT from the Se hyperaccumulator *A. bisulcatus*, it diverted the flow of Se from the Se amino acids that may otherwise be incorporated into protein, leading to alterations in enzyme structure, function, and toxicity. Transgenic plants overexpressing SMT show enhanced tolerance to Se, particularly selenite, and produced 3- to 7-fold more biomass than wild type and 3-fold longer root lengths (LeDuc et al. 2004). Indian mustard plants overexpressing the *A. bisulcatus smt* gene, exhibited a greatly increased accumulation of MetSeCys and tolerance to Se compounds, in particular selenite.

13.3.3 Tobacco

Scientists at the University of Cambridge expressed the bacterial gene encoding pentaerythritol tetranitrate reductase in transgenic tobacco, conferring the ability to survive on growth media containing otherwise toxic levels of the nitrate ester class of explosives. Further analysis also demonstrated an enhanced degradation of these compounds by transgenic tobacco plants relative to untransformed seedlings.

13.3.4 Tomato

When bacterial gene 1-aminocyclopropane-1-carboxylic acid (ACC) deaminase was expressed in tomato plants, it showed enhanced metal accumulation and tolerance levels for a range of heavy metals (Cd, Cu, Ni, Mg, Pb, and Zn) than untransformed plants (Grichko et al. 2000).

13.3.5 *Pteris vittata*

Pteris vittata can effectively remove As metalloid from soil. For example, in soil contaminated with As at a concentration of 97 ppm, the older fronds of the fern had As concentrations of up to 3894 μg per gram of tissue. Less than 168 μg As per gram was found in the root tissue. More than 95 % of the As removed from the soil by the fern was translocated to the aboveground biomass (Doty 2008). Unfortunately, this plant species grows well only in warm, humid environments.

13.3.6 Poplar and Willow

Although the research on several hyperaccumulating species is promising, the species themselves are too small and slow-growing for some phytoremediation applications. For this reason, high-biomass crops such as poplar and willow are being studied for phytoremediation of metals. Poplar and willow are not hyperaccumulators as they do not concentrate metals to high concentrations, but because of their greater biomass and deep root systems, they are also effective remediators of metal contamination. Poplar plant transformation was conducted using *Agrobacterium*-mediated transfer of genes. Transgenic poplar plantlets which express genes of interest were then multiplied by cuttings. After multiple-year field trials, it can be confirmed whether transgenic poplar plants are indeed improved in the capacity to remediate the polluted environment, thus suitable for actual phytoremediation. There is a possibility that plants with single gene insertion may not be sufficiently improved in the capacity to remediate the actual field. To make the process more effective, multiple genes with different and synergistic mechanisms of phytoremediation are expressed at the same time. To develop more and better plants for phytoremediation, the search for new genes with different functions and stacking them (introducing many genes into the same line of plants) will hopefully further improve the plants' capacity to clean up the environment.

Willow was specifically suggested for phytoremediation of heavy-metal contaminated lands because of its unique ability to re-grow readily after its shoots have been harvested, a distinctive trait of willow. Further there is substantial genetic diversity of willow, with over 450 *Salix* species. Willow plants (*Salix matsudana* × *Salix alba* NZ1040) grown in soil contaminated with Cd at concentrations commonly found in agricultural sites fertilized with Cd-containing fertilizers accumulated Cd in the aboveground tissues with BCFs of about 10 (Doty 2008).

13.4 Challenges

To improve the process of phytoremediation using genetic modifications success has been achieved, and these processes have great potential for field applications, the only constraint being public acceptance of genetically modified organisms.

The key concerns in remediating contaminated sites by transgenic plants include:

1. Characterization of vector system, its stability, and expression in the plant cell.
2. Avoidance of unwanted properties encoded by introduced DNA like toxic, pathogenic, or deleterious functions.

3. Potential risks to birds and insects that might feed on these plants.
4. Proper disposal of plant biomass containing high concentrations of hazardous substances, particularly metals.
5. Competitiveness of transgenic to wild type.
6. The ability to outcross with related species (particularly wild relatives).

13.4.1 Complexity of Phytoremediation in the Field

Uneven distribution of contaminants is the frequently encountered challenge in field studies, including ''hot spots'' across a site. In laboratory and greenhouse experiments, soils generally have uniform matrix. This may not be possible in the field even if the site is extensively tilled prior to planting; this spatial heterogeneity of initial contaminant levels results in data scatter, which can make it difficult to statistically show significant treatment effects for field trials. This is also problematic in terms of regulatory standards because remediation success is often judged on a point-by-point basis rather than an average of data points from across the site. Each and every point should meet the regulatory targets; otherwise, the whole site fails.

Other factors such as soil pH, moisture content, root structure, soil structure, organic composition of the soil, and microbial activity often exhibit spatial variability at a given site and can change over time. One way to mitigate the problem of spatial variability is to increase the number of samples analyzed per treatment plot. However, extensive sampling can lead to highly variable data that is difficult to interpret.

Agronomic techniques are often employed prior to planting. Tilling and addition of nutrients and organic matter are generally designed to improve soil texture and composition to facilitate plant and/or microbial growth. They also cause changes in soil pH and oxygen content, which can affect the bioavailability, hence degradation of contaminants even in bulk (non-rhizosphere) soil. Nutrients in the bulk soil, such as nitrogen and phosphorus, will positively affect the indigenous microbial populations, including contaminant degraders. Aeration of field soil increases its oxygen concentration, which promotes oxidative degradation of numerous organic contaminants. This may also lead to photo-oxidation of previously buried organics. Hence, there is often a decrease in contaminant concentrations in bulk soil (generally used as the negative control) as well as in treatment plots. As a result it is difficult to statistically show that phytoremediation is superior to natural attenuation, particularly in the initial stages of a field trial. Less remediation is thought to occur in the rhizosphere (2 mm zone extending from the roots), and much less in bulk soil. Determining the boundary between rhizosphere and bulk soil is almost impossible. Furthermore, it is impractical to definitively separate fine roots, soil directly in contact with the roots, and rhizosphere soil into different samples for analyses. This can present an analytical challenge for researchers endeavoring to show PHC and PCB remediation where rhizoremediation is thought to occur at the root–soil boundary (rhizoplane). There is evidence that some plants can provide the impetus for movement of hydrophobic organics such as PAHs from bulk soil into the rhizosphere (Liste and Alexander 2000). Mobilization of PAHs may be accelerated by the release of organic acids from roots, which putatively increases PAH solubility and bioavailability.

13.4.2 Physical Restrictions and Stress Factors

Despite increasing interest *in situ* remediation of contaminants, the number of reports pertaining to phytoremediation appears to be declining, suggesting a decrease in the activity for this area of research. Numerous unsuccessful and inconclusive field trial reports provide some insight into the challenges that researchers have encountered in this field (Gerhardt et al. 2009). The slower rate of remediation compared to other strategies, such as excavation and *ex situ* treatment, is a frequently cited disadvantage of phytoremediation. This may be probably because plant growth is hard to achieve in heavily impacted soils, thereby limiting catabolism of contaminants. Factors that affect phytoremediation in the field are not encountered in the laboratory or greenhouse. These include stressors like plant pathogens; herbivore (insects and/or animals); variations in temperature, nutrients and precipitation; and competition from weed species that are better adapted to the site. Any of these abiotic or biotic stressors can slow down or prevent plant growth in the field. For example, attempted phytoremediation of PHC and other organics at a hydrocarbon burn facility at NASA Kennedy space center failed due to drought and competition from weeds (http://www.cluin.org).

There are other physical challenges that limit the use of phytoremediation. For instance, increasing the moisture content of hydrophobic PHC-contaminated soils can be problematic once they have become dry, which can prevent germination and seedling growth. Another problem is faced when the contaminated soil is deeper than the rooting zone, e.g., average root depth of herbaceous plants is 50 cm. In that case, there is a requirement of excavation prior to phytoremediation. Trees have deeper roots (average 3 m, with much longer roots in some species) which can facilitate remediation at greater depths without excavation. Dendroremediation (phytoremediation using trees) of explosives, e.g., TNT and TCE from soil and groundwater has shown great promise (Van Aken et al. 2004). However, here the challenge is that it can be difficult to grow and establish trees in contaminated soils. The process is quite long and they take several years to attain sufficient biomass for efficient rates of phytoremediation. Further, in case of trees there may be a disposal problem if the roots and wood are deemed to be contaminated after completion of remediation.

13.4.3 In-field Performance of Genetically Modified Organisms

Genetically engineered microbial strains often fail to compete with native microbes in the rhizosphere, and their numbers diminish to levels that merely support remediation.

Degradation of organic contaminants generally requires the concerted action of numerous enzymes and it is generally impractical to introduce all the genes required for degradation of an organic contaminant into a single plant or microbial genome. It can be difficult to stably maintain even a single gene in transformed or recombinant organisms and the desired trait, such as enhanced degradative capacity, is often lost. For example in plants, silencing of transgenes, *via* mechanisms such as cytosine methylation, makes the use of this technology inherently unpredictable.

Though genetically modified organisms in the laboratory and greenhouse for phytoremediation has made considerable progress, regulatory restrictions for *in situ* applications have prevented any substantial accumulation of field data. There is the potential for inserted genetic material to be transferred to indigenous populations if genetically modified organisms are used without adequate containment systems. Further, recombinant strains that contain antibiotic resistance genes from the cloning procedures cannot be released in the environment. Besides, there has also been low public acceptance of genetic engineering.

13.4.4 Regulatory Acceptability

The issue of regulatory acceptability for phytoremediation of hazardous waste is to ensure public safety and to protect the environment. The Government regulators must be convinced that a given remedial strategy will diminish contaminant toxicity, mobility, and/or concentration before it is approved for use in the field. Bioaccumulation of hazardous compounds in plants is one consideration. In situations where plants do not catabolize the contaminants, provisions may be required for removal of contaminated plant materials as part of a remedial plan. Another consideration is the potential ecological risks of introducing non-native plant and microbial species into field sites. These species can move from the contaminated site by displacing, or hybridizing with native species.

Phytoremediation of organic compounds often takes place in the rhizosphere of fine roots that have a high turnover rate. These roots have a very short life and the decaying root tissue is converted into bulk soil during humification. Contributions of organic matter from the degraded root tissue can complicate sample analyses due to the structural similarities to compounds being remediated. Most of the regulatory agencies for example, United States Environmental Protection Agency (EPA) and Canadian Council of Ministers of the Environment (CCME), do not distinguish between petrogenic and phytogenic carbon compounds. This means that in the case of PHC remediation naturally occurring organic matter that is co-extracted with organic contaminants can easily lead to exaggerated PHC levels in soil samples.

13.5 Overcoming the Challenges

13.5.1 Chloroplast Engineering

Transformation of chloroplast prevents the escape of transgenes *via* pollen to related weeds and crops. This method was used to stably integrate the bacterial merAB operon into the chloroplast genome of tobacco. The resulting plants were substantially more resistant to highly toxic organic mercury (Bizily et al. 2000), in the form of phenyl mercuric acetate, than wild type. Other important advantages of chloroplast transformation include the fact that codon optimization is not required to improve expression of bacterial transgenes.

13.5.2 Minimizing Ecological Risk

To minimize the risks of introducing non-native biota (including genetically modified species) into an ecosystem, the best strategy is to use native species for phytoremediation. It can be applied to microbial species that are used to facilitate plant growth and/or degrade contaminants (Davison 2005). Using native plant species can serve the dual purpose of remediation as well as native habitat reconstruction/reclamation (for microbe-assisted phytoremediation), which may be required after successful remediation.

To minimize ecological risks from non-native (transgenic and nontransgenic) phytoremediation species, a biological containment system is often employed. For example, genes can be introduced to prevent propagation, or to render a species overly sensitive to abiotic stressors such as temperature changes or chemicals. Multiple transgenes are employed to prevent gene flow between the introduced species and other species in the environment. This containment system can be reinforced by adding mitigator genes linked to the primary transgene. Mitigators genes confer non-deleterious traits to the phytoremediation species, but are harmful to relate species in case of gene transfer. Alternatively, they can prevent the phytoremediation species from successfully competing outside the contaminated site.

13.5.3 Overcoming the Challenge of Plant Stress

Naturally occurring endophytic bacteria have been found to degrade organic contaminants such as nitro-substituted explosives. In contrast to root-colonizing microbes, endophytes colonize internal plant tissues, including root, leaf, and vascular tissues. Inoculating plants with endophytes can thwart competitive pressures in the rhizosphere as well as some of the challenges that limit effectiveness of root colonizers, including dependence on specific soil pH, temperature, and soil moisture content for optimum growth. Field trials will be required to assess the effectiveness of the use of endophytes.

13.5.4 Monitoring, Sampling, and Analyzing Experimental Data from the Field

To demonstrate adequate performance of remediation systems in the field, government agencies are focusing on issues that address the inconsistencies in experimental protocols, particularly regarding which analytical parameters need to be measured. Remediation Technologies Development Forum, a group with government, industry, and academic partners has developed the protocol of phytoremediation of PHCs in soil in the 1990s. This forum intends to develop protocols that would allow comparative tests of remedial strategies at numerous and varied geographical locations. These protocols recommend standards for plot size; number of replications; choice of plant species; plant and soil sampling procedures; microbial and hydrocarbon analyses; statistical treatment of data; time-points and end-point (3 years). A three-year end-point is considered more realistic than remediation in a single growing season. Establishing a longer time frame for field trials is always advantageous for researchers because ambiguous results are often observed in short-term field studies as a result of tilling and amending the soil at the onset of a field trial. The three-year frame duration also allows for assessment of the phytoremediation system and improvement of methods between field seasons. If any problems in the field are encountered, laboratory and greenhouse experiments can be conducted to resolve the issues prior to the onset of the next field season.

One way to facilitate statistical analysis of field data is to use particularly recalcitrant compounds in the soil samples to normalize the data. These are called internal markers (also called "conservative biomarkers"). They are not generally degraded to any appreciable degree during phytoremediation. Concentrations of individual contaminants can be directly compared (normalized) to the internal marker. The relative ratio of a specific compound to the internal marker decreases as remediation occurs. For example, Hopanes, compounds found in crude oil, can be used as biomarkers for PHC remediation, including PAHs.

The methods developed to distinguish biologically derived organic compounds from organic contaminants can be used for chemical fingerprinting to assess liability after accidental release of chemicals into the environment. These methods can also be used to distinguish plant-derived hydrocarbons from petrogenic hydrocarbons for assessing phytoremediation. Although these methods are not widely employed in the phytoremediation field, in case of PHC phytoremediation it becomes necessary to distinguish contamination from naturally occurring and anthropogenic hydrocarbons (i.e., background hydrocarbons). A number of hydrocarbon indices have been derived to assess the source of environmental hydrocarbons, including the carbon preference index, average chain length, and various n-alkane/acyclic isoprenoid ratios.

13.6 Future Outlook

The phytoremediation technology is still in its early development stages and full scale applications are still limited. The study and use of genetic modifications must be performed to determine the true costs and benefits of this technology to the ecosystem as a whole. Progress in the field of molecular genetics will allow the analysis of metal hyperaccumulator plants and should provide new insights into metabolic detoxification processes and identify tolerant genes, thus providing more information about the genomes of these model organisms used for this technology.

The genomics can accelerate the discovery of genes that confer key traits, allowing their modification. In addition, metabolomics can provide biochemical and physiological knowledge about network organization in plants subject to toxic metal stress, providing a much more detailed understanding of the molecular basis of hyperaccumulation.

Identification of many signaling pathways and proteins within the stress response network can contribute to the cellular stress response. The development of DNA and RNA microarray chip technologies in systematic genome mapping, sequencing, functioning, and experimentation may allow the identification and genotyping of mutations and polymorphisms. This will provide a better insight into structure–function interaction of genome complexity under toxic metal stress. For example, Mitogen-Activated Protein Kinase (MAPK) pathways are activated in response to metal stress, which encourages new strategies for improving plant tolerance to heavy metals and phytoremediation.

Molecular genetics approaches such as insertion mutagenesis can be used to identify genes involved in hyperaccumulation. Similarly, transposon tagged plants can be screened to identify mutants impaired in the ability to accumulate metals. Recently, considerable progress has been made in identifying plant genes encoding metal ion transporters with important functions in cation transport and homeostasis.

Modern molecular techniques, bioinformatics, and computational techniques are effective tools for detailed structure–function genome analysis. Approaches allowing recombination hotspots to be highlighted will further aid plant breeding efforts. It is clear that both fundamental and applied research must be carried out in association, since the lack of the basic understanding will make it difficult to exploit many of the recent advances in plant molecular biology.

13.7 Conclusion

It is evident that phytoremediation has benefits to restore balance to a stressed environment, but it is important to proceed with caution before it is applied on a larger scale. The results already obtained have indicated that the plants are effective and could be used in toxic metal/organic contaminants remediation. Although it appears to be a widely accepted technology amongst scientists, engineers, and regulators, it is important that public awareness about it is considered. Clear and precise information should be made available to the general public to enhance its acceptability as a global sustainable technology.

References

Baker AJM, McGrath SP, Reeves RD, Smith JAC (2000) Metal hyperaccumulator plants: a review of the ecology and physiology of a biological resource for phytoremediation of metal polluted soils. In: Terry N, Banuelos G (eds) Phytoremediation of contaminated soil and water. Lewis Publishers, Boca Raton, pp 85–107

Bizily SP, Ruch CL, Meagher RB (2000) Phytodetoxification of hazardous organo-mercurials by genetically engineered plants. Nat Biotechnol 18:213–217

Chaney RL, Malik M, Li YM, Brown S, Brewer EP, Angel JS, Baker AJ (1997) Phytoremediation of soil material. Curr Opin Biotechnol 8:279–284

Davison J (2005) Risk mitigation of genetically modified bacteria and plants designed for bioremediation. J Ind Microbiol Biotechnol 32:639–650

Dhankher OP, Li Y, Rosen BP, Shi J, Salt D, Senecoff JF, Sashti NA, Meagher RB (2002) Engineering tolerance and hyperaccumulation of arsenic in plants by combining arsenate reductase and γglutamylcysteine synthetase expression. J Nat Biotechnol 20:1140–1145

Doty SL, Shang QT, Wilson AM, Moore AL, Newman LA, Strand SE, Gordon MP (2000) Enhanced metabolism of halogenated hydrocarbons in transgenic plants contain mammalian P450 2E1. Proc Natal Acad Sci USA 97:6287–6291

Doty SL, James CA, Moore AL, Vajzovic A, Singleton GL, Ma C, Khan Z, Xin G, Kang JW, Park JY, Meilan R, Strauss SH, Wilkerson J, Farin F, Strand SE (2007) Enhanced phytoremediation of volatile environmental pollutants with transgenic trees. Proc Natl Acad Sci USA 104(43):16816–16821

Doty SL (2008) Tansley review: enhancing phytoremediation through the use of transgenics and endophytes. New Phytol 179:318–333

Dowling DN, Doty SL (2009) Improving phytoremediation through biotechnology. Curr Opin Biotechnol 20:1–3

Eide D, Broderius M, Fett JM, Guerinot ML (1996) A novel iron-regulated metal transporter from plants identified by functional expression in yeast. Proc Natl Acad Sci 93(11):5624–5628

Flocco CG, Lindblom SD, Smits EAHP (2004) Overexpression of enzymes involved in glutathione synthesis enhances tolerance to organic pollutants in *Brassica juncea*. Int J Phytoremediat 6:289–304

French CJ, Rosser SJ, Davies GJ, Nicklin S, Bruce NC (1999) Biodegradation of explosives by transgenic plants expressing pentaerythritol tetranitrate reductase. Nat Biotechnol 17:491–494

Fulekar MH, Singh A, Bhaduri AM (2009) Genetic engineering strategies for enhancing phytoremediation of heavy metals. Afr J Biotechnol 8(4):529–535

Gandia-Herrero F, Lorenz A, Larson T, Graham IA, Bowles J, Rylott EL et al (2008) Detoxification of the explosive 2,4,6- trinitrotoluene in *Arabidopsis*: discovery of bifunctional O and C-glucosyltransferases. Plant J 56:963–974. doi:10.1111/j.1365-313X.2008.03653.x

Gerhardt KE, Huang XD, Glick BR, Greenberg BM (2009) Phytoremediation and rhizoremediation of organic soil contaminants: potential and challenges. Plant Sci 176:20–30

Grichko VP, Filby B, Glick BR (2000) Increased ability of transgenic plants expressing the enzyme ACC deaminase to accumulate Cd, Co., Cu, Ni, Pb and Zn. J Biotechnol 81:45–53

Hamer DH (1986) Metallothioneins. Ann Rev Biochem 55:913–951

Hansen D, Duda P, Zayed AM, Terry N (1998) Selenium removal by constructed wetlands: role of biological volatilization. Environ Sci Technol 32:591–597

Hirose S, Kawahigashi H, Ozawa K, Shiota N, Inui H, Ohkawa H et al (2005) Transgenic rice containing human CYP2B6 detoxifies various classes of herbicides. J Agric Food Chem 53:3461–3467

Kawahigashi H, Hirose S, Inui H, Ohkawa H, Ohkawa Y (2003) Transgenic rice plants expressing human CYP1A1 exudes herbicide metabolites from their roots. Plant Sci 165:373–381

Kawahigashi H, Hirose S, Ohkawa H, Ohkawa Y (2006) Phytoremediation of herbicide atrazine and metolachlor by transgenic rice plants expressing human CYP1A1, CYP2B6 and CYP2C19. J Agric Food Chem 54:2985–2991

LeDuc DL, Tarun AS, Montes-Bayon M, Meija J et al (2004) Overexpression of selenocysteine methyltransferase in *Arabidopsis* and Indian mustard increases selenium tolerance and accumulation. Plant Physiol 135:377–383

Liste HH, Alexander M (2000) Accumulation of phenanthrene and pyrene in rhizosphere soil. Chemosphere 40:11–14

Lu YP, Li ZS, Rea PA (1997) AtMRP1 gene of Arabidopsis encodes a glutathione S-conjugate pump: isolation and functional definition of a plant ATP-binding cassette transporter gene. Proc Natl Acad Sci USA 94:8243–8248

Malin M, Bülow L (2001) Metal-binding proteins and peptides in bioremediation and phytoremediation of heavy metals. Trends Biotechnol 19:67–72

Monti MR, Smania AM, Fabro G, Alvarez ME, Argarana CE (2005) Engineering *Pseudomonas fluorescens* for biodegradation of 2,4-dinitrotoluene. Appl Environ Microbiol 71:8864–8872

Oller ALW, Agostini E, Talano MA, Capozucca C, Milrad SR, Tigier HA et al (2005) Overexpression of a basic peroxidase in transgenic tomato (*Lycopersicon esculentum* Mill. cv. Pera) hairy roots increases phytoremediation of phenol. Plant Sci 169:1102–1111

Phytoremediation at the hydrocarbon burn facility at NASA Kennedy Space Center in Florida http://www.cluin.org/products/phyto/search/phyto_details. cfm? ProjectID= 25 (July 22, 2008)

Rugh CL, Wilde D, Stack NM, Thompson DM, Summers AO, Meagher RB (1996) Mercuric ion reduction and resistancein transgenic *Arabidopsis thaliana* plants expressing a modified bacterial merA gene. Proc Natl Acad Sci USA 93:3182–3187

Schat H, Llugany M, Vooijs R, Hartley-Whitaker J, Bleeker PM (2002) The role of phytochelatins in constitutive and adaptive heavy metal tolerances in hyperaccumulator and non-hyperaccumulator metallophytes. J Exp Bot 53:2381–2392

Shah K, Nongkynrih J (2007) Metal hyperaccumulation and bioremediation. Biol Plant 51(4):618–634

Tong YP, Kneer R, Zhu YG (2004) Vacuolar compartmentalization: a second-generation approach to engineering plants for phytoremediation. Trends Plant Sci 9:7–9

Van Aken B, Yoon JM, Schnoor JL (2004) Biodegradation of nitro-substituted explosives 2,4,6-trinitrotoluene, hexahydro-1,3,5-trinitro-1,3,5-triazine, and octahydro-1,3,5,7-tetranitro-1,3,5-tetrazocine by a phytosymbiotic *Methylobacterium* sp. associated with poplar tissues (*Populus deltoides* × *nigra* DN34). Appl Environ Microbiol 70:508–517

Villacieros M, Whelan C, Mackova M, Molgaard J, Sánchez-Contreras M, Lloret J, Aguirre de Cárcer D, Oruezábal RI, Bolaños L, Macek T, Karlson U, Dowling DN, Martín M, Rivilla R (2005) Polychlorinated biphenyl rhizoremediation by *Pseudomonas fluorescens* F113 derivatives, using a *Sinorhizobium meliloti* nod system to drive bph gene expression. Appl Environ Microbiol 71:2687–2694

Wang GD, Li QJ, Luo B, Chen XY (2004) Ex planta phytoremediation of trichlorophenol andphenolic allelochemicals *via* engineered secretory laccase. Nat Biotechnol 22:893–897

Yamada T, Ishige T, Shiota N, Inui H, Ohkawa H, Ohkawa Y (2002) Enhancement of metabolizing herbicides in young tubers of transgenic potato plants with the rat CYP1A1 gene. Theor Appl Genet 105:515–520

Yee DC, Maynard JA, Wood TK (1998) Rhizoremediation of trichloroethylene by a recombinant, root-colonizing *Pseudomonas fluorescens* strain expressing toluene ortho-monooxygenase constitutively. Appl Environ Microbiol 64:112–118

Zhu YL, Pilon-Smits EAH, Tarun AS, Weber SU, Jouanin L, Terry N (1999) Cadmium tolerance and accumulation in Indian mustard is enhanced by overexpressing glutamylcysteine synthetase. Plant Physiol 121:1169–1177

14 Future of Biotechnology Companies: A Global Perspective

Sunita Chauhan and Pradeep Bhatnagar

Abstract

In the current scenario of environmental friendliness, need of sustainable development, globalization, and fast technological developments, biotechnology which has been used in a crude manner by mankind since thousands of years, has become quite relevant especially in its most modern form. Out of the three forms viz. white, red, and green biotechnology, white or industrial biotechnology has become the technology of the day. The present chapter briefs about the global and regional drivers of its developments all over the world. With an overview of the global biotechnology clusters, challenges involved in setting biotechnological companies are discussed. Finally, the great market potential involved for biotechnological companies has been highlighted with a suggestion of three groups of companies (http://www.oecd.org/sti/biotech/44776744.pdf) that can be set up for harnessing the vast potential involved in biotech products all over the world.

14.1 Introduction

The survival and well-being of any country depends on sustainable development. The sustainable development is a process of social and economic betterment that satisfies the needs and values of all interest groups without foreclosing future options. For this, we must ensure that the demand on the environment from which we derive our sustenance, does not exceed its carrying capacity for the present as well as future generations.

The Chemistry Nobel Laureate of 2000, Alan G. MacDiarmid during his lecture on "The World is Becoming Smaller" at the Fifth APEC Research and Development (R&D) Leaders' Forum on March 11, 2004 at New Zealand, brought out the fact that the world population in 2003 was 6.3 billion, and by 2050, it would be 10 billion. Further the humanity's top ten challenges for the next 50 years have been listed as: (1) Energy, (2) Water, (3) Food, (4) Environment, (5) Poverty, (6) Terrorism and war, (7) Disease, (8) Education (9) Democracy and

S. Chauhan
Kumarappa National Handmade Paper Institute (KNHPI), Jaipur, Rajasthan, India
e-mail: itsneeru@yahoo.com

P. Bhatnagar (✉)
Dean, Life Sciences, The IIS University, Mansarovar, Jaipur, India
e-mail: Pradeepbhatnagar1947@yahoo.com

I. Ravi et al. (eds.), *Advances in Biotechnology*,
DOI: 10.1007/978-81-322-1554-7_14,

(10) Population. Our best hope of overcoming these challenges is with new technologies like Biotechnology (Lim 2009).

14.2 Role of Biotechnology in Sustainable Development

Biotechnology has been used for more than 6,000 years for lots of interesting and practical purposes: making food such as bread and cheese, preserving dairy products, and fermenting beer. Although we do not always realize it, biotechnology is a huge part of our everyday lives, from the clothes we wear and how we wash them, the food we eat and the sources it comes from, the medicine we use to keep us healthy and even the fuel we use to take us where we need to go, biotech already plays, and must continue to play, an invaluable role in meeting our needs. No other industry is better placed to enhance quality of life and respond to society's *Grand Challenges* of tackling an aging and ever increasing population, healthcare choice and affordability, resource efficiency, food security, climate change, sustainability, and energy constraints.

The "new biotechnology" that was started in the early 1970s with direct manipulations of the cell's genetic machinery through recombinant DNA techniques has fundamentally expanded the utility of biological systems with its application on an industrial scale. Scientists and engineers can now confer new characteristics to microbes, plants, and animals by changing their genetic make-ups. Just like inanimate matters, some biological molecules can now be biologically manufactured or biofactured. The new biotechnology, combined with the existing industrial, government, and university infrastructure in biotechnology and the pervasive influence of biological substances in everyday life, has set the stage for unprecedented growth in products, markets, and expectations (Lim 2009).

Biotechnology which is divided into three sub fields—red, white (grey) and green biotechnology— has been very useful for the mankind. Red Biotechnology is used for manufacturing products like insulin and vaccines for medical uses and in reproductive technologies like *in vitro* fertilization, DNA profiling, forensics, and transplantation technologies. White Biotechnology that deals with creating useful chemicals for the industrial sector through molds or yeast has proven to be of immense benefit environmentally in cleaning oil spills and in storing DNA samples of endangered species for future research, besides being useful for removing excess nutrients in soil and water and for detection of landmines. Green Biotechnology deals with the use of environmental friendly solutions as an alternative to traditional agriculture, horticulture, and animal breeding processes.

Today, Biotechnology is perceived as a revolution throughout the world. Scientists, through R&D have already developed and are continuing to develop cures for diseases that have affected people for decades and even centuries. Similarly, certain crops have been developed that can withstand the brutalities of weather changes, helping poor farmers of the developing countries to retain their yield, and increase their output manifold. Biotechnology can make agriculture more competitive and sustainable by creating new non-food markets for crops. It can help industry increase its economic and environmental efficiency (eco-efficiency) and sustainability while maintaining or improving its competitive advantage and ability to generate growth. Biotechnology, considered a boon by some, provides great hope to many around the world, and its benefits will be realized by more and more people over the years.

14.3 Industrial Biotechnology

Industrial biotechnology has tremendous potential to improve industrial production along all the three dimensions of sustainable development viz. the Society, the Environment, and the Economy. The application of biotechnology to the eco-efficient production and processing of chemicals, materials and bioenergy, and Industrial biotechnology as such is a cornerstone of the knowledge-based bioeconomy. It utilizes the extraordinary capabilities of microorganisms and enzymes, their diversity, efficiency, and specificity to make products in sectors such as chemicals, food and feed, pulp and paper, textiles, automotives, electronics, and crucially energy. Biological processes are generally more environmentally benign than industrial chemical processes as they take place at low temperature and pressure, have lower energy input requirements and lower greenhouse gas (GHG) emissions. Also, the raw materials for production are renewable agricultural feed stocks. Thus, according to Organisation for Economic Co-operation and Development (OECD), Biotechnology, probably more than any other technology, offers full or partial solutions to major societal problems like healthcare, environmental degradation, food security and safety, and energy supply (OECD 2011).

14.4 Major Drivers of Industrial Biotechnology

The fast growth of Industrial Biotechnology all over the world is clearly linked to the global challenges of climate change, energy security, and the financial crisis. This "Green Growth" challenge is faced by developing and developed countries alike (OECD 2010). There are many global as well as regional drivers of industrial biotechnology as discussed below:

14.4.1 Global Drivers

14.4.1.1 The Fast Technological Development within Industrial Biotechnology

Progressing at an enormous rate include development of basic technologies (e.g., directed evolution, genetic modification tools), development of tailor made and high performance enzymes (especially in the number of available enzymes), improvement of efficiency through new developments in reactor and process design, implementation of new applications for existing and new enzymatic systems, and increasing production of high-value products, such as nutraceuticals, cosmaceuticals, and performance chemicals.

14.4.1.2 The Need for Sustainable Development

With the growing awareness all over the word, consumers are increasingly conscious about the impacts of their consumption and choosing products with low negative impact on the environment (e.g., low-carbon economy). Such environmental and social considerations have created the need for a switch to clean and sustainable processes using renewable resources. Added to the converging economic and environmental dilemma is the increase in global population and per capita income. Population explosions are resulting into a huge impact on land use and water resources, increased waste and waste water production, and impact on food prices through growing demand (OECD 2011). Therefore, need for a sustainable development is one of the major global drivers of industrial biotechnology.

14.4.1.3 Ongoing Globalization

Globalization is acting as a general driver. The challenge from the Asian chemical industry is causing the European Union (EU) and the United States to look at industrial biotechnology as means of sustaining competitiveness. Global competition drives companies to biotech routes, especially in Europe and the USA. Reduction of trade barriers toward global biotech markets is important aspect which cannot be ignored. Another aspect of globalization is that feed stocks and their costs vary across the globe, for example, feed stocks for bio fuel production include sugar cane in Brazil, maize in United States, wheat in Europe, and palm oil in Indonesia.

14.4.2 Regional Drivers

There are regional differences in the drivers for the development of industrial biotechnology.

14.4.2.1 The US Top-Down Approach

The widening ban on the use of the contaminating and potentially carcinogenic methyl tertiary- butyl ether (MTBE) as a gasoline oxygenate necessitates an alternative and ethanol is gaining share (LoGerfo 2005). Apart from the use of ethanol as a fuel, its use as a fuel additive is itself a billion-dollar market. But the much larger issue and driving force in the United States is the growing concern over energy security and independence which led them to make vast investments in bioethanol development. The US industrial biotechnology drive has been led centrally, initiated by Government and/or the administration with massive public funding (Lorenz and Zinke 2005). A further dimension of it is the need of regeneration of the rural environment because according to United states Department of Agriculture a huge number of agricultural jobs have been lost over the years owing to increased efficiency (USDA 2010).

14.4.2.2 The EU Chemical Industry and a Bottom-Up Approach

In the EU the drivers are different. The development of industrial biotechnology in the EU has been derived from the desire of the chemical industry to remain competitive. Data for the 10 years from 1999–2009 indicate that the EU has been the clear leader in terms of world chemical sales, but it has gradually lost ground to Asia, principally owing to the rise of China. Still, EU exports of chemicals in 2009 accounted for 46 % of global chemical exports. (Hadhri 2010). Nonetheless, the threat to its position from Asia is indubitable due to high production costs. The role of SusChem, the European Technology Platform for Sustainable Chemistry[1] is to enhance the European Chemical industry, and industrial biotechnology is one of its key strategic areas. Overall, the EU approach to industrial biotechnology has been bottom-up, motivated by the chemical industry. Europe has developed a vision for 2025 as they feel that development and use of industrial biotechnology is essential to the future competitiveness of European industry and provides a sound technological base for the sustainable society of the future (Moll and Tanda 2012). In general terms, the United States has focused on biomass-based energy supply and bulk chemicals, whereas the EU has focused more on the manufacture of novel, high-margin products. The EU has also been relatively more concerned with the environmental impacts, e.g., compliance with demands for Kyoto protocol.

14.4.2.3 Biomass Utilization in Japan

Japan is developing a national strategy for biotechnology, which it sees as a transformative technology of strategic importance (Lynskey 2006). In March 2006, the Japanese Government renewed the biomass Nippon Strategy of 2002, approved by the cabinet to promote biomass utilization and to implement new measures (Kiyoshi 2006). The emphasis is on the greater use of transport biofuels and acceleration of the creation of biomass towns, local municipalities with infrastructure for utilizing biomass. Owing to limited agricultural resources and the food versus fuel debate, Japan is to focus strongly on biofuels derived from cellulosic or other materials which do not compete with food supply (USDA Foreign Agricultural Service 2009). They have set a 2030 goal of producing biofuels equivalent to 10 % of domestic fuel consumption from cellulosic materials such as rice straw, wood, and resource crops such as sugar cane and sugar beet.

14.4.2.4 The Rise of Asia

The Asian chemical industry as a whole has overtaken the EU in terms of sales, and there are good prospects for the use of biofuels as transport fuel in developing countries. Most of these countries face severe energy insecurity and have large agricultural sectors able to support production of biofuels from energy crops (Liaquat et al. 2010). Population and GDP growth and environmental and social pressures in developing countries could be significant drivers for competiveness in industrial biotechnology. The literature reveals

[1] (www.suschem.org)

the depth and breadth of industrial biotechnology research and innovation in Asian countries such as China (Zhang et al. 2011); India (Gupta et al. 2008); Japan (Anazawa 2010); Malaysia (Hassan and Yaacob 2009); Chinese Taipei (Lin et al. 2010); Thailand (Hniman et al. 2011); and Vietnam (Thanh et al. 2008).

14.4.2.5 China as a World Force

China has clearly signaled its intention to be a world force in industrial biotechnology, with a focus on biofuel. Drivers include a long history of expertise in fermentation, a desire for energy security and rapidly increasing energy consumption, volatility of fossil fuel prices, environmental concerns, and providing farmers with an additional income stream to support rural areas (Nesbitt 2009). China is the world's third largest producer of ethanol. Existing biobased production includes vitamin C and citric acid, and the Chinese chemical industry is making increasing use of industrial biotechnology, particularly in biopolymers. China is mapping out a 5-year development plan (2011–2015) to further concentrate on its bioindustry and raise its international profile.

14.5 Endeavor of Setting Biotechnology Companies and Global Biotechnology Clusters

One of the greatest challenges for researchers is converting scientific discoveries and innovations into successful biotechnology companies. To succeed, the spin-off phase of biotechnology companies has to be crossed by bioentrepreneurs and venture capitalists, reluctant to invest in early stage biotechnology companies. Ukropcova and Sturdik (2011) have summarized biotechnology commercialization in the world and identified significant biotechnology and life sciences clusters on the world map.

There are many issues to be addressed when commercializing biotechnology research. The obvious lack of pre-seed capital and inadequate financial support from government are not always to blame (Pavlou 2003). In many cases, a lack of commercialization skills in the field of biotechnology and innovative financial tools can be the missing factors to capture the significant value from the biotechnology laboratories (Nagle et al. 2003). Although growth and development of biotechnology spin-offs heavily depends on financial resources, presence of encouraging environment is a necessary condition (Booth 2006). Bains (2009) have reported the world's leading biotechnology clusters and companies to summarize the results of major biotechnology commercialization in the world.

Commercialized biotechnology concentrates in biotechnology clusters surrounded by universities and life science research institutes (Friedman 2006). The idea of a cluster is geographic concentrations of interconnected sectors, building on strengths, and removing barriers to development. It requires actions and coordination between government departments, regional economic development agencies, universities, companies, and others (Sainsbury 1999). Effective technology transfer is also necessary with a formal legal infrastructure for university participation and sufficient funds to file patents. The formation of new companies requires a business infrastructure in the community, researchers, technology transfer professionals, entrepreneurial company founders, scientists, managers to staff the companies, and knowledgeable investors. It takes a whole community to build a biotechnology cluster. Once built, the cluster can achieve a sustaining life that strengthens itself. The world's biggest clusters are San Francisco and Boston area in USA, Cambridge, and Oxford area in Great Britain and lastly Medicon Valley and Bio Valley in the continental Europe. South Pacific Asian and Australian clusters are lately emerged but are fast growing areas (Table 14.1).

14.6 Barriers in Biotechnology and Challenges Ahead

It is becoming a reality in the world that biotechnology and life sciences are the frontier of a knowledge-based society. Biotechnology is a

Table 14.1 World's biotechnology and life sciences clusters with high growth competitiveness index 2004–2005 (EMBO Reports 2006)

Region	Cluster
Life sciences clusters of North America	Seattle, USA
	San Francisco, USA
	Los Angeles, USA
	San Diego, USA
	Saskatoon, Canada
	Minneapolis/St. Paul/Rochester USA
	Austin, USA
	Toronto, Canada
	Montreal, Canada
	Boston, USA
	New York/New Jersey, USA
	Philadelphia, USA
	Baltimore/Washington, DC, USA
	Research Triangle NC, USA
Central America/ South America	West Havana, Cuba
	Belo Horizonte/Rio de Janeiro, Brazil
	Sao Paulo, Brazil
Africa	Capetown, South Africa
United Kingdom/ Ireland	Glasgow-Edinburgh, Scotland
	Manchester-Liverpool, England
	London, England
	Cambridge-SE England
	Dublin, Republic of Ireland
Continental Europe	Brussels, Belgium
	Medicon Valley, Denmark/ Sweden
	Stockholm/Uppsala, Sweden
	Helsinki, Finland
	Paris, France
	Biovalley, France/Germany/ Switzerland
	BioAlps, France/Switzerland
	Sophia-Antipolis, France
	BioRhine, Germany
	BioTech Munich, Germany
	BioCon Valley, Germany
Middle East	Israel
Oceania	Brisbane, Australia
	Sydney, Australia
	Melbourne, Australia
	Dunedin, New Zealand
Asia	Beijing, China
	Shanghai, China
	Shenzhen, China
	Hong Kong, China
	Tokyo-Kanto, Japan
	Kansai, Japan
	Hokkaido, Japan
	Taipei, Taiwan
	Hsinchu, Taiwan
	Singapore
	Dengkil, Malaysia
	New Delhi, India
	Hyderabad, India
	Bangalore, India

unique industry sector where a high failure rate for companies is considered the norm. High priority for earlier stage companies is to secure funding with more dependence on external factors such as governmental support. The later stage companies, having access to product-derived funds, are more able to build internal resources and expanding to global markets (Vanderbyl and Kobelak 2008). Some of the major barriers for Biotechnology companies are as below:

1. **High investment and production costs as well as critical raw material prices**
   (a) Restrictions of biotechnological production processes from an economic viewpoint.
   (b) High investments in R&D and process development.
   (c) Massive new investments to build new production facilities.
   (d) Higher prices for products from biotechnology processes normally not achievable.
2. **Cyclical raw material prices and limited availability**
   (a) Almost 1:1 correlation of crude oil and biomass prices during the last years so no general cost advantages of biobased routes.
   (b) Rising food and feed demand as a critical driver and a significant limitation on the uptake of Industrial biotechnology (especially in debates around biofuels, land use has been a controversial issue).

3. **Complex innovation processes as well as critical social acceptance and regulations**
   (a) Industrial biotechnology know how mainly used in the early stages of the value chain.
   (b) Specialized companies normally covering only a small share of the value-added along the value chain.
   (c) Combination of "technology push" and "market pull" along the value chain.
4. **Critical social acceptance and regulations**
   (a) Social acceptance of industrial biotechnology is normally high but some regions still have rather a low acceptance of genetically modified organisms—in the field of genetic engineering considerably more bureaucracy and legislation have been seen.
   (b) Problem accepting green biotechnology, especially in Europe, has a direct impact on industrial biotechnology.

Growth depends very much on the development of green biotechnology. Green and industrial biotechnology often combines to an integrated value chain. The growth and success of biotechnology sector depends on a combination of good education, good science, and good business (Moses 2003). Biotechnology education and bioentrepreneurship is a long-term issue requiring a long-term view; it should not be constrained by short-term funding. The ability to take risks, prior work experience in private firms, and personal experience in cooperating with industry lead to a positive attitude toward switching to private sector employment or entrepreneurship (Fritsch and Krabel 2010). However, despite numerous initiatives to popularize and sell science, it seems the attitude and understanding of society toward science and scientists remain lower than expected. Scientists' communication in society comes forward as high priority and great importance (Baron 2010).

Today, the health care ecosystem and its constituents face historic challenges. At a time when key stakeholders—payers, pharma companies, biotech firms, and their investors—are increasingly resource-constrained, we need R&D paradigms that are several shades more efficient and productive. With aging populations and rapidly growing middle classes in emerging markets, societies need ways to accelerate cures for ailments that are expected to impose huge societal costs, such as neurodegenerative and chronic conditions. For life sciences companies, this will involve different ways of thinking about intellectual property and recognizing that in some situations, sharing information may create more value than protecting it. Regulators will need to adapt frameworks to allow for drug development paradigms that are flexible and learn in real time. And ultimately, patients will need to willingly share their personal health data, with the recognition that they might reap some of the biggest dividends from this approach: better health outcomes, better drugs and cures for long-intractable diseases.

14.7 Market Potential for Biotechnology Companies

Emerging industries such as the life sciences, animal health, agricultural biotechnology, and environmental products offer both a potential for economic growth and improvements in quality of life, the environment, and industrial productivity. Even governments in developing countries and investors are seeking to create and enhance biotech entrepreneurship face. Several enabling trends include increasing numbers of science graduates worldwide, accelerating pace of scientific advancement, dominating role of globalization enabling greater collaboration, and the relentless competitive pressure to innovate (Thorsteinsdótti et al. 2004). The largest market potential lies in the production of fine chemicals for the pharma and agro industry, biopolymers and biofuels (Table 14.2).

The potential involved with Biotechnology companies and their future prospects may be further understood from the following requirements.

14.7.1 Move from Fishing to Aquaculture

A 1997 United Nations estimate indicates that the supply of seafood products would have to increase sevenfold to meet the worldwide

Table 14.2 Market potential involved and the major trends

S.No.	Sector	Major trends
1.	Fine chemicals	• Growing importance of chiral-active pharmaceutical ingredients
		• Cost reductions through simplified synthesis paths (especially if molecules are complex)
2.	Polymers	• Stronger use of renewable raw materials due to cost reasons
		• Access to polymers with new properties (e.g., biodegradability)
		• Increasing regulatory pressure to realize sustainable solutions (e.g., for packaging applications)
3.	Specialty chemicals	• Strong growth in industrial enzymes for more and more applications
		• Increasing advantages of enzymatic processes especially in the food, cosmetic, textile, and leather industries due to customer requirements
4.	Base chemicals and intermediates	• Stronger use of renewable raw materials due to cost reasons
		• Cost advantages of biotech processes due to the stricter regulatory environment
5.	Biofuels	• Increasing advantage due to rising oil prices and climate change
		• Significant progress in production technologies with increasing cost competitiveness without subsidies

requirement for fish and other seafood by the year 2020. Given the rapid decline in world fish stocks, caused mainly by over fishing, it is clear that demand can only be met by aquaculture.

Worldwide aquaculture production of many fish species faces several challenges. Among them are the selection and supply of suitable broodstock, growth rate, feed conversion efficiency, feeding costs, control of the reproductive cycle, and disease protection. Traditionally, the broodstock is selected by cross breeding to enhance the fishes' desired traits. Cross breeding is generally time-consuming and traits are slow to emerge and unpredictable. Gene transfer technology—identifying genes responsible for desirable traits using molecular biology and then transferring them to the broodstock—is an improvement over traditional selection and breeding methods. New traits not present in a genome can also be genetically transferred into an unrelated species, enabling the production of new and beneficial phenotypes. Thus biotechnology companies have a great scope in Aquaculture.

14.7.2 From Green Revolution to Agricultural Biotech Revolution

The Green Revolution of the 1960s depended on agrochemicals and achieved a doubling of production with only a 10–20 % increase in the amount of land under cultivation. The world population is exploding, while one-third of all crops are still lost to pests and diseases. As the world population rises to 10 billion over the twenty-first century, the green revolution will soon reach its saturation point. Its impact on the environment is also increasingly unacceptable. Sir Robert May, Chief Scientific Adviser to the British government, warned, "…we will not be able to feed tomorrow's population with today's technology. We will now need to create crops that are shaped to the environment, with biotechnology, whereas before, with the Green Revolution, our environment was shaped by crops that were created with the use of chemicals derived from fossil fuels."

Metabolic or nutritional genomics is using genetic engineering to improve nutritional value of plants. Rice, though a staple food for half the world's population, is a relatively poor source of many essential nutrients, including vitamin A and iron. An estimated 124 million children worldwide are deficient in vitamin A, and a quarter million in Southeast Asia go blind each year because of the deficiency. Improved nutrition can prevent 1–2 million deaths a year.

The International Rice Research Institute (IRRI) have succeeded in splicing three genes into rice to make it iron-enriched and four other genes to make it rich in β-carotene, a source of vitamin A. The iron-enriched rice and vitamin A-enriched functional rice are combined by crossing. Thus biotechnology companies may play a very important role in the Agro-biotech revolution.

14.7.3 Increasing Role of Biotechnology in Pharmas

There are many other actively researched biotech areas. But a recurring confusion is the boundary between biotechnology and pharmaceutical industry. To be fair, the boundary is fuzzy. After all, of the more than 300 publicly traded biotechnology companies, Amgen ($69 billion in capitalization) and Genentech ($55 billion in capitalization), both in California, account for about a third of the market capitalization.

Genentech, which essentially started the biotechnology industry in 1976, introduced three drugs in 2003. Chief among them is Avastin for cancer treatment. Amgen, founded 4 years after Genentech, has relied on its two runaway multibillion-dollar blockbusters introduced around 1990: Epogen, a red blood cell booster to treat anemia; and Neupogen, a white blood cell booster to treat side effect chemotherapy. Generally speaking, biotechnology companies have concentrated on using genetic engineering to make proteins that can be used as drugs. Pharmaceutical companies, on the other hand, tend to rely more on drugs made from chemicals.

Stem cells also regarded as "magic seeds" since they possess the ability to replicate indefinitely and morph into any kind of tissue found in a human body. They are nature's blank slates, capable of developing into any of the nearly 230 cell types that make up the human body. Potential uses of stem cells are immense. Stem cells can be used in delineating the complex events during human development. Also, they can be used in the biotechnology and pharmaceutical industry to streamline drug safety tests and in cell and tissue therapies, and organ transplants. Scientists believe they will lead to cures for debilitating diseases once thought untreatable.

So, to exploit the vast potential involved in Biotechnology companies, following three groups of industrial companies can be targeted by policy makers with different approaches:

Dedicated start-ups and small and medium enterprises (SMEs): provide incentives to foster growth based on R&D based innovations.

Diversified SMEs: provide incentives to enable the usage of biotech technologies in established production processes.

Dedicated and diversified multinational enterprises (MNEs): provide incentives to support market introduction and penetration of biotech products.

14.8 Conclusion

Keeping in view of the estimated 10 billion population of the world by the year 2050 and humanity's top 10 challenges, biotechnology can be the most appropriate technology to help us in meeting those demands and facing all the challenges.

Although the biotechnology industry has endured several slumps during the last 2–3 decades, the overall trends have been positive by many important measures: number of companies, number of approved products, market capitalization and revenues. It is these positive factors that should help companies with the right scientific and commercial strategies to prosper in the future (Kermani and Bonacossa 2003). The number of biotechnology compounds has been increasing steadily over the past 20 years, reflecting the key contribution that biotechnology is now making to healthcare. Recombinant DNA technology has been used to develop a number of therapeutic proteins, including antibodies, cytokines, hormones, and vaccines for use in tackling and diagnosing a range of disorders. Worldwide there are more than 4,000 specialized biotechnology companies. The most well-known companies are located in the USA and Europe, but there are significant companies emerging in Canada, Australia, New Zealand, and throughout Asia.

Biotechnology companies are working everyday to solve the greatest challenges facing our society whether it is finding a cure for cancer, protecting against bioterror threats, or creating renewable energy sources, to feed the hungry or to clean our environment. Thus biotechnology companies are bound to grow and prosper all over the world. The need is to formulate suitable strategies by making national, regional, and internationally harmonized policies, smooth regulatory procedures, and to provide financial and government support so that the biotechnology companies may actually transform the innovative ideas of today into the realities of tomorrow.

References

Anazawa H (2010) Bioindustry in Japan: current status and challenges. J Biotechnol special abstracts. 150S: S1–S576 (special abstracts)

Bains W (2009) Bonsai biotech. Nat Biotechnol 27:233. doi:10.1038/nbt0309-233

Baron N (2010) Stand up for science. Nature 468: 1032–1033

Booth BL (2006) Early stage return. Nat Biotechnol 24(11):1335–1340

EMBO (European Molecular Biology Organization) Reports (2006) 7(2):133–136. doi:10.1038/sj.embor.7400633

Friedman Y (2006) Building biotechnology. Think Biotech LLC, Washington

Fritsch M, Krabel S (2010) Ready to leave the ivory tower?: Academic scientists' appeal to work in the private sector. J Tech Transfer. doi:10/1007/S10961-010-9174-7

Gupta S, Kapoor M, Sharma KK, Nair LM, Kuhad RC (2008) Production and recovery of an alkaline exo-polygalacturonase from Bacillus subtilis RCK under solid state fermentation using statistical approach. Bioresour Technol 99:937–945

Hadhri M (2010) The European chemical industry in a worldwide perspective. In: Proceedings of CEFIC facts and figures 2010

Hassan MA, Yaacob WARW (2009) Current status of industrial biotechnology in Malaysia. J Biosci Bioeng 108:S1–S3

Hniman AO, Thang S, Prasertsan P (2011) Developing a thermophillic hydrogen producing microbial consortia from geothermal spring for efficient utilization of xylose and glucose mixed substrates and oil palm trunk hydrolysate. Int J Hydrogen Energy 36:8785–8793

Kermani F, Bonacossa P (2003) Current and future prospects for the global biotechnology industry. J Commercial Biotechnol 10(2):154–161

Kiyoshi F (2006) New biomass Nippon strategy. Environ Res Q 142:13–19

Liaquat AM, Kalam MA, Masjuki HH, Jayed MH (2010) Potential emissions reduction in road transport sector using biofuel in developing countries. Atmos Environ 44:3869–3877

Lim HA (2009) Biotechnology: past, present and future. In: Proceedings of contribution to symbiosis-biotechnology, pp 1–8. (www.dtrends.com/publication/biotech_symbiosis.pdf)

Lin CY, Wu SY, Lin PJ, Chang JS, Hing CH, Lee KS, Chang FY, Chu CY, Chang CH, Lay CH, Chang AC (2010) Pilot scale hydrogen fermentation system start up performance. Int J Hyd Energy 35:13452–13457

LoGerfo J (2005) Fermentation funding fervor. Ind Biotechnol 1:20–22

Lorenz P, Zinke H (2005) White biotechnology: differences in US and EU approaches? Trends Biotechnol 23:570–574

Lynskey MJ (2006) Transformative technology and institutional transformation: coevolution of biotechnology venture firms and the institutional framework in Japan. Res Policy 35:1389–1422

Moses V (2003) Biotechnology education in Europe. J Commercial Biotechnol 9(3):219–230

Moll N and Tanda S (2012) Innovating to secure Europe's future: biotechnology as a cornerstone of the EU 2020 agenda. Annual report of Europabio (the association for bioindustries), pp 1–30

Nagle T, Berg C, Nassr R, Pang K (2003) The further evolution of biotech. Nat Rev Drug Recovery 2:75–79

Nesbitt E (2009) Industrial biotechnology in China amidst changing market conditions. Ind Biotechnol 5: 232–236

OECD (2010) Interim report of the Green growth strategy: Implementing our commitment for a sustainable future. Meeting of the OECD Council at Ministerial level held at Paris from 27-28th May, 2010, 1–94. www.oecd.org/greengrowth

OECD (2011) Introduction: scope and drivers for industrial biotechnology. In: Proceedings of future prospects for industrial biotechnology, OECD Publishing, pp 11–21. http://dx.doi.org/10.1787/9789264126633-en

Pavlou AK (2003) Market place leading biotechnology players, 2002–2008, key products, indications and technologies. J Commercial Biotechnol 10:167–176

Sainsbury Lord of Turville (1999) Biotechnology clusters. Department of Trade and Industry, London, UK

Thanh VN, Mai LT, Tuan DA (2008) Microbial diversity of traditional Vietnamese alcohol fermentation starters (banh men) as determined by PCR-mediated DGGE. Int J Food Microbiol 128:268–273

Thoresteinsdotti HT, Saenz W, QuachU DaarAS, Singer PA (2004) Cuba: innovation through synergies. Nat Biotechnol 22:19–24

Ukropcova D, Sturdik E (2011) Biotechnology commercialization in the world. Acta Chimica Slovaca 4(1):115–125

USDA Foreign Agricultural Service (2009) Japan biofuels annual. Japan to focus on next generation biofuels. Global agricultural information network (GAIN) report prepared by Midori Iijima

USDA (2010) A USDA regional roadmap to meeting the biofuels goals of the renewable fuels standard by 2022. USDA biofuels strategic production report, 23 June

Vanderbyl S, Kobelak S (2008) Risk management for the biotechnology industry: a Canadian perspective. J Commercial Biotechnol 14:128–140

Zhang L, Zhao H, Gan M, Jin Y, Gao X, Chan Q, Guan J, Wang Z (2011) Application of simultaneous saccharification and fermentation (SSF) from viscosity reducing of raw sweet potato for bioethanol production at laboratory, pilot and industrial scales. Bioresour Technol 102:4573–4579

About the Editors

Dr. Indu Ravi

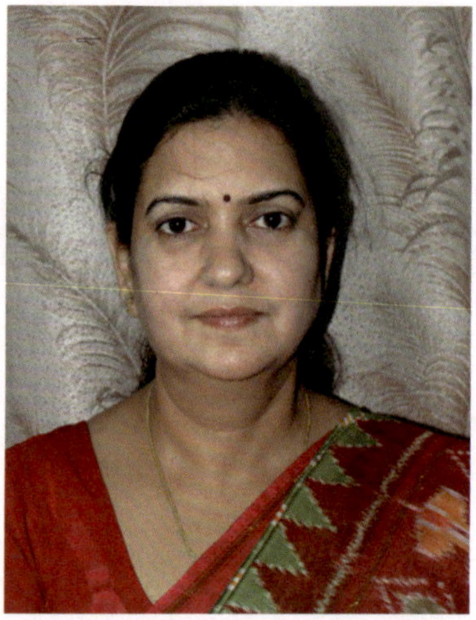

Dr. Indu Ravi is presently working as Assistant Regional Director, IGNOU Regional Centre, Jaipur. She has worked as Associate professor in the Dept. of Bioscience and Biotechnology, Banasthali University, Banasthali (Rajasthan) and has 14 years of teaching experience of undergraduate and postgraduate classes in the field of biotechnology. She acquired her post graduation in Biotechnology from Indian Veterinary Research Institute (IVRI), Izatnagar, and her Ph.D. in Biotechnology from Banasthali University. Her area of specialization is Plant Molecular Biology. She has written one book and published research papers in National and International journals. Email: induravi11@yahoo.com.

Dr. Mamta Baunthiyal

Dr. Mamta Baunthiyal has completed her Master's degree in Biochemistry from Kurukshetra University, India and her Ph.D. in Biotechnology from Banasthali University, Rajasthan, India. She has 17 years of teaching experience in undergraduate and postgraduate classes in Biotechnology. Her research is focused on the phytoremediation of fluoride and she has several research articles published in reputed journals. Presently, she is working as Head and Associate Professor in the Department of Biotechnology, GB Pant Engineering College, Pauri, Uttarakhand, India. Email: mamtabaunthiyal@yahoo.co.in.

I. Ravi et al. (eds.), *Advances in Biotechnology*,
DOI: 10.1007/978-81-322-1554-7, © Springer India 2014

Dr. Jyoti Saxena

Dr Jyoti Saxena is a Professor and Head in the Biochemical Engineering Department at Bipin Tripathi Kumaon Institute of Technology, Dwarahat (Uttarakhand). She has completed her M.Sc. and Ph.D. from Kumaun University, Nainital and her postdoc studies from UNL, USA. She has 19 years of teaching experience in undergraduate and postgraduate classes in Bioscience, Biotechnology, and Biochemical Engineering fields. Dr. Saxena has 51 research papers in reputed national and international journals and 4 books to her credit. Email: jyotisaxena2000@yahoo.co.in.

Index

I. Ravi et al. (eds.), *Advances in Biotechnology*,
DOI: 10.1007/978-81-322-1554-7, © Springer India 2014